THE COMPLETE ILLUSTRATED WORLD ENCYCLOPEDIA OF

INSECTS

THE COMPLETE ILLUSTRATED WORLD ENCYCLOPEDIA OF
INSECTS

A natural history and identification guide to beetles, flies, bees, wasps, mayflies, dragonflies, cockroaches, mantids, earwigs, ants and many more

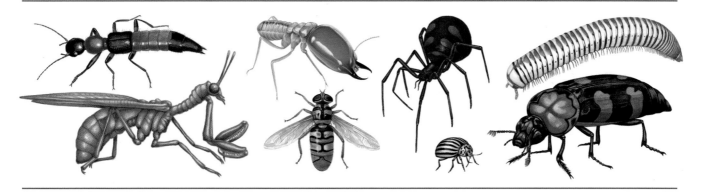

Featuring 650 arthropods, including common insect and spider species, illustrated with 750 specially commissioned illustrations and photographs

MARTIN WALTERS

HERMES
HOUSE

This edition is published by Hermes House, an imprint of Anness Publishing Ltd, Hermes House, 88–89 Blackfriars Road, London SE1 8HA; tel. 020 7401 2077; fax 020 7633 9499

www.hermeshouse.com;
www.annesspublishing.com

Publisher: Joanna Lorenz
Editorial Director: Helen Sudell
Project Editor: Simona Hill
Book and cover design: Nigel Partridge
Illustrators: Andrey Atuchin, Penny Brown, Peter Bull, Stuart Jackson-Carter, Felicity Cole, Joanne Glover, Paul Jones, Jonathan Latimer, Carol Mullin, Fiona Osbaldstone, Denys Ovenden
Production Controller: Pirong Wang

ACKNOWLEDGEMENTS
The author would like to thank the following people for their help with researching information: Yvonne Barnett, Sarah Hart, Virginie Mellot, Jane Parker, Steve Parker, Leanne Scott.

ETHICAL TRADING POLICY
At Anness Publishing we believe that business should be conducted in an ethical and ecologically sustainable way, with respect for the environment and a proper regard to the replacement of the natural resources we employ.
As a publisher, we use a lot of wood pulp in high-quality paper for printing, and that wood commonly comes from spruce trees. We are therefore currently growing more than 750,000 trees in three Scottish forest plantations: Berrymoss (130 hectares/ 320 acres), West Touxhill (125 hectares/ 305 acres) and Deveron Forest (75 hectares/ 185 acres). The forests we manage contain more than 3.5 times the number of trees employed each year in making paper for the books we manufacture.
Because of this ongoing ecological investment programme, you, as our customer, can have the pleasure and reassurance of knowing that a tree is being cultivated on your behalf to naturally replace the materials used to make the book you are holding.
Our forestry programme is run in accordance with the UK Woodland Assurance Scheme (UKWAS) and will be certified by the internationally recognized Forest Stewardship Council (FSC). The FSC is a non-government organization dedicated to promoting responsible management of the world's forests. Certification ensures forests are managed in an environmentally sustainable and socially responsible way. For further information, go to www.annesspublishing.com/trees

© Anness Publishing Ltd 2010

A CIP catalogue record for this book is available from the British Library.

PUBLISHER'S NOTE
Although the advice and information in this book are believed to be accurate and true at the time of going to press, neither the authors nor the publisher can accept any legal responsibility or liability for any errors or omissions that may have been made.

CONTENTS

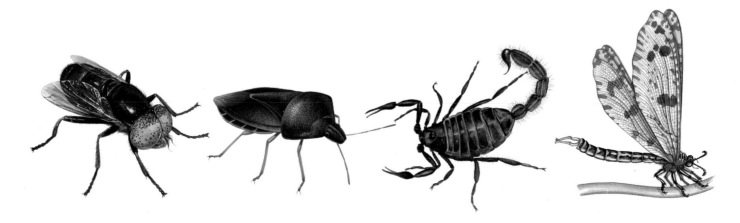

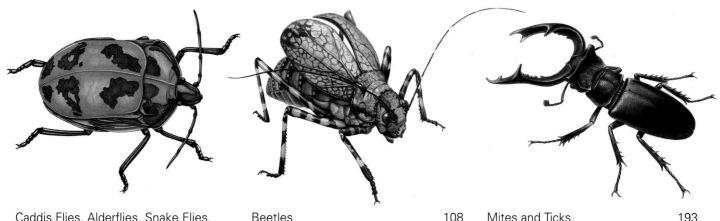

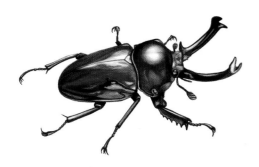

INTRODUCTION

Insects in all their diverse forms are all around us, in our immediate landscapes, our homes and gardens and even in microscopic form on our bodies. They are the most successful group of living creatures on Earth, and have colonized almost every habitat known, with the exception of the seas, where very few are found. From cold mountain terrain to humid riverbanks and every altitude and temperature zone in between, every landscape plays host to a myriad of insect species, each playing a vital role in the ecological balance of the area. Each species has adapted to survive in its landscape as well as alongside other species from the animal kingdom. It is these adaptations that make insects such a fascinating group of creatures to study.

It is easy to dismiss insects as unpleasant creepy-crawlies, with irritating buzzing, and a feeding habit or life cycle that causes damage to crops and even furniture. But insects in all their variety are collectively more beneficial than harmful to our human way of life. In fact, there are plenty of insects that play a vital role in our food chain, as well as others that add to the aesthetic appeal of the great outdoors. For these reasons, it is worth paying closer attention to these small creatures that share our world.

Insects have their small size in common, and while a scorpion may look dissimilar to a butterfly, these insects manage to share some characteristics. The

Above: This beetle has a formidable set of antlers that act as its primary defence against other insects. It uses them to jostle with other beetles.

typical adult insect has three pairs of legs and normally two pairs of wings. The insect's body is usually divided into three distinct sections: head, thorax and abdomen. The insect head has mouthparts (which may be specialized for chewing, biting or sucking), eyes (which may be simple or compound) and antennae. The insect uses antennae to sense vibrations or chemicals in the environment.

Because insects are so visible and common in the landscape, they are easy to recognize. Everyone is familiar with flies, bees, dragonflies, beetles, butterflies, centipedes, grasshoppers and bugs, as well as a whole host of other common creatures, and many are easy to identify. Sit a while in a small area of land such as a flower meadow, a railway embankment, or near a garden pond and before long the presence of insects will become apparent. Being able to identify exact species takes time, and many

Left: This hoverfly shows the typical features of an insect's body: three sets of legs, two sets of wings, and a body divided into three parts, made up of the head, upper body or thorax and lower body or abdomen.

species will not stay around long enough for you to get your identification guide book turned to the correct page before they have moved on. Having a camera to be able to take a photograph helps the process of identification, as does a notepad in which to make any obvious notes of body patterns, markings, approximate size and colouring. Knowing what species to expect in a given area helps too. That way you will know when you have seen something truly remarkable in the area that you are visiting. Some insects, for example, are attracted to specific plants for the pollen that they produce, while others will be sought as a tasty meal by a predator, whose presence in the vicinity may indicate that the species you are looking for is likely to be close by.

It is the sheer diversity of species of insects that has fascinated scientists for centuries. With one million known insect species identified, and many more awaiting identification, it would be impossible to do justice to the sheer variety of insect types in one volume. Instead this book aims to present a selection of the most common, as well as unusual species from around the globe. There should be plenty of insects presented here that you recognize without too much difficulty. Other insects are fascinating because of their unusual behaviour, whether that is the way in which they mimic other insects in order to avoid detection by predators, or are themselves the predator. Some are beautiful to look at, such as butterflies with their delicate scales arranged in attractive colours, while others, such as beetles, are breathtaking because of their sheer size. Included too are some of the common members of the Arthropoda phylum, such as spiders, mites, ticks and millipedes.

Above: Dragonflies' wings are held open when at rest.

Below: Butterflies can reach up to 30cm (12in) across in some tropical countries and appear in a limitless range of colours. They have the most dramatic life cycle of all the insects.

UNDERSTANDING INSECTS

Insects form the most diverse group of animals on Earth. They are extremely successful and, despite their small size, they affect many aspects of our lives. Insects may live on land or in water, although only a small number are known to inhabit the oceans. Insects in all their varied forms belong, along with crustaceans (crabs, lobsters, shrimps, woodlice etc), arachnids (spiders, scorpions etc) and myriapods (millipedes and centipedes), to a group known as arthropods. This group is the largest phylum of the animal kingdom. The word comes from the Greek and means 'jointed-limbed', a feature common to all members of this phylum.

This chapter aims to give an overview of insects, their evolution and development, and the way scientists have classified them into different orders. Insects have a common anatomical make-up and this is described and illustrated, with particular adaptations, such as antennae, highlighted. The life stages of insects are featured and shown too, including the metamorphosis from egg to larva, pupa and adult, where appropriate. There is discussion on how insects organize themselves, living in colonies and working for the collective group, or living solitary lives. Each order has adapted different mechanisms for self-defence, as well as methods of camouflage against predation, and in order to fool its prey, and these are all looked at in detail. The power of flight is another key feature of most insects and has enabled them to live active and mobile lives.

Left: A bumblebee feeds on nectar from a thistle. In return for this sugary meal it carries pollen to another thistle flower.

WHAT ARE INSECTS?

There are five features that are common to all insects, and which may be the key to the success of these amazing animals. These features are a tough outer skeleton, small size, adaptability, the power of flight (in some species) and metamorphosis (changing body shape).

Insects are fascinating creatures and account for more than three-quarters of all known animals. For example there are about 90,000 species of insect in North America, and a similar number in Europe. Some of the most common insects are flies, wasps, beetles, ants, crickets and grasshoppers. They are often persecuted as pests or dismissed as ordinary 'creepy-crawlies'. However, many people love them and there is a great deal left to be learned about them. Many insects exist in enormous numbers, for example ants, gnats and termites. Being small, they are able to colonize areas quickly, and also reproduce at a fast rate.

Insects also play an important role in the economy of nature. They pollinate plants, serve as food for other animals and dispose of dead organisms, as well as doing many other ecologically essential tasks, such as the recycling of organic matter. On the downside, many insects, such as mosquitoes and locusts, transmit diseases and damage crops, but the good services that most insects offer greatly outweigh the harm caused by the few destructive ones.

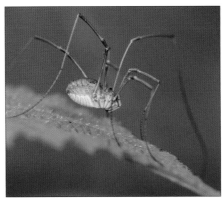

Insects may have gregarious or solitary social lifestyles and may be conspicuous, mimics of other objects, such as leaves or other forms of natural life, or they live concealed and camouflaged. They may also be active by day or night. Due to their life cycle, insects are able to survive under a wide range of conditions, such as extreme heat or cold, and have adapted to survive in all habitats except the ocean. The following five features are characteristic of insects.

The exoskeleton

Insects, like other arthropods, possess a tough outer skeleton made of a special protein called chitin. This layer

Left: The harvestman is an arachnid, rather than an insect. Unlike insects it has four pairs of legs. Spiders and scorpions are also arachnids, a subphylum of arthropods.

protects the body and enables them to survive in extreme environments. The exoskeleton also restricts water loss and because of this it has freed insects from the necessity to live in damp places and allowed them to move into a wide range of habitats on land.

Size

Compared with most other animals, most insects are fairly small, with many having body lengths of just 1–10mm (0.04–0.39in) (although some, such as stick insects, can reach 30cm (12in) in length. Wingspans range between 0.5mm (0.02in) and 30cm (12in). Insects that are unusually long or have a greater wingspan than most still have quite slender bodies. The largest insects are usually found in tropical areas.

The small size of insects can be attributed mainly to their breathing mechanism. Insects use a different breathing system from that of vertebrates and most other groups of invertebrate animals. Insect respiration is accomplished without lungs but instead via a system of internal tubes and sacs through which gases either diffuse or are actively pumped, delivering oxygen to all the parts of the body. Since the circulatory system of an insect is not used to carry oxygen, it is small and simplified. It consists of little more than a single, perforated dorsal tube, which circulates the liquid inside the body through muscular pumping movements.

Small size is a great advantage to insects when it comes to finding places to live. It enables them to colonize tiny places that larger creatures are totally unable to access. Tough, but also light, many insects fly or drift long distances.

Below: Insects have six legs, a body divided into three parts, and often have wings and the ability to fly. This is a hoverfly.

Below: Unlike insects, myriapods have many pairs of legs. They include centipedes and millipedes.

Above: Some insects spend the time between larval and adult stages as a pupa.

Adaptability

There are very few places on Earth that have not been inhabited by insects. They survive in a range of habitats, from mountain-tops and hot deserts to lakes and rivers. Their adaptability on land seems almost unlimited. Some of the most important adaptations in insects are related to their feeding habits. For example, their jaws can deal with a wide range of foods, both solid and liquid. In fact, there are few organic materials that are immune to insect attack. Most plants play host to one or more species.

Below: Woodlice are crustaceans, not insects, and are classified together with crabs and lobsters.

Insects can feed off plants that are poisonous to humans and other animals. Some feed off animals, eating smaller creatures and sucking the blood of larger ones, while others survive by living off unlikely foods such as dung, furniture or clothes.

Flight

Insects are the only group of invertebrates to have developed flight. Being able to fly allows insects to escape from predators more effectively, find mates more easily and move long distances to find their feeding grounds.

Metamorphosis

As they mature, most insects change their body shape, a process known as metamorphosis. The most abundant and successful insects are those which undergo complete metamorphosis. This

Above: Damselflies are among the many winged insects.

is when the larvae are totally different in appearance from the adults. The reason for this success is because while the insects are at differing stages of development, they survive on different kinds of food. For example, a caterpillar will live on a diet of leaves, but as an adult butterfly it sips nectar.

These five factors of toughness, small size, adaptability, flight and metamorphosis, along with others, such as the ability to quickly reproduce, all contribute to the abundance and success of the members of this amazing group.

Below: Moths use their powerful wings to fly from flower to flower collecting nectar and spreading pollen.

EVOLUTION OF INSECTS

Evolution has produced an astonishing variety of insects. Most of the major orders of insects alive today were already distinguishable 250 million years ago. They first appeared in the fossil record 400 million years ago and have continued to evolve and adapt ever since.

The oldest known insect fossil is *Rhyniognatha hirsti*, believed to be about 400 million years old. This fossil was found in red sandstone in Scotland in rocks dated at between 396 and 407 million years old. Its existence suggests that insects probably evolved in the Silurian Period (412–438 million years ago). *R. hirsti* possessed insect-like mandibles rather like those of winged insects today, suggesting that wings may have already evolved by this time.

By the Carboniferous Period (290–355 million years ago), different kinds of insects had evolved. The most ancient winged insects probably included primitive cockroaches, whose fossils date back to the late Devonian Period (370 million years ago). Their early appearance shows the significance of a scavenging lifestyle in terms of adaptability. By about 300 million years ago these insects were diverse. The members of this order varied in size and morphology, most notably in mouthparts, wing articulation and the pattern of veins in the wings.

An early origin of plant feeding is indicated by the beak-like, piercing mouthparts of some Carboniferous insects. It was in the Permian Period (250–290 million years ago) that conifers became abundant in a flora that was previously dominated by

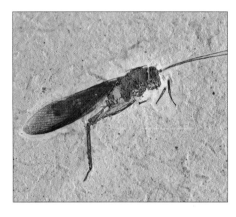

Above: Fossils show that many insect groups that thrive today, such as this cricket, lived millions of years ago.

ferns. A dramatic increase in insect diversity is seen in this period, with 30 orders known to exist by that time. Newly available plants may have had something to do with the evolution of the plant-sucking true bugs (Hemiptera). Other insects in the Permian included those that fed on pollen, another resource that had previously been unavailable.

Giant insects

Some Carboniferous and Permian insects were large, and included giant dragonflies and mayflies. Wingspans of up to 71cm (28in) have been recorded on fossils of these gigantic insects. A possible explanation for this large size is that there may have been higher

levels of oxygen in the atmosphere at that time and that this made for more efficient diffusion in the insects' breathing tubes. The size of modern insects is restricted by the difficulties of getting enough oxygen to their muscles. If other gases were unchanged, the extra atmospheric oxygen would have made the air denser than it is today, making flight for larger creatures easier.

Some early medium to large insects disappeared at the end of the Permian, while others including the mayflies (Ephemeroptera), dragonflies and damselflies (Odonata), stoneflies (Plecoptera), wingless insects (Grylloblattodea), crickets and grass-hoppers (Orthoptera), and cockroaches and mantids (Dictyoptera) survived. The phase between the Permian and Triassic Periods was one of major extinction that dramatically reduced diversity within the insect orders.

Insects of the dinosaur era

The Triassic (205–250 million years ago) is best known for the appearance of dinosaurs and mammals. By the beginning of the period, the major orders of modern insects, apart from the sawflies, wasps, bees, ants, and moths and butterflies, were established. The order Hymenoptera was only weakly represented in this period by sawflies and wood wasps. The appearance of the oldest still-living families occurred in this period, including modern dragonflies and damselflies, true bugs and true flies.

In the Jurassic Period (135–205 million years ago), bees, wasps and ants and many forms of true flies appeared. In the Cretaceous Period (65–135 million years ago), many insects and other arthropods were

Left: Many modern-day insects can be traced back to the era of the dinosaurs.

preserved in exuded tree resin that occasionally trapped insects and then turned into a clear fossilized substance known as amber. In comparison to stone fossils, which may have consisted of little more than a crumpled body or wing, this process preserved whole insects.

The rapid increase and diversification of insects in the Cretaceous Period coincides with the diversification of flowering plant species (angiosperms).

However, the evolution of the major mouthpart types seen in insects today occurred before the appearance of flowering plants.

Modern insects

Some fossils from the Cretaceous Period are so similar to modern insects that they have been able to be classified in existing genera. Much of our knowledge of more modern

insects, from the Tertiary Period (65–1.8 million years ago), comes from amber fossils.

All northern temperate, subarctic and arctic zone fossil insects, many of which are beetles, dating from the last million years, appear to be almost identical to existing species. Climatic variations in the last 1.8 million years have seemingly caused many changes in the ranges of existing species.

Key

1 Collembola (Springtails)
2 Diplura (Diplurans)
3 Protura (Proturans)
4 Archaeognatha (Bristletails) and
 Thysanura (Silverfish)
5 Ephemeroptera (Mayflies)
6 Odonata (Dragonflies and
 Damselflies)
7 Grylloblattodea (Rock Crawlers)
8 Plecoptera (Stoneflies)
9 Zoraptera (Zorapterans)
10 Embioptera (Web Spinners)
11 Dermaptera (Earwigs)
12 Isoptera (Termites)
13 Blattodea (Cockroaches)
14 Mantodea (Mantids)
15 Phasmatodea (Stick and
 Leaf Insects)
16 Orthoptera (Crickets and
 Grasshoppers)
17 Psocoptera (Booklice and Bark-lice)
18 Phthiraptera (Parasitic Lice)
19 Thysanoptera (Thrips)
20 Hemiptera (True Bugs)
21 Neuroptera (Lacewings),
 Megaloptera (Alder-flies and
 Dobsonflies), Raphidioptera
 (Snake Flies)
22 Mecoptera (Scorpion Flies)
23 Trichoptera (Caddis Flies)
24 Lepidoptera (Butterflies and
 Moths)
25 Siphonaptera (Fleas)
26 Diptera (Flies)
27 Strepsiptera (Strepsipterans)
28 Hymenoptera (Bees, Wasps,
 Ants and Sawflies)
29 Coleoptera (Beetles)

Insect evolution

This diagram shows the possible relationships of the major insect orders and the approximate geological times during which they are thought to have arisen. Note, however, that insect classification is complex and the relationships of the groups and their ages are constantly being modified in the light of new research.

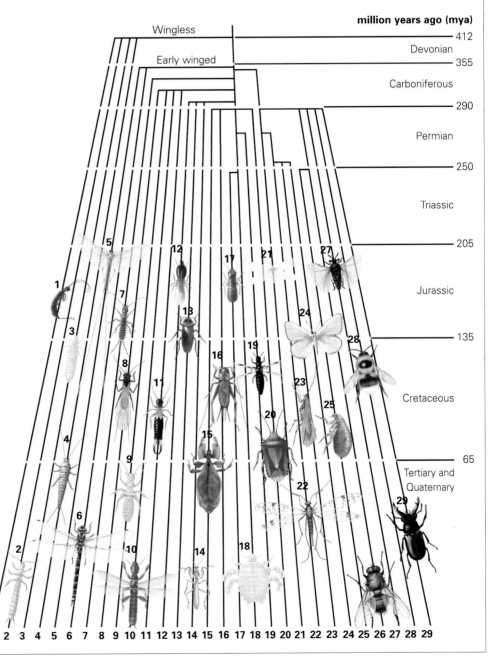

CLASSIFICATION OF INSECTS

The creatures that make up the animal kingdom are arranged into phyla according to their characteristics. Arthropoda is the largest phylum. It can be divided into four classes: Insecta, Crustacea (crabs and relatives), Arachnida (spiders and relatives), and Myriapoda (centipedes and millipedes).

The class Insecta contains almost one million species, split into two subclasses: wingless insects and winged insects. These two groups are further subdivided into different orders.

The orders that make up the class Insecta are primarily the focus of our interest in this book and make up the majority of entries in the regional directory. However, the most common creatures from the other three classes of Arthropoda are also included, such as the wingless near-insects. These creatures share some features with true insects, such as number of legs, and are often thought of as being insects, such as spiders and millipedes.

THE CLASS INSECTA

Bristletails (Order Archaeognatha)
Small, wingless insects with flexible bodies, and thread-like bristles at the tail.

Silverfish (Order Thysanura)
Similar to bristletails but often with a shiny, silvery sheen and flattened, tapering body.

Mayflies (Order Ephemeroptera)
Primitive insects with two pairs of wings, a long body with a number of 'tails' and a brief adult lifespan.

Dragonflies and Damselflies (Order Odonata)
Primitive slim-bodied adults with two pairs of wings, biting mouthparts and large compound eyes for hunting aerial prey.

Gladiators (Order Mantophasmatodea)
Only discovered in 2002, this small group of wingless insects live in southern and eastern Africa.

Cockroaches (Order Blattodea)
Adults (and most nymph stages) have a flattened body with an enlarged pronotum body, compound eyes and strong biting jaws.

Termites (Order Isoptera)
The body consists of a soft cuticle with a harder area on the head. Highly social. Reproductively active adults have wings while workers and soldiers do not.

Mantids (Order Mantodea)
Characteristic triangular shaped, downward-facing head with large eyes. Predatory habits aided by modified forelegs and cryptic coloration.

Rock Crawlers (Order Grylloblattodea)
Another small group of wingless insects, related to stick insects. They live in high mountains in western North America and eastern Asia.

Web Spinners (Order Embioptera)
Brown, soft-bodied insects with biting jaws on a broad head; males have two pairs of wings; females are wingless.

Stick and Leaf Insects (Order Phasmatodea)
Long, slender or flattened insects with biting mouthparts. May or may not have wings. Usually shaped in the form of vegetation (leaf or twigs).

Earwigs (Order Dermaptera)
Long antennae, a slender, flattened body (some winged) and enlarged pincer-shaped appendages (cerci) on the rear end (though absent in some).

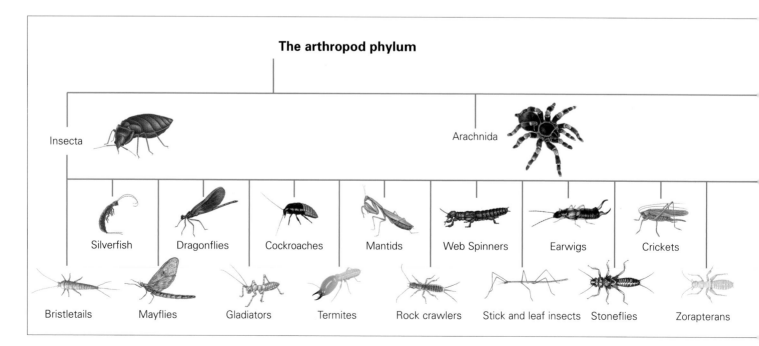

The arthropod phylum

Insecta — Arachnida

Silverfish — Dragonflies — Cockroaches — Mantids — Web Spinners — Earwigs — Crickets

Bristletails — Mayflies — Gladiators — Termites — Rock crawlers — Stick and leaf insects — Stoneflies — Zorapterans

Stoneflies (Order Plecoptera)
Primitive insects with two pairs of wings, a cylindrical body and segmented, thread-like tails.

Crickets and Grasshoppers (Order Orthoptera)
Stout, rigid-bodied insects with large modified hind legs.

Zorapterans (Order Zoraptera)
Tiny insects with soft bodies, long antennae, biting mouthparts, an enlarged prothorax and, in some species, wings.

Booklice (Order Psocoptera)
Small, soft-bodied insects which hold their wings in a tent shape at rest (although some are wingless). Usually 1–2mm (0.04–0.08in) long.

Thrips (Order Thysanoptera)
Tiny insects (most less than 2mm (0.08in) long) with long, thin, soft bodies, piercing and sucking mouthparts, sensitive antennae, and some species also bearing two pairs of hair-fringed wings.

Parasitic Lice (Order Phthiraptera)
Flattened, wingless, often with strong claws, one pair of spiracles (the opening through which oxygen enters the body) on the side surface, short antennae and a range of mouthparts.

Bugs (Order Hemiptera)
Insects with specialized piercing and sucking mouthparts (rostrum and stylets); body shape and size vary.

Snake Flies (Order Raphidioptera)
Long-necked (enlarged prothorax) predatory insects with four wings and bristle-like antennae.

Alder-flies (Order Megaloptera)
Biting mouthparts and four wings which join together.

Lacewings (Order Neuroptera)
A diverse order of insects with sucking mouthparts, long antennae, four clearly veined wings, large compound eyes and an often 'hairy' appearance.

Beetles (Order Coleoptera)
The dorsal area of body has adapted forewings creating hard wing cases (elytra) for the protection of underlying wings. Biting mouthparts.

Strepsipterans (Order Strepsiptera)
Minute insects with biting mouthparts and adult males (when present, as some species are parthenogenetic) with broad membranous hindwings.

Fleas (Order Siphonaptera)
Wingless, with laterally flattened bodies, bloodsucking mouthparts and powerful hind legs.

Scorpion Flies (Order Mecoptera)
Front of head projects down into a pointed beak with biting parts at the end; two pairs of wings; abdomen tip in males often curved upward creating the appearance of a scorpion sting.

Flies (Order Diptera)
Most species usually have two membranous wings for flight and a second reduced pair of 'halteres' for balance. Mouthparts adapted for sucking or piercing; varied appearance with many examples of mimicry.

Caddis Flies (Order Trichoptera)
These insects have two pairs of membranous wings (less veined than lacewings, though superficially similar to some species).

Butterflies (Order Lepidoptera)
Adult mouthparts are usually reduced to a proboscis which coils away neatly when not in use. The wings are covered in a wide variety of distinct overlapping, often very colourful, scales.

Bees, Wasps, Ants and Sawflies (Order Hymenoptera)
These species usually have two pairs of membranous wings (with the front pair being larger) joined by hooks; chewing mouthparts. Ants are mainly wingless, except for reproductives.

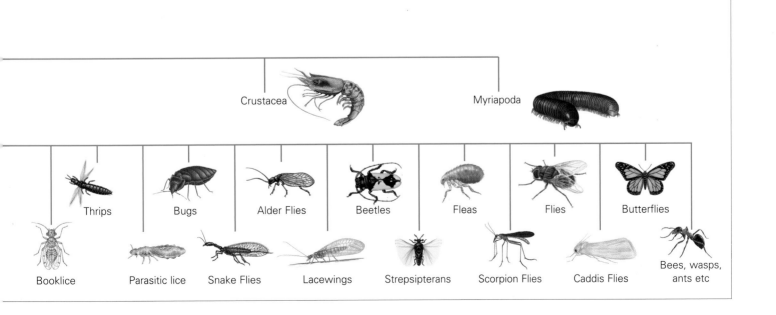

INSECT ANATOMY

The body of an adult insect consists of three main parts – the head, thorax (upper body) and abdomen (lower body). Even though insect species can look vastly different to each other in scale and make up, they all have a common body configuration.

An insect's body is segmented with usually 20 similar rings or segments. There are six segments in the head, three in the thorax and eleven in the abdomen. Some may be fused together. The main part of the body wall is the cuticle, a major component of which is chitin. Chitin is a protein that forms up to 60 per cent of the cuticle's dry weight. A process called tanning brings about hardening of the cuticle. Tough plates called sclerites are formed in most segments as a result. The cuticle remains soft and flexible between the segments, forming joints that enable the body to move.

The head

An insect's head is a tough capsule. It supports a pair of sensory antennae, a pair of compound (complex) eyes, one to three simple eyes and three sets of variously modified appendages that form the mouthparts, usually specialized for sucking or biting.

The antennae are mainly concerned with the senses of smell and touch. They are composed of numerous separate segments. Most insects have compound and simple eyes, but in some species, one set of these may be missing. Compound eyes are larger and more conspicuous than simple eyes, the latter merely distinguishing light from dark.

The thorax

An insect's thorax is the middle section of the body. The three segments that make up the thorax are named, from front to back, the prothorax, mesothorax and metathorax. A pair of legs is present on each segment while the mesothorax and metathorax carry the wings. The forewings are

Below: A cockroach is a typical insect. Some species have wings, although this species lacks them. As with all insects, the body is divided into three parts: the head, the thorax and the abdomen. These, in turn, are subdivided into smaller sections.

normally larger than the hindwings and are found on the mesothorax, which in consequence is normally larger than the metathorax. Almost all adult insects have six legs. The basic structure of the leg comprises the coxa (which articulates with the thorax), the trochanter (small and movable on the coxa but firmly fixed to the femur), the femur (the largest segment), the tibia (which often carries a number of spines), and the tarsus.

The wings

Functional insect wings are membranous, transparent projections that are supported and strengthened by a network of veins. The same basic wing pattern of veins is consistent within families and orders and, as such, is useful in insect classification and identification. The textures of both pairs of wings are generally quite similar; however, in some insects, such as beetles and bugs, they are sometimes different, with the forewings being tougher in order to protect the softer hindwings.

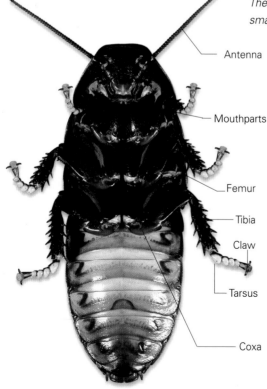

Antenna

Mouthparts

Femur

Tibia

Claw

Tarsus

Coxa

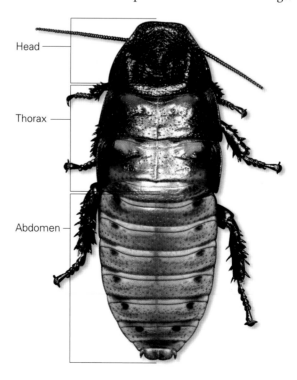

Head

Thorax

Abdomen

Above: The eyes of a damselfly bulge out like beads from the side of its head; giving it almost 360-degree vision. Dragonfly eyes have 30,000 lenses. Most insects have compound eyes made up of many tiny facets.

The abdomen

An insect's abdomen may appear to have fewer segments than are actually present because some may be fused together. The first segment may be reduced or incorporated into the thorax and the eleventh segment may be absent in certain species. The pregenital segments (the first seven abdominal segments) of most adults are similar in arrangement and lack appendages. In many insects, the abdomen is brightly coloured and patterned, often as a warning signal. In others, such as the stick and leaf insects, it has spines or flanges that help protect the insect or conceal it from predators.

The genitalia are formed from modified appendages on segments eight and nine of the abdomen. These appendages are usually small and concealed inside the body, with certain exceptions such as female crickets, which have highly conspicuous ovipositors that they use to lay their eggs deep in soil.

Apart from the ovipositors, the most obvious abdominal appendages are the cerci (paired appendages), which articulate from the last abdominal segment. They are typically long and slender (as in mayflies) but may be modified (for example, the forceps of earwigs). Not all insects possess cerci.

The senses

Insects owe much of their adaptability to their acute senses, especially their ability to sense vibrations, scents and light, including ultraviolet. Many insects have large eyes, providing all-round vision and colour discrimination, used notably by those species that feed from flowers. Some can even detect the plane of polarization of light. In most insects, there is a trade-off between visual acuity and chemical or tactile acuity: those insects with well-developed eyes usually have reduced or simple antennae, and vice-versa. Touch sensors are concentrated on the antennae and legs, but also occur elsewhere on the body. Many insects are very sensitive to vibrations.

In many insects, the sense of taste is well developed and in some, for example the housefly, the taste receptors are found not in the mouth region but on the feet.

There are a variety of different mechanisms by which insects can pick up sound. Crickets' ears occur on their legs and some insects have membranous organs similar to our own eardrums while others have hairs or bristles to detect sound. The range of frequencies insects can hear is often quite narrow. Many insects rely on sound for effective communication.

Insects have also been shown to be able to detect and respond to gravity and the Earth's magnetic field. In summary, insects are mostly highly responsive and often react quickly.

Different antennae

Insects use their antennae as feelers and scent detectors, learning about their surroundings through touch and smell. Those which rely on vision, such as bees, have relatively simple antennae. Those of some beetles are long and used to investigate the area immediately in front of the insect's head. Insects for which smell is important for finding food or mates often have quite complex antennae, adapted to increase the surface area for sensory cells.

Above: Male moths use their feathery antennae to detect the pheromones produced by potential mates.

Below: Some beetles have long, branched antennae.

Above: The antennae of a cockchafer end in leaf-like extensions. Males have seven leaves and females have six.

Below: Insect antennae are composed of many segments as shown below.

INSECT LIFE CYCLES

The vast majority of insects lay eggs: only a few give birth to active young. A tough shell and one or more internal membranes protect these eggs, enabling them to survive a wide range of conditions. A period of dormancy often interrupts the progression from egg to newly hatched larva or nymph.

The wingless insects and some secondarily wingless species are similar in form to the adult when they hatch, but they lack reproductive organs. As the young insect grows up, apart from an increase in size, there is little or no change in its physical appearance.

Winged insects have a more complex development. When the young insect hatches it is often visibly quite different from the adult and must undergo a series of changes before reaching the adult state. This series of changes is known as metamorphosis. Insects possess a tough, non-living external skeleton, which must be shed periodically to enable them to grow in stages. This shedding process is called moulting or ecdysis. Most insects moult between four and ten times. The stages of life between moults are called instars. Newly moulted insects are soft and vulnerable at first.

Partial or incomplete metamorphosis

Winged insects fall into two groups according to the way in which the wings and bodies of the young develop. In the first group, which includes cockroaches, grasshoppers and bugs, the wings of the insect develop gradually on the outside of the body and get larger at each moult until they are fully formed. Nymphs, as the larval insects are usually called at these young stages, often resemble the adult in appearance and behaviour, inhabiting the same places and eating the same types of food. This group, called the Exopterygota in reference to the external development of the wings, are said to undergo a partial (or incomplete) metamorphosis since there are no dramatic changes in body form during development. The young insect gradually changes as it grows, eventually attaining full adult appearance. Some, however, such as lice, are wingless.

Complete metamorphosis

In the second group of winged insects, which includes butterflies, moths, beetles and flies, the young are very different from the adult. Larvae, as the insects of this group are known in the young stages, often occupy a completely different niche and exist on completely different diets from those of the adult. Larvae are generally worm-like in appearance and can be divided into five different forms; eruciform (caterpillar-like), scarabaeiform (grub-like), campodeiform (elongated, flattened and active), elateriform (wireworm-like) and vermiform (maggot-like). The young of this group must undergo one very dramatic change to reach the adult form as opposed to a series of small changes. Because of this, a resting stage is needed, in which the young larva does not feed. This resting stage is the pupa or chrysalis.

The resting phase

There are three types of pupae; obtect (the pupa is compact, with the legs and other appendages enclosed), exarate (the pupa has the legs and other appendages free and extended) and coarctate (the pupa develops inside the larval skin).

Emergence from the pupa

Although most insects need some time for their crumpled wings to open out, certain species of fly such as the mayfly and caddis fly, are able to fly as soon as they emerge from their pupal skins. Those that must wait usually try to find a place that can hold them up, enabling their wings to open out freely. Blood pumped through the veins helps the wings to open out. An insect's

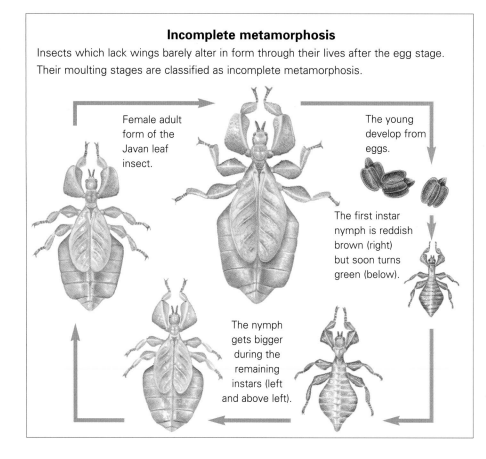

Incomplete metamorphosis

Insects which lack wings barely alter in form through their lives after the egg stage. Their moulting stages are classified as incomplete metamorphosis.

Female adult form of the Javan leaf insect.

The young develop from eggs.

The first instar nymph is reddish brown (right) but soon turns green (below).

The nymph gets bigger during the remaining instars (left and above left).

wings usually need a couple of hours to harden before it is ready to fly. In order for their wings to be ready for flight at a certain time of day, some insect species time their emergence from the pupa very efficiently. Most insects will usually stop growing once the wings are fully developed or when sexual maturity has been reached. A fully developed adult is known as an imago.

Adult life stage

The adult stage has a reproductive role and in insects with relatively inactive larvae, it is often the stage during which they disperse. The adult may be able to reproduce as soon as it has emerged from the pupa or there may be a maturation period before sperm transfer or oviposition can occur. The number of reproductive cycles an adult insect has is dependent upon the species and food availability. It can range between one and many. The adults of certain species are very short-lived, for example, mayflies, midges and male scale insects. They fly for only a few hours, or at the most a day or two, before dying. They need to mate within this time. Most adult insects live for a few weeks or months and some are particularly long-lived – queen bees and ants, and queen termites, for instance. Many large, winged insects, such as butterflies, may fly long distances, even migrating to regions with a more favourable climate before reproducing. Many also hibernate to survive the winter, often in the pupal stage, emerging when food is available the next season.

Above: A dragonfly perches next to its empty larval skin. Dragonfly larvae live in water and must climb out before they can break free from their skin and emerge as adults. These ghost-like empty cases may be found clinging to waterside stems.

Complete metamorphosis

Butterflies undergo complete metamorphosis. Their bodies and habits change dramatically from the many-legged caterpillar borne of the egg, to the winged adult form, via a 'resting' pupa (chrysalis) stage, which may last for weeks.

1 *A butterfly egg waits to hatch. Butterflies lay on the stems or leaves of plants on which their caterpillars feed.*

2 *The newly hatched caterpillar is small but completely independent. Its job now is to feed and grow.*

3 *As the caterpillar grows it changes colour. The bright yellow indicates that it is unpleasant to eat.*

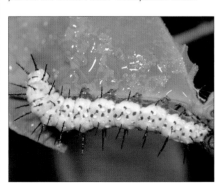

4 *After several weeks of feeding the caterpillar has grown to hundreds of times its original size. It is now ready for metamorphosis.*

5 *Following a week or more as a pupa the adult butterfly emerges from its chrysalis (case). Next it will unfurl its wings and pump them up with blood.*

6 *After extending and drying its fully formed wings for several hours, this butterfly can take to the air and search for a mate and food.*

INSECT FLIGHT

The insects are the only group of invertebrates to have developed flight. This development has allowed insects greater mobility, which helps enormously when searching for food and mates. This ability has also allowed insects to exploit many new environments and habitats.

Insects began flying 350 million years ago. The earliest flying insects had two sets of wings, and were unable to fold them over their abdomen. Most insects today have either one pair of wings or two pairs functioning as a single pair. Natural selection has played a huge role in refining the wings, and the control and sensory systems, as well as everything else that affects aerodynamics and movement through the air. One significant trait is wing twist. Most insect wings are twisted, between 10 and 20 degrees, giving them a higher angle relative to the ground at the base. As well as this, the wing membrane is distorted and angled between the veins in such a way that the cross-section of the wing acts as an airfoil. The wing's basic shape is therefore very efficient at producing lift.

Below: Dragonflies have two pairs of long, rather rigid, membranous wings. Unlike their smaller relatives, the damselflies, they hold their wings open when at rest (damselflies close their wings over their long abdomens).

Most insects use tiny muscles in the thorax to adjust the tilt, stiffness and flapping frequency of the wings, thus keeping them in control. The capabilities of flying insects vary but some are able to hover and even fly backward.

Take-off

In order to fly, an insect must overcome the forces of gravity and air resistance to movement. In gliding flight, in which the wings are held rigidly outstretched, the use of passive air movements helps overcome these forces. By adjusting the angle of the leading edge of the wing when orientated into the wind, an insect achieves lift. As this angle (the angle of attack) increases, lift also increases. The angle of attack of insects can be raised to more than 30 degrees, giving them great manoeuvrability. There are two considerably different insect flight mechanisms, and each has its own advantages and disadvantages. Most insects glide; some a little, for example the flies (Diptera), and some a lot, for instance the dragonflies (Odonata).

Above: Grasshoppers and locusts only have fully formed wings as adults. When immature nymphs, as here, their wings are visible but only partially formed and useless for flight. Their enlarged legs provide the power for leaping instead.

Getting airborne: method one

It is by beating their wings that most winged insects fly. A single wing beat comprises three interlinked movements. The first is a sequence of downward, forward movements followed by upward and backward movements using the whole wing, fully extended. Second, during the sequence each wing is rotated around its base for extra manoeuvrability. Third, in response to local variations in air pressure, various parts of the wing flex. In true flight, the movement of the wings produces the relative wind whereas in gliding it comes from passive air movement.

Wing muscles

The different flight mechanisms used by flying insects are brought about by two kinds of arrangements of muscles powering their flight. These are: direct flight muscles connected to the wings, and indirect muscle action, whereby internal body muscles cause wing movements by changing the shape of the thorax. The wings have rubber-like hinges that add extra elasticity.

Above: Butterflies and moths have wings covered with numerous tiny scales – it is these which give them their colour and pattern. The wings of butterflies and moths are the largest of any insects relative to body size.

Getting airborne: method two

Unlike those of most other insects, the wing muscles of insects belonging to the orders Odonata (the dragonflies and damselflies) and Blattodea (the cockroaches) insert directly at the wing bases, which are hinged so that a small movement of the wing base downward lifts the wing itself upward. In mayflies, the hind wings are reduced, sometimes absent, and play little role in their flight. The primitive flight mechanism of the Odonata and Blattodea does not necessarily mean they are poorer fliers. More advanced insects use indirect muscles for flight. However, they do also retain direct muscles which are used for making fine adjustments to wing orientation during flight. When the muscles attached to the wing base inside the pivotal point contract, an upward stroke is produced. When the muscles that extend from the sternum (the upper half of the exoskeleton) to the wing base outside the pivot point contract, a downward stroke is produced. These are direct flight muscle movements. The indirect flight muscles are attached to the tergum (the lower half of the exoskeleton) and the sternum. The tergum connects to the base of the wing. When the vertical muscles inside the thorax contract, the tergum is pulled down, thus raising the wings in the upstroke, and stretching the horizontal muscles. When the horizontal muscles contract, the wings are lowered in the down-stroke, stretching the vertical muscles.

Synchronized wings

Although the beating of the wings may be controlled independently, it is more common for them to be harmonized, as in butterflies, bugs and bees, for example. This is accomplished by locking the fore- and hindwings together and also through neural control, synchronizing the nerve patterns going to each from the brain. Insects that have slower wing-beat frequencies (for example, dragonflies) possess synchronous muscles, which help to maintain one nerve impulse for each beat. Insects with faster beating wings (such as wasps, flies and beetles) have asynchronous muscles whereby a single nerve impulse causes a muscle fibre to contract multiple times allowing the frequency of wing-beats to exceed the rate at which the nervous system can send impulses. Some small flies can flap their wings at more than 1,000 times per second.

As well as powered flight, many of the world's smaller insects are dispersed on the wind. For example, aphids are often transported long distances by low-level jet-stream winds and many migrant species make use of prevailing winds and air currents.

Below: Hoverflies are master fliers. As their name suggests, they can stay in one spot in midair. They can also fly backward. It is the venation in the wings that allows the species to be told apart from bees and wasps, all of which are common visitors to garden flowers.

INSECT SOCIAL ORGANIZATION

The vast majority of insect species are solitary. Any interaction within these species is usually confined to mating or competition. However, some insects spend their whole lives in huge colonies, while others gather together in large numbers at certain times of the year.

Bees, some wasps, ants and termites are all social insects, living in caste systems. Their social organization is known as eusociality. Other insects such as beetles, cockroaches, bugs, butterflies, hymenopterans and thysanopterans have less developed social habits and group together for specific reasons. Their social interaction is known as subsocial.

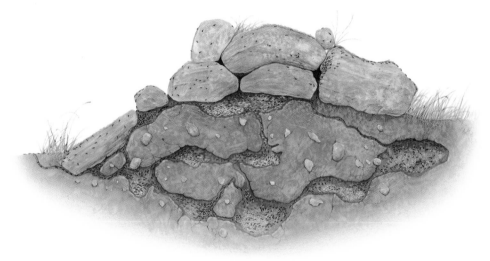

Eusocial insects

In a caste system there is a clearly defined division of labour among members. The members co-operate in order to ensure the life of the colony. Many sterile members of a colony care for a few reproductive members. The sterile members carry out specialized tasks, such as feeding reproductive members and caring for the young. Workers maintain the hive and defend it from attack. An overlap of generations assists the functioning of the colony.

Caste systems

The caste system is made up of one or several reproductive individuals called queens who are aided by the numerous non-reproductive workers. In termite

Below: Many ant species build their nests underground, excavating numerous chambers. An entrance is shown here.

and ant colonies there is often an additional defensive group of individuals called soldiers. In some species the queens and soldiers are unable to feed themselves and it is the job of the workers to bring them food.

In eusocial hymenopterans the queens control the sex of their offspring by fertilizing the eggs with stored sperm. These eggs then develop into diploid females. These sisters are more related to each other than queens are to their offspring, sharing 75 per cent of their genes on average. Male

Below: Unlike most ants, termites are exclusively herbivorous insects, eating only plant matter. They are particularly common in grassland habitats, where they tend to live in mounds. This is a queen termite.

Above: Areas beneath surface stones are often chosen for ants' nests as they soak up the sun's warmth by day and radiate it out at night.

offspring are produced when eggs are unfertilized. The production of males is infrequent and they die after mating.

The queen of a colony is generally larger than any worker and has an extended abdomen. In her first brood, she will only produce workers. Once these have hatched she will cease to forage and devote herself entirely to reproduction. The workers carry out tasks such as foraging and food distribution, cleaning, ventilating and guarding the nest.

In an ant colony there are two major female castes, the queen and the workers, which usually look completely different. Some ant workers are divided into subcastes known as minor, media or major, according to their size. In contrast to the female-only castes of the eusocial hymenoptera, termite castes include both male and female representatives. All termite colonies have a king as well as a queen who act as the primary reproductives. The workers and soldiers in ant colonies also consist of males and females. The colonies of some termites are huge, with up to a million individuals.

Subsocial insects

Insects with subsocial habits are more common than those with eusocial habits. An example of a subsocial interaction is the non-reproductive gathering of monarch butterflies at overwintering sites in Mexico and California. Many tropical butterflies gather together to roost in safety, particularly those species that have warning coloration on their bodies, or odours and an unpleasant taste. A solitary insect of this kind is more at risk of being predated upon than a single member of a conspicuous group.

Parental roles

Among subsocial insects, parents do take care of their young. For those insects that do not nest, it is predominantly the adult female that attends to the eggs and early instars, although paternal guarding is known in some tropical assassin bugs and giant water bugs. The parent protects the eggs against parasitization and predation, keeping eggs free from fungi and maintaining the correct conditions for egg development.

Nesting

Only five social insect orders build nests in which to reproduce, four of which are subsocial and one that is eusocial. Insects such as earwigs and mole crickets, which are solitary insects, exhibit similar parental care. After overwintering together in the same nest, the female of the species ejects the male in the spring as she begins to tend to her eggs.

Below: Queen bees are larger than the male bees, or drones, which, in turn, are larger than the workers.

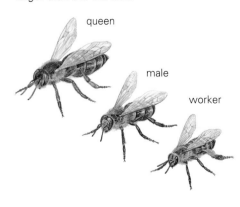

queen

male

worker

Above: A termite mound is a complex structure built from soil particles held together with sticky saliva. The large central chamber acts like a chimney, so that air is drawn in at the bottom, ventilating the mound.

Among dung and carrion beetles, there is competition for nutrient-rich dung in which to lay eggs. The female is mainly responsible for burrowing and preparing the dung, which is sometimes rolled away from its source or else coated in clay. Some species of beetle take no further interest in the eggs once they have been laid.

Subsocial parasitoid wasps, which attack and immobilize arthropod prey upon which their young feed, exhibit numerous prey-handling and nesting strategies that can be quite complex. Some use the prey's own burrow for nesting while others either build a simple burrow after prey is captured or make their nest burrow before they capture their prey.

Parental care with communal nesting most commonly occurs among subsocial bees and wasps. The females of these species remain in the nest after laying and often until the next generation have emerged as adults. They maintain nest hygiene by removing faeces and guard the nest against specialized nest parasites.

Above: Tropical butterflies with distasteful flesh often roost together. Although most carry warning colours for potential predators, these are enhanced and the effect amplified by the insects' occurrence in groups.

Some aphids have a soldier caste in which some first- or second-instar nymphs never develop into adults. These nymphs exhibit aggressive behaviour and are physically modified so that they are larger than non-soldier nymphs. Using either their frontal horns or their mouthparts as piercing weapons, they attack intruders and protect the aphid colony. Potter wasps make neat cup-shaped cells of mud, each containing a paralysed prey and an egg laid by the wasp.

Below: Termite queens are so large that they are unable to move much. They rely entirely on the other members of the colony for survival.

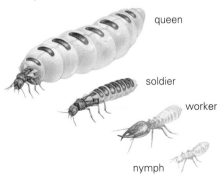

queen

soldier

worker

nymph

INSECT DEFENCES

Insects use a range of defence mechanisms, and for a variety of reasons. They may be used to ward off competition during mating and feeding, or to avoid or deal with predators. Mimicry, chemicals, weapons and camouflage comprise some of an insect's arsenal, often coupled with specialized behaviour.

An insect's in-built weaponry includes stings, biting mouthparts and adapted claws for nipping. An insect's sting evolved with the specific purpose of delivering venom. In the case of bees, using the sting leads to the insect's death, as crucial parts of organs are removed as the sting is torn away. Bees sting to defend their colonies from attack by large predators.

Some insects, such as those with adapted claws or mandibles, indulge in ritualized fighting, which avoids real harm to either party. The impressive mandibles of the male stag beetle *Lucanus cervus* are used as an indication of strength and to wrestle opponents (unlike the nipping mouthparts of the smaller female of the species).

Speed as defence

The ability to react swiftly and move away quickly is a good form of defence. Ground beetles can run fast, while winged insects such as flies and dragonflies can fly away from danger and are also equipped with excellent vision. Other forms of locomotion for escape include physical propulsion, as in the click beetles, which, due to a mobile joint in the thorax, can flick themselves high into the air. Click beetles make a loud noise when

Above: A tough, spiky carapace offers protection from some predators, although it is not always effective against larger creatures, such as birds.

Above: Powerful mandibles can give a nasty nip. Perhaps surprisingly, short mandibles like these bite with more force than longer ones.

flipping themselves, which may also be used as an alarm to distract or confuse enemies. Grasshoppers and crickets have excellent strong jumping legs, as do fleas, which can jump great distances relative to their size.

Swift movements combined with a flash of bright coloration can also confuse a predator. Some tropical locusts have brightly coloured wing patches which they flash at their predators as they flee from them. If the bright colour is only viewed briefly and the insect swiftly drops into grass, the predator will try to follow the

insect in the direction of the flash, which tends to guide it away from the potential prey hidden safely below. Conversely, staying still can also be an effective defence. Fierce, predatory mantids will tend to overlook a motionless insect, not recognizing it as prey.

Warning colours

Some insects use the toxic qualities of their food to generate similar compounds within themselves and thus become unpalatable to predators. They advertise the fact that they taste bad (or are toxic) by using bright warning colours such as red, black and yellow on their body parts. The black and yellow striped cinnabar moth caterpillar which feeds on ragwort, a plant poisonous to many other animals, is just one example of an insect which does this. Insects generating their own chemical protection often group together to

Left: Although it looks dangerous, the male scorpion fly's 'sting' is actually fake. It is thought that the fly mimics a scorpion's sting for protection from predators.

increase the effect of this defence should they come under attack. Parasites of such insects may also utilize these chemicals further for their own defences.

Mimicry

Many non-poisonous and harmless insect species mimic the warning coloration and patterns of more dangerous ones to take advantage of the effects of the predators' learned response to those creatures without having to eat the food plant or synthesize chemicals themselves. To be convincing, these mimic insects also adopt the postures of the species they are pretending to be. Thus the European bee beetle (*Trichius fasciatus*) mimics bumblebees (such as *Bombus lucorum*) by appearing to feed on flower-heads in the same way as a bumblebee (plunging its whole body into the pollen of composite flowers). It also has a large yellow-and-black body with hairs on and around its elytra. Its general bulk and appearance is similar to a bee when glimpsed in flight.

However, it is not as convincing as some of the clearwing moths, from the family Sesiidae, for instance, which mimic the appearance of paper wasps. They are active in the daytime and move in the same way as the wasps. Another mimic is the European bee fly *Bombylius major,* which looks rather like a fluffy bee with its hairy round abdomen.

The male scorpion fly, on the other hand, superficially resembles a scorpion. It has a bulbous projection on the end of its abdomen, which it curls up like a scorpion's sting to ward off predators.

Chemical warfare

Many other insects use chemicals as defence. These are often produced within the insect iself, sometimes by eating particular food plants. Chemicals are synthesized by the insects internally to squirt or inflict on enemies by a variety of methods. Particularly noxious chemicals are thought to be made swiftly on demand

Above: Bright colours and bold patterns are used by many insects to warn predators that they are poisonous or taste bad.

to reduce any toxic effects on the insect's own body. The bombardier beetle (*Brachinus crepitans*) dramatically produces a blister-forming volatile liquid, which is squirted from its abdomen. This quickly turns to gas with a dramatic pop to ward off unwanted attention. Many ants squirt formic acid, and worker termites are capable of squirting noxious secretions from specialized head glands to ward off predators. Stink bugs gained their common name from their ability to release a foul-smelling fluid from special glands when alarmed. Sometimes defence mechanisms combine and mimics may even go as far as producing identical chemical defences to those of the insect they mimic (as in the case of ant mimics).

Alarm pheromones are chemicals synthesized within an insect and which are used to communicate danger to other insects of the same species. They act by triggering an immediate response from the recipient of the chemical message. Such chemicals are highly specific and even very similar molecules create a different reaction. Alarm pheromones take effect quickly then soon fade away after their task

has been accomplished. Interestingly this method is also used by predatory insects, which in effect raise a false alarm. By releasing massive quantities of simulated alarm pheromone an invading species is able to attack a colony under the cover of the almost certain chaos the false alarm will create.

Social defence mechanisms

Termites and ants have developed co-operative social responses to defending vulnerable offspring, using armoured soldiers that repel invaders of their colonies. Hard chitinous processes have evolved to provide weapons on the specially adapted heads of soldiers and, among ants, strong biting jaws also dissuade hungry visitors. In soldier termites the jaws are purely for defence. Their fellow workers feed them as they do not search for food themselves.

Certain species of honey bee have been shown to co-operate in defending their hives by attacking enemies as a group. Where their stings are unable to penetrate the hard surface of a hornet, for example, attacking their nest, the honey bees have been observed surrounding the enemy until it is completely smothered by many individuals. In some cases, the intruder is killed through suffocation or even by the heat generated by the mass of attacking bees.

Below: Camouflage is used by some insects to hide from both predators and prey. This praying mantis has a wide, flat carapace which helps disguise it as a leaf.

INSECT CAMOUFLAGE

Most insects are vulnerable to attack from a wide range of predators. Camouflage is one of their favoured methods of escaping detection and many insects closely resemble their backgrounds, some having strange spines or projections to enhance this effect.

Many insects display cryptic coloration, whereby their body colour is used to help them blend into the background and effectively disappear from view. Cryptic coloration indicates that an insect is well adapted to the habitat or niche in which it lives. It allows an insect to become part of its immediate background – a plant or other surface. It is an approach used both by predators who lie in wait, and by potential prey species trying to avoid being eaten by making themselves less conspicuous to the creatures which hunt them.

Many moths rest by day against the bark of trees and can then be extremely hard to spot. Some have cryptic patterns on their upper wings. Butterflies that are brightly coloured on the upperside frequently have well-camouflaged underwings. Red admirals and peacock butterflies, for example, can be very hard to see when at rest or when hibernating with their wings closed, as their hindwings are brown, and speckled and patterned like bark.

Below: Many crickets hunt from leaves. Most mimic leaf colour and texture well, being green and brown.

Above: Most moths rest by day and are beautifully camouflaged. The true number of moths around us is often under-estimated, as they are so difficult to see.

Chameleon insects

Some species have the ability to change colour with seasonal changes in their environment. Predatory African mantid species, for example, can be very green in the wet season in the presence of lush vegetation but change to a brown coloration during the dry season when most of the vegetation has dried out and died back. In Australia, a similar mechanism, known as fire melanism, occurs whereby Australian mantid species are able to camouflage themselves in the black landscape that appears after a bush fire.

Melanism also occurs in the cryptic camouflage of the peppered moth (*Biston betularia*), which closely resembles the pattern and colours of the tree bark on which it rests. As far back as the 19th century it was noted that moths resting on blackened surfaces resulting from industrial pollution tended to be darker than those found in less polluted areas, where the tree trunks had a lighter covering caused by active growth of lichens. In this case, predation by birds removed more of the less well-camouflaged moths in each area.

Body shape

Insect exoskeletons vary in shape and provide a disguise with another dimension, as in the case of leaf and stick insects. They enable insects to take on not only the colour of their surroundings but also the shape of objects in it. Many mantids also have bodies resembling parts of plants, with very convincing leaf-like or petal-like projections or textures. Stick and leaf

Below: When at rest, orthopterans often pass unnoticed. They are recognizable by their large hind legs, adapted for jumping.

Left: A slender mantis hides very well on the delicate stem of a flower.

Above: This mantis blends into its background seamlessly.

insects provide very convincing imitations of twigs and leaves, even swaying to resemble foliage in the breeze. They also assume a pose with their front legs facing forward in line with the twig they are emulating. Mantids, too, often sway gently from side to side in order to match the movements of the leaves or flowers on which they sit.

Imitation

Insect eggs often closely resemble plant seeds and are dropped to ground level or glued to plant leaves. Those at ground level may be carried away by ants mistaking them for seeds. Young stick insects hatching in ant nests suffer less predation than those left in the open, so this helps their survival. The small dark young quickly travel upward once hatched to the relative safety of the underside of a leaf until, after a few moults, they too take on the colour and camouflage of the host plant. Stick insects, as well as camouflaging themselves to resemble plants, try to make their overall presence less obvious by propelling their droppings beyond food plants with some vigour. This avoids obvious signs of their presence.

Patterns and textures of mimics, though very convincing, will often also involve behaviour that enhances their effect, such as the closed-wing posture of the Indian leaf butterfly (*Kallima*

inachus) whose wing undersides are shaped and patterned to appear similar to the fallen leaves it perches among. When this butterfly opens its wings it becomes much more obvious, due to the bright colours of its upper wings. Some caterpillars appear very like the twigs of their food plants and adopt an angled posture on a twig, often causing them to be overlooked.

The colours and patterning of cryptic insects and plant or habitat mimics are designed to help them blend in with the backgrounds of the places where they live. However, these insects may have hidden areas of bright colour, which they flash to startle predators when they find themselves attacked. An insect such as a stick insect or cricket may suddenly take flight to avoid danger and as it does so it will display normally hidden wing patches or bright areas on legs. This serves to shock the predator, stopping it in its tracks and giving the insect time to make an attempt at escape. It may also confuse the predator into losing the trail of its prey. Flash coloration can also take the form of false eye markings, as on the normally hidden wings of the

otherwise cryptic peanut-headed bug (*Fulgora laternaria*). These markings are at the opposite end to its actual head, which also aids its survival, as predators often strike at the head when attacking.

It is possible that there are forms of camouflage in nature that human eyes do not register, since we see a different range of wavelengths from insects. In ultraviolet light, for example, other patterns may be revealed.

Below: This scorpion has coloration that blends well with the gravel background on which it is resting.

Below: This leaf insect gives the appearance of a dead leaf. Its colouring helps to camouflage it from predators.

INSECT FEEDING METHODS

Insects generally have two methods of feeding. They either bite off and chew their food or they drink it in liquid form. The types of food consumed by different insect groups are many and varied. They include sap, leaves, vertebrate blood, dry wood, bacteria and algae, and the internal tissues of other insects.

Insects have various different mouthparts, which correlate with the sort of diet they have. For example, termites have tough biting jaws with which they feed on wood, and butterflies have a long proboscis with which they suck nectar from flowers. The gut structure and function of each insect family varies too depending on the nutrient composition of the food they eat. Insects that eat solid food have a broad, straight, short gut, while liquid-feeding insects have a longer, narrower gut that allows more contact with the ingested liquid.

Feeding categories

Insects generally specialize in one of four different feeding categories, depending on whether the food they eat is solid or liquid, or from a plant or an animal source. However, those with a more generalized diet will fall between two or more of these categories, for example bees feed both on pollen (a plant solid) and nectar (a plant liquid), and most endopterygotes (insects with larval stages) will change from one category to another depending on their life stage (for

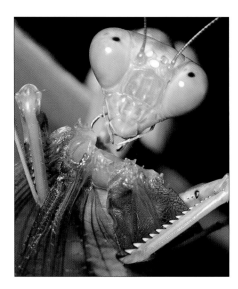

Above: A praying mantis feeds on prey. Its mouthparts are small but sharp, for slicing flesh.

instance moths and butterflies switch from a diet of leaves as larvae to a diet of nectar as adults).

Mandibulates

Insects that bite and chew their food are known as mandibulates. Examples of these include cockroaches, crickets and earwigs. The mouthparts of these insects are the more basic design and are composed of five main parts. The mandibles (or jaws), which can be extremely hard, help to cut and grind food. They do this with tooth-like

Above: Bumblebees have both a long 'tongue' for collecting nectar and mandibles for crushing pollen.

ridges for cutting, and crushing surfaces for grinding. The maxillae lie behind the mandibles and assist them in processing food. The maxillary palp is found behind the maxillae and bears sensory hairs. The upper lip forms the roof of the preoral cavity and mouth, while the lower lip forms the floor of the preoral cavity. The labium has one pair of palps also bearing mechanoreceptors. The final mouthpart is a tongue-like structure which divides the preoral cavity into a dorsal food pouch and a ventral salivarium from which saliva enters the mouth.

Haustellates

Evolution has led to the development of an array of different mouthpart types. There has been such diversification in different orders that in addition to the basic chewing method, lapping, sucking, biting, piercing and filter-feeding are all also employed by various species. Bees use

Below: Moths and butterflies suck liquid food from plants by means of a proboscis which is coiled out of the way until needed.

Left: Blister beetles are voracious feeders, contaminating animal feed crops with toxins at the same time.

their mouthparts for both chewing and lapping. Lapping occurs when an insect uses a protrusible organ to acquire a liquid or semi-liquid food, which is then transferred to the mouth. The honey bee, *Apis mellifera*, does this by dipping its hair-covered tongue into nectar or honey, which sticks to the hairs and is then retracted and carried up the bee's food canal.

Insects that obtain their food solely by sucking up liquids are called haustellates. Most moths and butterflies, and many adult flies fall into this category. The sucking mouthparts of these insects are elongated. Muscles help pump the liquid food. The proboscis of moths and butterflies is loosely coiled until the insect wishes to feed, whereupon it is extended. The upper surface of the proboscis is more or less flat when coiled, but contraction of the muscles causes the upper surface to become slightly domed, which automatically causes the proboscis to uncoil.

Combination feeders

A combination of sucking with either piercing or biting is used by a few moths and many flies. For example, bloodsucking flies such as mosquitoes and horseflies possess sharp, piercing mandibles and maxillae. Lobes carried on the tip of the labium or proboscis, are used to suck up the liquid food after piercing. In contrast, tsetse flies

Below: Leaf cutter ants return to their colony. These ants use leaf sections to grow fungi, on which they feed.

Differing mouthparts

As a group, insects have incredibly varied mouthparts, although in most these are formed from the same basic components: the mandibles, the maxillary palps, the labrum and the labium. Most insects either lap up food, crush or chew it, or suck it up through a tube. In the latter, the tube like mouthparts can either be stiff for piercing or flexible for probing.

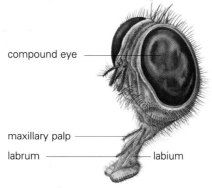

compound eye —
maxillary palp —
labrum — labium

Above: A fly's mouthparts are adapted for lapping up tiny particles or liquid food.

Below: Predatory beetles have long spiked mandibles for grabbing prey.

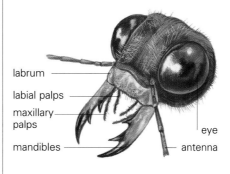

labrum —
labial palps —
maxillary palps —
mandibles —
eye
antenna

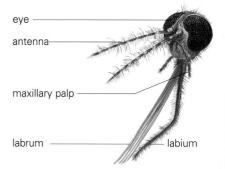

eye —
antenna —
maxillary palp —
labrum — labium

Above: The female mosquito feeds on the blood of mammals and other large animals.

Below: The tube like mouthparts of butterflies are adapted for sucking up nectar.

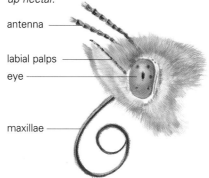

antenna —
labial palps —
eye —
maxillae —

and stable flies use their highly modified labium for piercing.

True bugs, thrips, fleas and sucking lice have modified mouthparts for sucking and piercing. The labium of nymphal damselflies and dragonflies is

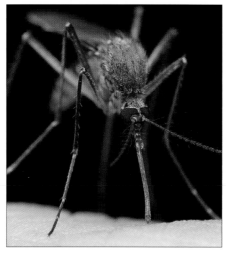

uniquely modified. This mask-like structure is hinged and can be folded so that it covers most of the underside of the head. Some insects are filter-feeders (for example larval mosquitoes, black flies and caddis flies) and obtain their food by filtering particles from the water in which they live. This food includes bacteria, microscopic algae and detritus. The mouthparts of such filter-feeders include a number of setal 'brushes' or 'fans' which generate feeding currents or trap tiny food particles and move them to the mouth.

The mouthparts of some adult insects, such as mayflies and warble flies, are greatly reduced and non-functional.

Left: Mosquitoes use their sharp mouthparts to pierce flesh and suck blood.

USEFUL INSECTS

Among the vast range of insects alive today there are many that are useful to people, notably as pollinators of crops that humans eat. Honey bees also produce honey and wax. Many insects are beneficial to the gardener, killing insect pests.

Insects, especially bees, butterflies, flies and moths, play a valuable role in the pollination of flowering plants. This is of economic importance for some crops and of vital importance for food chains and webs.

Insects break down matter at the lower end of the food chain, improving soil quality by affecting its structure and fertility. This in turn aids the plants that grow in it and benefits everything up the food chain. The effects of insects recycling decaying matter in rainforests may be as important as the effect of worms on the soil structure in such habitats.

Beneficial bees

Some insects provide us with food. Honey is the delicious and nutritious liquid we harvest from honey-bee hives. This practice has gone on for centuries, honey often being collected from wild bee nests by native people or cultivated by providing hives for bees and in effect farming them.

Wax is a another natural product derived from beehives and people have found many uses for its numerous qualities. It burns (making it a useful alternative to oil and tallow), melts and acts as a sealant (for dyeing processes such as batik), and is used as a traditional material in woodworking and crafts (as a wood sealant or lubricant), as well as being added to some skin products.

In the realms of alternative medicine, bees also figure prominently. Propolis is a substance synthesized in beehives as part of bees' own natural defences to bacteria. It is made available to humans in the form of a herbal remedy for coughs and sore throats and thought generally to boost immunity. Herbal medicines also use royal jelly (made to feed the queen bee larvae) as a health supplement. Many alternative health practitioners believe that eating a locally produced honey will help some individuals who suffer from hay fever to overcome their allergy, as a result of exposure to a product derived from local pollen.

Nutritious foodstuff

Whole insects are consumed by some indigenous people as part of their normal diet, being considered both tasty and nutritious, due to their high protein content. In Africa some people collect midges and squash them into a

Above: Dung beetles help keep the African savanna clean, collecting dung and making it into balls in which they lay their eggs.

flat patty to eat when they appear in huge numbers at certain times. The abundance of locusts means they too are often caught (and sometimes cooked) as an easily available meal in a number of places around the globe. In parts of Asia the nymphs of dragonflies are caught and fried like fish, and even the larvae of some tropical wasps are eaten.

In recent times people have come to value the role of insects in the wider food web. Various companies now tempt an increasingly conservation-minded public with a variety of insect-studded bird foods and larvae for the garden bird table as part of supplementary all-year-round feeding.

Beneficial insects

Cochineal insects (*Dactylopius coccus*) have been gathered for at least the past 500 years to produce a red dye (derived from the carminic acid they contain) still used in some cooking and dye products today.

Left: Honey bees are perhaps the most useful of all insects to humans. They not only pollinate crops but also make honey, beeswax and other products we use.

Insects are increasingly being researched and used as biological control agents to deal with outbreaks of species threatening to affect crops or plant collections. These are alternatives to chemically derived and often non-specific pesticides, and they are increasingly being seen as an organic alternative to more destructive chemical approaches. A predatory insect species is introduced to deal with outbreaks of 'harmful' insect species that may otherwise destroy the plants. One example is the tiny hymenopteran parasitic wasp (*Encarsia formosa*) which is introduced to plants to eliminate whitefly *Trialeurodes vaporariorum* (actually a member of the bug order). *Encarsia* acts by feeding on the nymph stages of the bug.

Since 1200BC, shellac has been derived from insects (notably from the hemipteran *Kerria lacca*) for varied uses including varnish, food and hair dye. Shellac is still economically important enough to be harvested in India, Thailand and China, and is used in jewellery and ornaments as well as the aforementioned products.

Forensic science

Forensic entomology is a useful if somewhat unusual application of the study of insects which can provide crucial evidence in unsolved murder

Below: Flies are often perceived to be pests, but may be useful to criminologists.

cases. An in-depth knowledge of fly species (Diptera) is often used, especially the larval stages or maggots which may be found on a corpse. By analysing details of the stage of the life cycle relative to the species observed, estimations can be made of the actual time of death of a victim when no other evidence is available. Clues for reconstructing a crime scene may result.

We have made use of our knowledge of insect pheromones by synthesizing them to place in traps to act as insect lures. Pests can be captured before they get the chance to raid or contaminate food (for example cockroaches or beetles in grain stores). Traps have also

Above: Ladybirds are beneficial insects to have in the garden – they eat aphids, which damage plants.

found an unexpected use in museums, helping to protect antiquities and precious items made from natural fibres which are attacked by dermestid beetles. One such pest is so prevalent that it has become known as the 'museum beetle' due to the problems it has caused in collections it perceives as edible. Traps are set in a similar way to those for food store pests, particularly at certain times of year when the beetles may be more mobile than normal, flying some distance to find mates and lay their eggs.

Insect sex pheromones are also used by humans to interrupt mating cycles where a species causes considerable harm to economically important crops. Artificially created sex pheronomes are used to fool a species into believing mating is imminent before the crops break through the soil, thus saving the crop from insect damage.

One example of a species being dealt with in this way is the pink bollworm moth (*Pectinophora gossypiella*) which causes damage to cotton crops in the USA, Pakistan and Egypt. Pheromone traps are used, as is biological control using a parasitic wasp.

THE INSECT ORDERS

There are 30 orders of insect included within the class Insecta. These range from tiny microscopic creatures to tropical stick insects that can measure up to 30cm (12in), and beetles with a similar wingspan. Many of these insects are around us in our homes and gardens and will be familiar creatures.

This chapter provides a brief look at the insects that are included within each order and describes their typical features. The general appearance of each insect group is detailed, with dimensions, body shape and colouring all described. Aspects such as feeding, defence, reproduction and characteristic behaviour are also covered. The life cycle of each insect type is explained, together with any identifying features for each life stage. Finally information is provided about the type of habitat in which typical members of the group are found. Some insects are cosmopolitan, while others have very specific requirements, with a need to be near water, or in damp leaf litter, for example.

Included too are spiders and relatives, millipedes and centipedes, and the wingless near-insects. These groups, like the insects themselves, are (mainly terrestrial) arthropods; they are sometimes confused with insects and may be found in similar habitats.

Left: A damselfly coming in to land shows the net-veined, transparent wings typical of many flying insects. Damselflies are recognizable by their long tapering bodies, which are usually brightly coloured.

WINGLESS NEAR-INSECTS AND WINGLESS INSECTS

The orders Collembola, Diplura and Protura make up the wingless near-insects. These small arthropods have six legs, and simple mouthparts that are enclosed inside a cavity. The wingless true insects are the bristletails and silverfish (Archaeognatha and Thysanura).

Springtails (Collembola)

These wingless insects are one of the most successful groups of hexapods and are found in amazing numbers. Commonly known as springtails, they range in size from 0.2–10mm (0.01–0.39in).

The name 'springtail' refers to the furcula, a specially adapted appendage underneath the abdomen (not present in all species). It is held in place by a catch. When this is released the furcula will spring back suddenly, launching the animal into the air. Collembolans can be identified by the fact that they have six abdominal segments, four antennal segments, four leg segments and lack terminal cerci. The 6,000 living species of springtail are divided

Above: Diplurans are primitive insects that are wingless and not easy to spot.

into three suborders: the elongate bodies (Arthropleona) containing two superfamilies; and the globular bodies, divided into Neelipleona, which has one family, and Symphypleona, which contains five superfamilies.

Above: Springtails can leap by releasing a powerful spring underneath their bodies.

Reproduction is indirect, with a spermatophore deposited by the male and either left for the female to pick up or placed in her genital opening. The eggs hatch into nymphs that only differ from the adults in size and lack sexual maturity. These nymphs undergo five to eight instar moults, but once adult they continue moulting. Depending on the species, springtails may produce one generation or several. In a lifetime (12 months) a female may lay 90–150 eggs.

The arctic species, *Cryptopygus antarcticus* can tolerate temperatures as low as -60°C (-76°F).

Diplura

The name diplura is derived from 'diploos' (double) and 'oura' (tail), referring to the cerci. Diplurans are slender and small, 2–5mm (0.08–0.19in) long, eyeless and mostly white, with long slender antennae, biting mouthparts and two prominent cerci that are either short and forceps-shaped or long and thread-like.

There are 800 species in nine families. Diplurans may be either herbivorous or carnivorous (predacious hunters), using their pincers to catch prey. The lifespan is approximately one year. Reproduction involves the males leaving spermatophores for females to

---

Life cycle of a springtail

The Lucerne flea, *Sminthurus viridis*, is a primitive wingless insect, whose young look like miniature versions of the adult, once free of the egg.

The female walks over the spermatophore and it comes into contact with her genital opening.

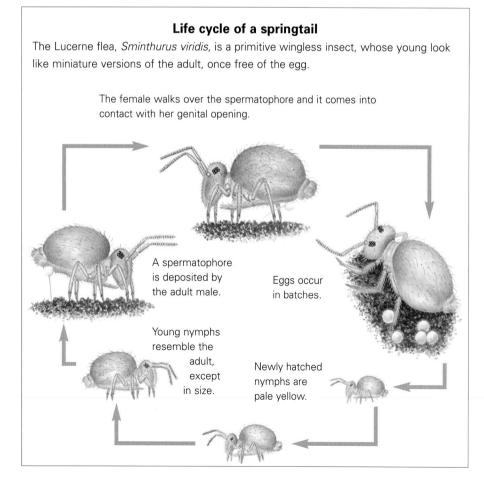

A spermatophore is deposited by the adult male.

Eggs occur in batches.

Young nymphs resemble the adult, except in size.

Newly hatched nymphs are pale yellow.

collect. Their life cycle is simple. The eggs, which are laid in burrows, hatch into nymphs that resemble small adults but lack reproductive organs. Moulting occurs both in the nymphs and adults and may take place over 30 times in a lifetime.

Diplurans can regenerate a lost appendage and may drop their cerci when under attack.

Protura

Less than 2mm (0.08in) long, eyeless and lacking antennae, proturans hold their enlarged front legs forward antennae-like and have sensory abilities. They also lack cerci. The cylindrical abdomen has 12 segments and the legs have five segments.

Between four and eight families have been recognized, with about 500 species described to date.

The more long-legged species appear to have one generation a year, as opposed to the shorter-legged species, which tend to reproduce continuously all year round. Little is known about their reproductive processes, but it is believed that from the egg they have five stages before adult: a pre-larva, two larval types, a junior stage, and a pre-imago (pre-adult).

Bristletails (Archaeognatha)

With elongated cylindrical bodies up to 20mm (0.79in) long, bristletails are shaped rather like a turnip. They have

Below: Silverfish have a silver body and scuttle across damp surfaces.

Habitats of wingless near-insects and wingless insects

Up to 200,000 springtails can live in a single square metre/yard of land, depending on the habitat. Springtails are found from mountain-tops to seashores, and occur on all of the world's continents including Antarctica. The majority of species inhabit leaf litter or damp soil in temperate and tropical climes, where they feed on decaying plant material, fungi, bacteria, excrement and dead insects.

Diplurans have a worldwide distribution, occurring in dark, humid places such as soil or under bark.

Proturans occupy moist leaf litter and soil, commonly to depths of 23cm (9in), and feed on fungus and decaying plant matter.

Bristletails and silverfish are found in varied environments throughout the world. They feed on algae, lichen and plant debris.

long antennae, six legs and two long cerci with a long central filament, giving them the appearance of having three bristles for tails.

These insects comprise about 350 species in two families, Meinertellidae and Machilidae. Bristletails can be either nocturnal or diurnal. Little is known of their breeding habits. On hatching, the young look like miniature versions of the adults.

Silverfish (Thysanura)

They are 2–22mm (0.08–0.87in) long, with flat elongate shiny silver bodies. Their antennae are long and they have two cerci and a tail-like telson at their rear.

There are 370 species in four families: Lepidotrichidae, Nicoletiidae, Lepismatidae and Maindroniidae.

Most species are nocturnal omnivores, scavenging under bark, in leaf litter. Males secrete silk that they suspend sperm droplets on for the female to take up. The eggs hatch into nymphs that, apart from size, are the same as the adults in appearance.

Like bristletails, silverfish can run fast, but unlike them they do not jump.

MAYFLIES AND STONEFLIES

The mayflies belong to the order Ephemeroptera and have been present since the Carboniferous.
Stoneflies are in the order Plecoptera, present since the Permian. Both are found near water and have
aquatic nymphs that are important components of the food chain, especially for fish.

Mayflies (Ephemeroptera)

The word Ephemeroptera is derived from the Greek 'ephemera' (fleeting), and 'pteron' (wing) – they are short-lived winged insects. They are small to medium-sized, soft-bodied insects with two pairs of wings: two forewings and two smaller hindwings, all of which are closed vertically above the body. They have short antennae, large compound eyes and three ocelli (light sensors), and at the rear end have two long cerci and a long 'tail' filament.

Their common names include shadfly, day fly, Canadian 'soldier fly' and fishfly. The adults do not feed, but the larvae feed on diatoms, algae and detritus, or are carnivorous.

The order is split into two suborders, the Schistonota (split-backs) and the Pannota (fused-backs) with approximately 2,500 species in 23 families. The two suborders can be distinguished by looking at the nymphs; the Schistonota nymphs have wing pads (immature wings) that are separate on each side of their thorax (next section after head), while the Pannota nymphs have their wing pads stuck together on their thorax (back). For adults identification is by the differing wing venation (lines on wing surface) of species.

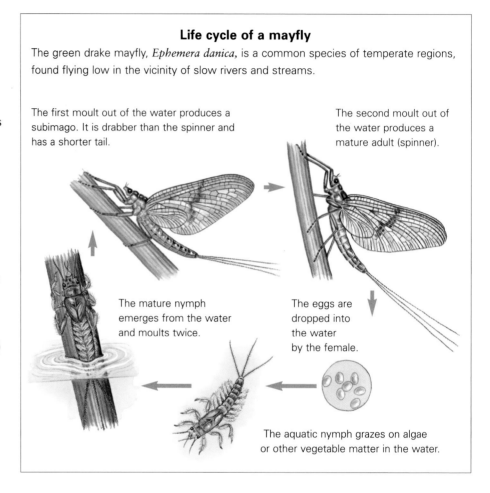

Life cycle of a mayfly

The green drake mayfly, *Ephemera danica*, is a common species of temperate regions, found flying low in the vicinity of slow rivers and streams.

The first moult out of the water produces a subimago. It is drabber than the spinner and has a shorter tail.

The second moult out of the water produces a mature adult (spinner).

The mature nymph emerges from the water and moults twice.

The eggs are dropped into the water by the female.

The aquatic nymph grazes on algae or other vegetable matter in the water.

The nymphs in the families have different feeding habits. Some are carnivorous, or eat decaying matter or a mixture of the two. Others are herbivores that feed on decaying plant matter.

The newly emerged adults swarm above the water surface seeking out a potential mate. The male's long front legs hold on to their chosen female and mating takes place in the air. The female then, depending on species, drops eggs singly or in batches into the

Left: The dun is the newly emerged winged form of the mayfly; the spinner is the adult form following the second out-of-water moult. The spent are dead or dying mayflies, often seen on the water surface. Emergence of adult mayflies is usually from May to August depending on species.

water, or places them purposefully on submerged vegetation. The adult lives for approximately one day to one week, and dies soon after mating.

The young that hatch, though, can live up to three years and are known as nymphs, but they do not resemble the adult form. Instead, they are wingless and have chewing mouth-parts. Breathing is by external gills along the sides of the body and in certain species there are three feathery tails as well. Insects have a tracheal system for allowing oxygenated air to passively reach their body tissues via a series of holes along their abdomens, but in these nymphs the system is sealed. The nymphs will undergo as many as 50 moults before maturing.

When a nymph has reached maturity and emerges from the water, it moults twice before it is fully adult. The first moult out of water produces a subadult that can fly, but the second moult produces the fully active adult.

Stoneflies (Plecoptera)

The name of the stonefly order Plecoptera is derived from Greek; 'plectos' (pleated) and 'pteron' (wing). They range in body size from 3–48mm (0.12–1.89in) with a maximum membranous wingspan of 10cm (4in). Their bodies are soft, dark brown or black, and slender, and they have bulging eyes, three ocelli and long antennae. The front wings are long and narrow compared to the hindwings, which are pleated when closed and lie flat against the body. The rear end has two long cerci.

Freshwater fishermen model artificial stoneflies (nymph or adult form) that are also referred to as salmon fly, trout fly or willow fly as lures for trout. All stages/ages of nymphal instars are present year-round in water bodies and are prey items for fish. When depositing their eggs the females disturb the water surface, attracting the predatory fish.

There are 2,000 species in 15 families divided between two suborders. They have worldwide distribution mainly in temperate cool regions.

The eggs when deposited undergo diapause (rest) from two weeks to two months (dependent on species) and when water temperature and the day length are suitable, will hatch.

The adults are recognized by their flattened bodies and their wings, which in most species extend beyond the end of their body. As adults they are nocturnal and by day seek out dark places to hide, such as under stones or bark. Nymphs with an omnivorous diet will not feed on emerging as adults, but the herbivorous nymphs tend to produce adults that do feed.

Males attract a mate by drumming or tapping on a terrestrial surface – single taps or more complex 'songs' that in turn are replied to by a female, and the two search each other out. The slimy egg mass, of up to 1,000 eggs, is deposited and adheres to plant or rock

Habitats of mayflies and stoneflies

Mayfly nymphs live in an aquatic habitat, which may be fast/slow running rivers, ditches, ponds, streams or lakes. Some species breed in water bodies with a gravelly bottom, while others prefer a silt or muddy substrate.

These insects have a worldwide distribution, but are found particularly in temperate regions where they require clean unpolluted and well-oxygenated water to complete their life cycle.

Stoneflies like well-oxygenated, pollution-free, flowing water in habitats such as streams and rivers with a rocky bottom. Lakes, sandy areas and damp areas also support certain species but less is known about these. These insects are well adapted to cold water conditions and, as such, a higher diversity of species are found in temperate zones, where they are mostly nocturnal.

surfaces or lodges in crevices within the water column. On hatching, the nymphs have the appearance of an earwig with two tails. The nymphs undergo up to 30 moults over one to four years, before finally emerging from the water as adults. They are eagerly eaten by many birds.

Below: A stonefly nymph lies on the bottom of a river, clinging tightly in the current.

Stonefly nymphs have an array of defence mechanisms to prevent being eaten, such as feigning death, reflex bleeding and using known 'boltholes' within their area of foraging, such as under pebbles and rocks.

Below: Stoneflies emerge and moult at the edges of watercourses, usually when dark, leaving behind their shed skin or 'shuck'. They survive for around three to four weeks before dying.

DRAGONFLIES AND DAMSELFLIES

The order Odonata is divided into two suborders – the dragonflies (Anisoptera) and the damselflies (Zygoptera). A third suborder, the Anisozygoptera, is sometimes recognized, but these are mainly extinct forms. Odonata is derived from the Greek 'odontos' (toothed), referring to the mandibles.

Dragonflies (Anisoptera)

The suborder name Anisoptera means 'unequal wings', which refers to the broader hindwings compared to the forewings. There are five families, with 3,000 living species, told apart by their adult colours and their body shapes.

Dragonflies are large, colourful, long-bodied insects with amazing flying abilities. Their two pairs of wings are independent and capable of changing their angle, wing beat frequency and depth of wing beat, allowing dragonflies to fly backward, if necessary. Their thorax is packed with flight muscles and at rest the wings lie straight out sideways.

Dragonflies have large heads and compound eyes that have near 360-degree stereoscopic vision. A dragonfly's eye has 30,000 lenses (compared to a human's single lens) and they have been seen to home in on prey up to 12m (40ft) away. They also have strong biting mouthparts and paired claspers at their rear end. Their legs are used for standing or grabbing prey on the wing, but they are unable to walk well on them. All dragonflies are predators using their excellent sight to hunt prey.

Outside the polar regions they are found worldwide but more than half live in the tropics.

Dragonflies are active by day, and hunt other flying insects in the air. The pairing and position of the male and female during mating is known

Above: Dragonflies are robust. Their wings are held open when they are at rest.

as the 'copulation wheel'. This describes the coupling, where the male grasps the female behind her head with claspers while she raises her abdomen under herself and forward. Mating can occur while perched or in flight. The eggs are laid either directly into water, attached to weeds, or in marginal plants. The female uses her ovipositor to cut a slit in the vegetation and then lay her eggs within it.

On hatching, the nymphs are active hunters of tadpoles, fish fry and other

Below: Dragonfly nymphs are voracious hunters, capable of killing small fish.

Life cycle of a a dragonfly

The emperor dragonfly, *Anax imperator*. This widespread species is also one of the largest. It is found over much of Europe and Asia, and also in North Africa.

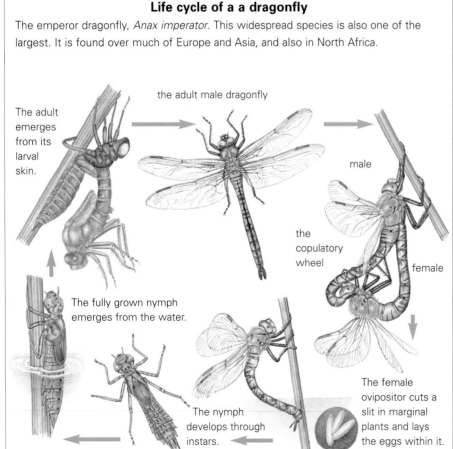

the adult male dragonfly

The adult emerges from its larval skin.

male

the copulatory wheel

female

The fully grown nymph emerges from the water.

The nymph develops through instars.

The female ovipositor cuts a slit in marginal plants and lays the eggs within it.

small creatures, including their own kind. They capture their prey with an elongated, hinged lower lip (labium) that shoots forward, grabbing the victim with clawed paired palps located on the end. This feature is known as the 'mask'. At rest the mask is held folded underneath the head and thorax, sometimes extending back beyond the front legs. The dragonfly nymph expands and contracts its abdomen to move water over its gills within the abdomen, and can squeeze the water out rapidly for a short burst of underwater jet propulsion.

The nymphs undergo between ten and 20 moults, and live from three months up to ten years. When mature, the nymph emerges from the water on to vegetation where it perches and moults into the winged adult form.

Damselflies (Zygoptera)

Zygoptera means 'equal-sized wings'. The hindwings are similar in shape to the front wings.

Damselflies are weaker fliers than dragonflies, with colourful but delicate and much smaller bodies. They have large heads, strong biting mouthparts and widely separated compound eyes, with more than 80 per cent of the brain devoted to visual processing – the same as dragonflies. They have paired claspers at their rear end and their legs are used for standing or grabbing prey on the wing, but they are unable to walk well on them. Damselflies are all carnivorous insects, using their excellent sight to hunt prey.

Damselflies have two pairs of elongated membranous wings with a strong cross vein and many small veins that criss-cross, as well as a small coloured patch, adding strength and flexibility. Damselflies can close their wings over their body at rest due to a hinge, whereas dragonflies cannot. They are not as restricted to sunny days as dragonflies, with males often defending their territories even when it is overcast.

All damselfly females possess an ovipositor. They are all predators, including the nymphs, which hunt by stalking in submerged vegetation.

Habitats of dragonflies and damselflies

Dragonflies are commonly found near fresh water but are in no way restricted to it; as strong fliers (50kmh/30mph) they can travel many miles to forage and some species even migrate over the sea.

Damselflies tend to be rather fluttery in flight. They are found around water and breed in varying freshwater flows, from still and stagnant to fast-moving, depending on the species.

When mating, damselflies form a 'copulatory wheel' like dragonflies. Mating may take from a few minutes to several hours, and when laying eggs, they may stay in the tandem position, the male guarding his mate. The eggs of damselflies are cylindrical, whereas dragonfly eggs are ovoid in shape. The places and manner in which they lay their eggs are similar to dragonflies. In the tropics, development of the nymph may be complete in 60 days, whereas in colder temperate climes it may take up to ten years. Development of damselfly nymphs, though, tends to be quicker than dragonflies, with fewer moults necessary, and usually takes approximately one year. Depending on the species, they may breed any number of times in a year, but this is dependent on food availability and water temperature. The eggs may enter a resting stage until the water temperature rises. Adults may live from weeks to a few months. Damselfly nymphs seize prey in the same way as dragonflies, with the elongate extensible, hinged labium or

'mask'. In the damselfly nymph the abdomen is longer and narrower with three fin-like gills projecting from the end, whereas the shorter, stockier dragonfly nymph has gills located inside the abdomen on plates.

Below: Damselflies are very slender and hold their wings closed over their bodies.

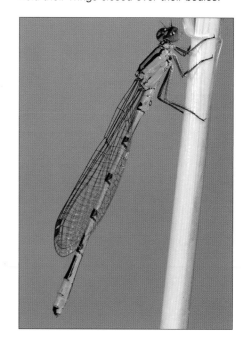

COCKROACHES AND TERMITES, MANTIDS AND EARWIGS

The four orders on these pages contain a wide range of insects. Cockroaches and termites feed mainly on plant matter. Most earwigs are omnivores, while mantids are strictly carnivorous, waiting patiently among the foliage that they mimic for their prey.

Cockroaches (Blattodea)

This order contains cockroaches, of which only one per cent of the 4,000 species are pests to man, the vast majority being harmless. The one per cent that are associated with man are attracted to food and dirty living conditions, and may carry germs on their feet or excrete dangerous viruses and protozoans. Of the other 99 per cent of harmless cockroaches, many are kept as pets, such as the Madagascan hissing cockroach *Gromphadorhina portentosa*, which when threatened expels air from the spiracles (breathing holes on body sides) to produce its characteristic hissing sound.

Cockroaches have flattened, oval-shaped, leathery bodies, with long thin antennae and strong spiny legs suitable for running. The forewings are toughened, protecting the membranous hind pair, and at rest the wings lie flat across the body, overlapping. Cockroaches have chewing and biting mouthparts and they are also able to lap up liquid food. They have large compound eyes and a large flattened plate attached to the thorax that covers part of the head.

Most female cockroaches produce a sex pheromone to attract a mate, and the males often produce 'aphrodisiac' secretions to prepare the female for mating. Courtship can vary from slight

body movements to production of sounds, or even head butting in the Madagascan hissing cockroach. The sperm is transferred in a spermatophore and the female either lays her eggs in an egg case and buries them, or carries them under her abdomen.

Termites (Isoptera)

Sometimes called 'white ants' due to their coloration, termites are social insects that live in colonies, with reproductives, workers and soldiers of both sexes. They have soft bodies, biting mouthparts, short cerci and simple antennae with 9–30 segments. Termites are mainly tropical or subtropical insects that feed on wood, fresh leaves, leaf litter or soil. They are grouped, according to their feeding or habitat preferences, as subterranean, soil-feeding, dry-wood, damp-wood or grass-eating. Dry-wood termites include some that are pest species of buildings.

Left: The Madagascan hissing cockroach plays an important role as a decomposer in the forests where it lives, helping to break down fallen fruits and other plant matter.

Above: Mantids are also known as praying mantises. This common name derives from the 'praying' pose they adopt when waiting for prey.

To maintain and guard their nests, termites have evolved a reproductive strategy that provides both a workforce and soldiers. The workers are sterile and make up most of the colony, carrying out nest building and repair duties, foraging, taking care of eggs and nymphs and caring for the queen. In lower termites, there may not be a distinct caste system, and nymphs often undertake these duties.

The 'primary reproductives' possess two pairs of wings, and will become the founders of a new colony. After the 'nuptial' flight, they shed their wings and mate, thus becoming king and queen. A colony can take up to two years to establish. The queen termite can grow to 90mm (3.54in), and in some species can produce an egg every three seconds.

In Amazonian forests the total biomass of termites can be ten per cent of the soil fauna and 80 per cent of the fauna of dead wood.

Above: Earwigs like confined, dark spaces.

Above: Termites are among the most common insects on Earth, with a global population measured in the trillions.

Mantids (Mantodea)

Found mainly in warm regions of the world, mantids are elongate, medium to large sized insects, growing up to 25cm (10in) in some South American species. Many are winged, with two leathery forewings and two hindwings. The head is triangular with large forward-facing compound eyes that have binocular vision. The two front legs are adapted to catch prey. Mantids are predators, feeding mainly on insects and spiders. There are about 2,000 species.

The male is generally smaller than the female, and in some species may be eaten by the female after or during mating. The female may lay 6–22 ootheca (egg cases), with 30–300 eggs in each, through her lifetime. The female will attach the ootheca to twigs or stones, and in some species will guard the eggs and even the nymphs for a few days after hatching.

Most mantids fly at night, when they may be preyed on by bats. Bats use ultrasonic clicks to locate prey, and mantids have an ultrasonic detecting 'ear' between their front legs. When harmful frequencies emitted by bats are detected, the mantid takes immediate evasive action, entering into a 'nosedive' if the bat is very close.

Earwigs (Dermaptera)

Earwigs have leathery short forewings and hindwings which are large and membranous, though they rarely fly. They are common, slender, brown or black, elongate 4–80mm (0.16–3.15in) and slightly flattened insects with biting mouthparts and two forceps-like cerci at their rear end, which are their most distinguishing feature. The cerci are curved in males and straight in females. There are about 1,900 species of earwig. Most earwigs are nocturnal and omnivorous, feeding on dead plant material and slow-moving invertebrates. Female earwigs show maternal care.

In temperate climates spring triggers females into constructing a brood chamber under a stone, in a burrow or within rotting vegetation. Mating is performed rear end to rear end, usually with each holding the other's pincers, and at night. The female tends the eggs, turning and cleaning them until they hatch, when she will leave to forage for food. The nymphs pass through four or five instars, but after the first two they disperse, to avoid the risk of the female eating them.

Some earwigs have defensive glands on their abdomens, which produce a noxious liquid that in some can be squirted 10cm (4in).

Most are considered beneficial insects since they eat pest species.

Habitats of mantids, earwigs and termites

Mantids live among tree foliage, flowers or grasses where they are expert ambush predators. They are often the same colour as their surroundings, and many have body outgrowths mimicking flower petals, leaves or twigs to help disguise them.

Most earwigs are free-living but some are parasitic or semi-parasitic on bats or rodents. Earwigs prefer confined humid habitats. They are found worldwide, except in the polar regions, and are most common in tropical areas.

Termites build nests, in fallen wood or above or below ground, in order to shelter the colony. Termite mounds are constructed from soil excavated in the course of digging underground tunnels, or from soil and sand collected on the surface, mixed with saliva and faeces. The interior has an intricate network of tunnels and galleries for movement, and in some species, for the cultivation of fungi. To maintain airflow and an even temperature, vents lead off from some tunnels.

GRASSHOPPERS, LOCUSTS AND CRICKETS

These insects belong to the order Orthoptera, which contains 22,000 species and includes some of the world's most familiar insects. The order name is derived from the Greek words 'orthos' (straight) and 'pteron' (wing). The order has a worldwide distribution with a high diversity in the tropics.

Grasshoppers (Caelifera)

This suborder consists of about 34 families of locusts, grasshoppers, pygmy locusts, bush-hoppers and bush locusts. Caeliferous insects have fewer than 30 segments in their antennae (termed short-horned), a reduced ovipositor and a hearing organ (if present) sited in the abdominal area. The production of sound is usually from the hind legs rubbing against the wings, or in some species, the clicking of the wings in flight.

Body size ranges from 5–11.5cm (2–4.5in) with wingspans up to 22cm (8.5in) and they possess compound eyes. The majority are plant eaters, but a few are omnivorous. Crickets and grasshoppers have stout bodies with large blunt heads and large rear legs adapted for jumping. Wings if present are two tough leathery forewings and a pair of membranous hindwings.

They have relatively short antennae and their chirpy song fills the air in warmer months. Most members of this family are grassland species but some are found in marshy areas. Sound production is either from pegs on the inner hind legs which are rubbed against a hardened vein on the forewing, or the other way around with the pegs on the forewing.

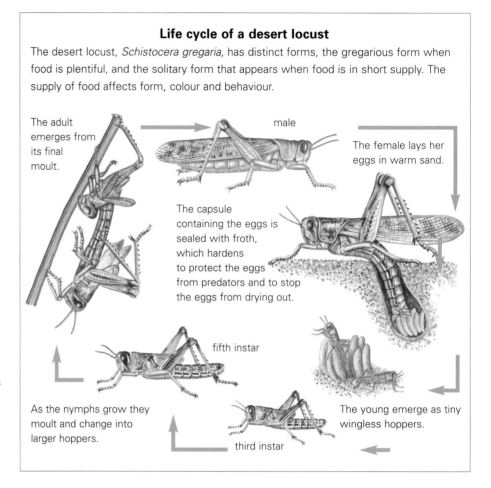

Life cycle of a desert locust

The desert locust, *Schistocera gregaria*, has distinct forms, the gregarious form when food is plentiful, and the solitary form that appears when food is in short supply. The supply of food affects form, colour and behaviour.

The adult emerges from its final moult.

male

The female lays her eggs in warm sand.

The capsule containing the eggs is sealed with froth, which hardens to protect the eggs from predators and to stop the eggs from drying out.

fifth instar

As the nymphs grow they moult and change into larger hoppers.

The young emerge as tiny wingless hoppers.

third instar

The members of the family Tetrigidae include the pygmy locusts, groundhoppers and pygmy grasshoppers. There are approximately 1,400 species. In Australia, a small family, the Cylindrachetidae, known as sandgropers, are an important economic pest. The females remain wingless as adults and resemble larvae, with a very elongated, cylindrical body, up to 70mm (2.76in) long, short, flattened front legs held beside the head, and simple eyes, all of which are adaptations for burrowing. Two species are also found in New Guinea and Argentina.

These are all predominantly tropical species, with many still undescribed.

The males sing to attract a mate. When mating occurs the male deposits a sperm package into the female's

Left: Grasshoppers have long bodies and many have large wings. Most are well camouflaged but a few are surprisingly colourful.

Habitats of crickets, locusts and grasshoppers

Few grass-dominated habitats are without crickets, grasshoppers or locusts, and grass and other plant matter makes up a large part of their diet. Most are rather secretive and may be heard rather than seen.

In the tropics, species are also found in the tree canopy or living among lichens and mosses, but the main families are well represented around the world in all habitats that support plants. These include deserts, bogs, marshes, grassland and woodland. Generally, they are ground-dwelling insects that feed mainly on grass during the day, but many grasshoppers can fly well. Locusts in particular are known for their long migrations in search of food.

ovipositor, and when the eggs have been fertilized she deposits them 20–50mm (0.8–2in) underground. In temperate regions, the eggs overwinter and can stay up to nine months underground until conditions are warm enough for hatching. They hatch into worm-like creatures, but on reaching the surface they discard this outer coating to reveal a nymph that will then undergo four to six nymphal instars. Life expectancy of an adult is around three months – they die in autumn with the next generation remaining as eggs underground.

Locusts change colour and behaviour at high population densities. Locusts are an important economic pest species capable, in their enormous swarms, of destroying vast areas of vegetation and crops, in places such as Africa, South America and the Middle East. The species *Schistocerca gregaria* is a significant pest, able to fly great distances in huge swarms, producing between two and five generations in a year and seriously affecting harvests and often devastating crops.

Crickets (Ensifera)

Ensifera is Latin for 'sword-bearer' and refers to the shape of the ovipositor. Crickets are distinguished from grasshoppers by their long antennae (over 30 segments) and ovipositor, and the fact that they have their hearing organ on their front legs. Some females

Below: Crickets are most easily identified by their long antennae. Unlike grasshoppers they tend to walk rather than fly or hop.

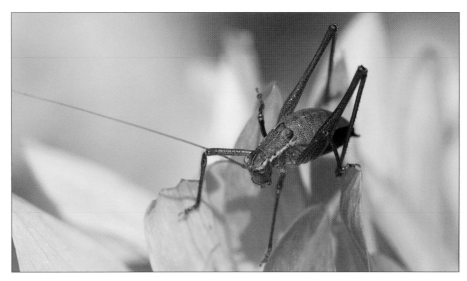

show maternal care. They are mainly nocturnal insects with a lifespan of over one year. Many crickets make loud rasping and repetitive calls to attract a mate.

Within the suborder are between four and six superfamilies (depending on classification used) and 12 families. The family Tettigoniidae are commonly known in the USA as katydids and elsewhere as bush crickets (or confusingly as long-horned grasshoppers). There are 5,000 species that are herbivorous, saprophagous or predacious, with some pest species. Their antennae can be as long as their body and most inhabit plant foliage. The majority of species are found in the tropics and they sing by rubbing their forewings together; the left wing, having the pegs, is rubbed over the right wing. Gryllidae is the family of 'true crickets'. It has approximately 1,500 species with moderately flattened bodies and long antennae. They are mostly nocturnal. Sound is produced in this family by the forewings where the right wing is toothed and is drawn across the left wing.

The Gryllotalpidae family are more commonly known as 'mole crickets'. They are capable of clumsy flying – males manage up to 8km (5 miles) in the mating season – and are thick-bodied with enormous shovel-like front legs for digging. They excavate tunnels in the soil to feed on plant roots and invertebrates, and to live in; males adapt a tunnel with a bell chamber to amplify their song.

STICK AND LEAF INSECTS

The order Phasmatodea contains the stick and leaf insects, strange creatures often kept as pets. They resemble grasshoppers and crickets in some respects but are adapted for camouflage with often bizarre body shapes. The name of the order is derived from the Greek 'phasma' meaning spectre or apparition.

Division of the order

There are three families in the order: Phylliidae, Phasmatidae and Timematidae. The Phylliidae contains the 'true' leaf insects which are found from South and South-east Asia to Australia. The Phasmatidae contain the vast majority of stick insects, while the Timematidae contains just a single genus of stick insect which has three segments in the tarsi (the end section of the leg) rather than five.

The key defining feature of their order is that all members are superb at camouflage and mimicking their habitat. Through the course of evolution they have adopted the appearance of twigs, twigs with lichens, mosses, and even bird faeces. Some of the leaf insects are remarkably similar to the plants they live among.

Stick insects (Phasmatidae and Timematidae)

Sometimes known as ghost sticks or walking sticks, stick insects are elongate 10–30cm (4–12in), slender and cylindrical with a variable body shape. They have long legs, short, leathery forewings and large membranous hindwings (in some species the wings are absent), thread-like antennae, small compound eyes and biting mouthparts.

Leaf insects (Phylliidae)

Leaf insects tend to be broad, flat and leaf-shaped. The leaf insects are impressive at mimicking leaves, often including the appearance of mildew, insect feeding damage and leaf veins. When seen they are hard to mistake for any other insect.

Above: This stick insect mimics the colours of its background, which effectively camouflage it from predators.

Common features

In many species males are rare or absent and the females produce viable eggs without fertilization (a process known as parthenogenesis). Males, if required for sexual reproduction, are generally smaller than females. They may be attracted by pheromones. In some, such as the North American species, *Diapheromera veliei*, the males compete by kicking out legs with spines at one another in an attempt to win a female.

The unmated female lays eggs that have just her genetic material, to produce female clones (which are genetically identical to her). The eggs resemble plant seeds, and depending on phasmid species the female may just drop them from where she is and lightly cover them in sand or soil, or stick them to foliage. Some species have eggs with an extra bit on the top which juts out to attract ants. When ants collect these from the ground and return to the nest, they eat the nutrient-rich top, leaving the egg untouched. This adaptation gives ants a rich food parcel and the phasmid egg a safe environment in which to develop. The egg, depending on the

---

Life cycle of a stick insect

The male of the species has stronger wings and it is he who flies to find a mate. Stick and leaf insects lay eggs that can take weeks or months to hatch depending on species. The eggs may be carried by ants far from the point where they were deposited.

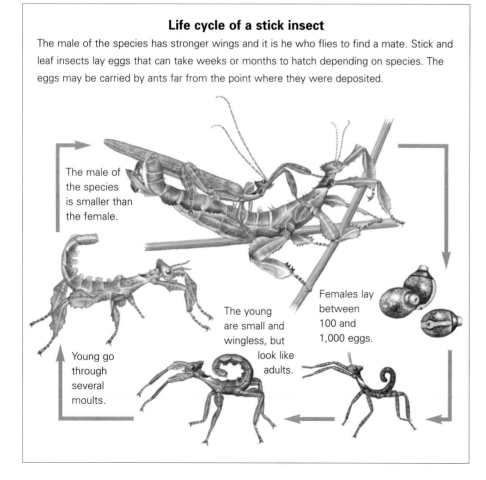

The male of the species is smaller than the female.

Young go through several moults.

The young are small and wingless, but look like adults.

Females lay between 100 and 1,000 eggs.

species, takes from one month to a year to develop and hatch. The hatched nymphs look like miniature adults and go through between two and eight moults before maturity.

Phasmids have the ability to feign death when disturbed, falling to the ground and remaining motionless until the danger has passed. Another strategy is to willingly shed a limb in an effort to escape. Although this may seem extreme to us, by losing a limb they are able to save their life. Many winged species flash open brightly coloured wings, or rustle them to startle predators, while others simply fly away and are lost in the vegetation. The nymphs of some species (*Extatosoma tiaratum*) are thought to mimic ants to prevent attack. Other species, such as *Eurycnema* from Australia, use spines on their spiky legs to strike out. *Anisomorpha buprestoides* from the south-east USA has warning colours on its body, and dispenses a foul-smelling liquid from glands on the upper part of its thorax and around its mouthparts. Stick insects are popular as pets and many species may be kept in a vivarium. Most will eat a range of food plants and will reproduce readily, producing many tiny young.

Below: This phasmid has flanges along its abdomen that resemble a leaf.

Habitats of stick and leaf insects

These are all herbivorous slow-moving insects, living and feeding within foliage and distributed mainly in tropical and subtropical regions. They are found in a variety of habitats including wet and dry forests and grassland. During the day, they tend to hide on the forest floor, under leaves or among twigs where they have adapted to sway gently like the surrounding foliage in a breeze, but at night, they come out to feed.

Some species have the ability to change colour to match their surroundings. Some are like twigs, many even with sharp spines, while others are flattened and flanged, like leaves.

Below: The variety of shapes among phasmids surprises many people. The intricate patterns on the body of this species help to give it camouflage.

Below: Leaf insects mimic the foliage of the plants on which they feed. This disguise works as effective camouflage from predators.

BOOKLICE, BARK-LICE, WEB SPINNERS, ZORAPTERANS, THRIPS AND PARASITIC LICE

The orders described on these pages all contain tiny insects. Some are so small that they may live on our bodies or share our homes in significant numbers without us ever noticing them. Others are irritating parasites of other animals, including humans.

Booklice and bark-lice (Psocoptera)

This order includes booklice and bark lice, which are active, fast-running, small 1–10mm (0.04–0.39in), soft-bodied insects whose coloration matches their surroundings. They have protruding compound eyes, thin antennae and simple, chewing mouthparts. The wings, if present (booklice are wingless), are two fore and two hind, with the forewings slightly larger.

There are about 3,000 species distributed worldwide. These insects feed on plant material, algae, lichens or fungi.

A limited number of species reproduce by parthenogenesis. The eggs are laid singly or in groups covered with silk. They may be left exposed or placed under bark or vegetation. The nymphs usually pass through six instars, and stay in nymphal groups until adulthood.

Web spinners (Embioptera)

The common name of these insects comes from their ability to spin silk galleries from threads exuded from a gland in the front leg. They are small- to medium-sized elongate, brown insects with kidney-shaped eyes, biting mouthparts and short legs. The front

Life cycle of a booklouse

The booklouse, *Ectopsocus briggsi*, is a common species found on bark and also on books and paper. Booklice live for up to six months. They lay eggs that hatch after 11 days. The eggs become mature adults within 15 days, so infestations can occur quickly.

the female

The wings develop as the nymph matures.

The female lays oblong eggs in clumps on leaves or bark.

The appearance of the insect in the nymphal stages (right and below) is similar to the adult.

legs have a swollen area where the silk gland is located. Males possess two pairs of hind and forewings, but the females are wingless.

There are between eight and ten families (depending on the system of classification used) with 170 species.

The female lays her eggs within the gallery and covers them with silk. She remains guarding them in the egg stage and in some species will even feed the first instars with pre-chewed food. Web spinners look rather like miniature earwigs, to which they may be related.

Left: A bird-louse grips the feathers of a peacock. Almost all birds and fowl are affected by some kind of lice.

Zorapterans (Zoraptera)

A number of this order are tiny termite-like, delicate, white to brown coloured insects. The dominant forms are blind and wingless, but they can produce winged, eyed offspring that can disperse to new areas.

There is a single family containing 30 species, mostly from tropical regions with four species known in warm temperate regions.

Zorapterans feed on fungi and predate and scavenge nematodes, mites and other small invertebrates in damp bark or logs.

Males offer 'gifts' of secretions from a gland on their heads to induce females to mate with them. Pairs mate two or three times to produce eggs.

Thrips (Thysanoptera)

This order contains the thrips. They are small 0.5–12mm (0.02–0.47in) yellow-brown or black-bodied, winged insects. There are approximately 5,000 species worldwide.

Thrips have only one mandible in their mouthparts, no cerci, and their wings have fringes of hair along the edges. The eyes are compound and their bodies are slender.

In some species, males fight for the right to mate, sometimes to the death, and mating occurs as the females lay their egg mass. The development of the offspring involves three nymphal stages and two pupal stages before adulthood and sexual maturity.

Parasitic lice (Phthiraptera)

There are approximately 3,150 species of lice distributed worldwide. Parasitic lice feed on the skin debris, secretions or blood of larger hosts, mainly birds and mammals, including humans.

Parasitic lice are small to minute 1–10mm (0.04–0.39in) insects with flattened bodies, and their mouthparts are either chewing mandibles or sucking stylets. Eyes are small or absent, antennae are short and stout with between three and five segments. Their body surface has many sensory hairs, heads are broad in biting lice and conical in sucking lice, and their legs are short with claws for gripping the feathers or hairs of their hosts. Many lice are host-specific, either to a group of similar birds or mammals, or to one single species.

The sucking lice suborder Anoplura feed on rodents. Two species feed on humans: *Phthirus pubis*, the pubic louse which is broad, squat, and moves little, and *Pediculus humanus*, which moves by its short front legs. It has two subspecies, one of which is the head louse and the other is the body louse, which lives on or in clothes and on body hair.

Above: Thrips are often found on flowers, sometimes in large numbers. They feed by piercing and sucking the plant tissues.

A female louse typically lays up to 100 eggs, which are fastened to hair or feathers with fast-acting waterproof glue. The nymphs undergo three nymphal stages which may last from two weeks to three months depending on species. The lice then continue the rest of their lives on that host. If removed from the host the louse will die within a few days. The human body louse attaches her eggs to clothing, especially at the seams. The adults will use the clothing as a hiding place between feeds. Lice can cause skin complaints and severe irritation and sores if persistent or present in numbers. Head lice in particular are a recurring problem in children, especially where many children meet, as at school. Human lice can also spread microbes such as that which causes typhoid fever.

There are also two suborders of chewing lice. More than 2,500 species parasitize birds or mammals feeding on skin, feathers, secretions or blood, some causing irritation.

The order Rhynchophthirina consists of only two species, which feed exclusively on either elephants or warthogs. They have specialized biting mouthparts on the end of a snout, which are used to bore through the host's tough skin. Other chewing lice attack sheep, dogs and various species of bird.

Habitats of booklice, bark-lice, web spinners, zorapterans, thrips and parasitic lice

Most thrips are found on plants where they feed on sap by making a hole with their mandibles then sucking the juices out with other mouthparts. A large number also feed on pollen and are useful flower pollinators. Some species live in leaf litter where they feed on fungal spores, and other species are predacious on other insects such as mites. Some thrips live in colonies with soldiers that protect other individuals.

Web spinners live in small communities. The silken tunnels are expanded to find new food sources as the need arises. Only females and nymphs feed, mainly on lichens, litter, mosses and dead plant material.

Booklice and bark-lice form gregarious groups under bark, leaf litter or wood dust. They are considered beneficial to the trees on which they live. Zorapterans live under bark or in rotting wood.

BUGS

Hemiptera is the largest order of insects with incomplete metamorphosis. There are estimated to be 82,000 species of bug worldwide, comprising about eight to ten per cent of all known insect species. The order includes pond skaters, cicadas, leaf-hoppers, aphids, scale insects and whiteflies.

Common features

The term 'bug' is often used for all insects, but should strictly apply only to the Hemiptera. All Hemiptera possess specialized mouthparts modified into a proboscis, which can appear like a beak. It typically forms a slender, piercing tube used for stabbing the host, either plant or animal, and sucking liquids. In herbivorous bugs the saliva injected into the host's tissue generally causes local necrosis in plants, whereas in predatory species, the saliva is highly toxic and can paralyse relatively large prey.

The majority of bugs are also characterized by the structure of their wings, which are hardened near the base but membranous at the ends, hence their order's name, which literally means half wing in Greek ('hemysis' – half; 'pteron' – wing).

Hemipterans have special organs which enable them to produce sounds. Some (for example cicadas) produce sounds by vibration of a pair of parchment-like tymbals in the base of the abdomen. In other families,

Above: Leaf-hoppers have piercing mouthparts and feed on plant sap.

sound is produced by different mechanisms, such as rubbing the wings over a striate area at the base of the abdomen. Some species are also well known for the secretion of aromatic compounds as a defence against predators.

Bugs range from minute, wingless scales to the fish-eating giant water bug (*Lethocerus maximus*) which can attain a length of 11.6cm (4.5in). Large bugs are more slender and thus less heavy than most other insects of a similar length, such as the beetles.

Many bug species are significant pests of crops and gardens. Among the worst are the many species of aphids, scale-insects and whiteflies. Injection of

saliva may be an important factor in the transmission of micro-organisms, especially plant viruses, which make these bugs serious vectors of plant diseases. Blood-feeding species, on the other hand, often carry potentially deadly infections. Some species, however, have a positive use, such as the scale insects, which are used in the production of dyes.

In general, hemipterans have simple or incomplete metamorphosis, in which the young, called nymphs, are similar to adults in shape. The wings develop and increase in size at each moult and become functional after the last moult, as do the reproductive organs. Many species among the aphids can lay unfertilized eggs that are genetically identical to their mother, a reproductive strategy called parthenogenesis. There can be one to several generations per year. Most bugs produce eggs which are laid inside plant tissue or on plants, or other substrates.

The following suborders, each containing species which share morphological and/or ecological similarities, are often recognized.

Below: Pond skaters are predatory bugs which hunt on the surface of freshwater habitats, catching other creatures which tumble in.

Below: Stink bugs, or shield bugs, produce foul-smelling liquid to aid their defence.

Stink bugs, bed bugs, leaf bugs and water bugs (Heteroptera)

This is the largest of the suborders and contains many familiar bugs. There are approximately 50,000 species, commonly called 'true bugs'. It includes well-known species such as water striders, pond skaters, water and marine bugs, bed bugs and stink bugs.

Species of Heteroptera are characterized by having their wings divided into two areas, hence their name, which means 'different wings', one part thickened and opaque and the other part membranous and usually transparent. Some species, such as the water striders, have lost their wings, while others have inefficient wings and are unable to fly.

The group is also characterized by the variety of form and colours displayed by the different species. These are often used as a defence mechanism. Some have adopted a camouflage technique in which the insects appear similar to their surroundings, while others mimic features of other organisms. Many species also have bright coloration, often used as a warning for predators. Shield bugs are among the most spectacular of all bugs due to their iridescent or metallic colours.

Nearly all species possess scent glands in the thorax, and stink bugs produce an irritant, smelly defensive secretion when disturbed.

About 60 per cent are diurnal, active plant feeders with well-developed eyes and wings, although many variations and exceptions exist.

Left: A bed bug feeds on human skin.

Habitats of bugs

Although less well known than beetles or butterflies, bugs are a diverse group of insects which very often live in association with humans and occur in a wide variety of habitats. Several species, such as the water boatmen and water scorpions, are aquatic.

The well-known pond skaters or water striders are also associated with water but only use the water surface. Another group is truly marine and some have been found on the surface of the ocean hundreds of miles from land. All aquatic species are predatory but the majority of bugs are plant feeders. A minority feed on other insects, while others are external parasites, feeding on the blood of large vertebrates.

They feed on various plants and suck the juices of fruits. Other species are predatory or external parasites.

The typical life cycle of a heteropteran bug consists of the egg, five nymph stages and the adult stage. They usually live in the same habitat as the adult and have identical lifestyles.

Some species are of agricultural importance, feeding on the reproductive parts of the plants such as flowers, ovaries or fruits. Others damage the plants by feeding on stems and roots, for example the chinch bugs *Blissus*, which occasionally wipe out entire lawns. Many species, such as lace bugs, attack and damage ornamental plants.

Moss bugs (Coleorrhyncha)

Forming the suborder Coleorrhyncha, moss bugs are limited to South America, Australia and New Zealand. There are 25 known species, most of which look like plant hoppers. They vary from 2–4mm (0.08–0.16in) in length. They are generally flattened, greenish or brownish with a broad head and lateral extensions. They feed mainly on mosses and usually live in moist habitats such as damp leaf litter on the forest floor; a few species have been found living in caves. Moss bugs are flightless.

Below: Aphids damage plants, slowing their growth by feeding on their sap. They also make a tasty meal for ladybirds.

Jumping plant lice, aphids, whiteflies, scale insects and (Sternorrhyncha)

These common insects are found in many different ecosystems throughout the world. They range in size from 0.1–10cm (0.04–4in) long, and also vary greatly in shape and colour. All species are plant feeders and many are considered as major pests. This group of insects is well known for the cotton-like wax produced by some of the species. The eggs are typically deposited on plant surfaces. Some species jump or fly, but they are predominantly quite sedentary, specialized for rapid feeding and reproduction. Parthenogenesis is widespread and can be alternated with sexual reproduction. Very often the juveniles differ morphologically from adults.

Jumping plant lice

These are found in most regions with the majority in the tropics. They are brightly coloured and range from 1–8mm (0.04–0.31in). Their common name refers to their ability to jump when disturbed. They all feed on woody plants. Many plant pests cause damage in the form of galls or poor plant growth.

Aphids

Also known as greenflies or plant lice, aphids are probably the most universally recognized members of this group, with about 250 species considered as serious pests of crops

and ornamental plants. They are minute insects ranging from 1–10mm (0.04–0.39in) in length. They are distributed throughout the world but are most common in temperate zones. Aphids typically travel short distances by walking or by being transported by ants, but they can disperse over great distances if carried by winds.

Many members of this order feed only on one species of plant; others are more opportunistic and feed on hundreds of species.

Aphids and their relatives tend to live in aggregations on their host plants. Many species form specialized relationships with ants, which are attracted to their excretions of honeydew and in return protect them from predators and move them about their host plants.

Aphids can reproduce asexually as well as sexually. Some lay eggs while others give birth to live young. Typically there are six stages of development: the egg, four nymph stages and the adult stage. Aphids live from 20 to 40 days but they are able to reproduce rapidly and some can produce over 40 generations of offspring. There are about 4,500 aphid species known.

Whiteflies

Adult whiteflies have a wingspan of 4mm (0.16in) or less. Roughly 1,200 species have been described. Whiteflies typically reproduce sexually, with unfertilized eggs becoming males. There are six stages of development: the egg, three nymph stages, the pupa stage and the adult. Many species are considered economically important, as pests of crop plants. They feed on plant tissues, typically on the underside of plant leaves, and all feeding stages produce honeydew. The most recognized pest of this group is the sweet potato whitefly. This pest is thought to carry more than 70 different viruses.

Left: Frog-hoppers, or spittlebugs, surround themselves with froth as nymphs. This is known as cuckoo-spit and can be highly visible on foliage.

Above: Aphids are significant plant pests.

Scale insects

These insects are the most diverse group among the Sternorrhyncha. There are about 8,000 species found worldwide on a wide range of plants. Most of them are plant parasites, feeding on sap, though a few species feed on fungi. They can vary greatly in appearance, from shiny pearl-like insects to insects covered with wax. They vary from 1–5mm (0.04–0.19in) in length. Adult females are not very mobile and mostly stay attached to the plant they have parasitized. They secrete a waxy substance for protection which makes them look like scales,

Below: Scale insects are a major problem for gardeners and commercial growers.

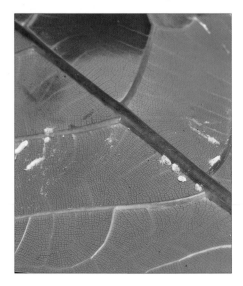

hence the name. Scale insects can be hermaphrodite, reproducing asexually or sexually. Eggs can be laid outside the body, in a protective sac, or withheld in the body, in which case live birth occurs.

Cicadas, frog-hoppers and leaf-hoppers (Auchenorrhyncha)

This is a diverse order, including familiar species such as cicadas as well as many less well-known groups. In this group, the rostrum (beak) arises from under the back of the head.

These species range in size from 0.2–10cm (0.08–4in) with wingspans up to 20cm (8in). There are about 17,000 species described, all terrestrial plant feeders.

Cicadas

About 4,000 species of cicadas are known. They reach lengths of 10cm (4in) although some are only 10mm (0.39in) long. Their most distinctive features are a stout body, large head and large eyes, as well as long and usually transparent wings. They are best known for their loud, high-pitched buzzing songs, produced by the males to attract the females.

Frog-hoppers

Spittlebugs or frog-hoppers are characterized by the frothy masses of 'spittle' produced by the nymphs. About 3,000 species have been described, most of them tropical. Their hind legs are long and adapted for leaping, hence their common name. Most species feed on sap from herbaceous plants or trees, and commonly live in grasslands with the nymphs feeding on roots.

Leaf-hoppers and tree hoppers

Found mainly in tropical America leaf-hoppers and tree hoppers occur in rainforests, savannas and deserts. They are extremely diverse. Many species mimic the substrate they live on in both shape and colour, some resembling thorns with spiny projections, or other plant parts. Many members of this group are gregarious, with young and adults feeding together. They tend to be active during

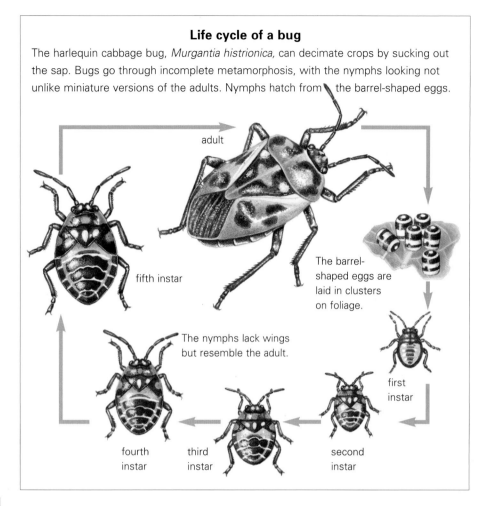

Life cycle of a bug

The harlequin cabbage bug, *Murgantia histrionica,* can decimate crops by sucking out the sap. Bugs go through incomplete metamorphosis, with the nymphs looking not unlike miniature versions of the adults. Nymphs hatch from the barrel-shaped eggs.

adult

fifth instar

The barrel-shaped eggs are laid in clusters on foliage.

The nymphs lack wings but resemble the adult.

first instar

fourth instar

third instar

second instar

the day. Some nymphs are tended by ants and parental care is also common. They feed on a wide range of host plants, including herbs, grasses, shrubs and trees. The adults are often highly mobile and can migrate long distances, which makes them especially effective as transmitters of plant diseases, especially viruses.

Plant-hoppers

Fulgorids, as plant-hoppers are known, are common in nearly all habitats worldwide. They range from 0.4–10cm (0.16–4in) in length, with wingspans of up to 15cm (6in). Species are diverse and sometimes they have bizarre body forms, with a bulging, disproportionate head. Some species have large colourful eye-spots on their hindwings to startle predators. All species are plant-sap feeders: some feed

on trees or shrubs, others on fungi. Most plant-hoppers spend their entire adult life on foliage, but the majority live underground as nymphs, feeding on grass roots, or in ant nests. Some species are of agricultural significance as pests, because of the plant pathogens they transmit when feeding, or because of the sticky honeydew they excrete. About 10,000 species have been described.

Right: A cicada has left behind its shell in its final moult. Cicadas are more often heard than seen. The sounds they make can travel more than a mile.

CADDIS FLIES, ALDER-FLIES, SNAKE FLIES, LACEWINGS AND SCORPION FLIES

The insects on these pages are a diverse group. All are winged adults and most are quite small, but they include a variety of herbivores, predators and scavengers. The lacewings and relatives are very varied in appearance, and some are quite large.

Caddis flies (Trichoptera)

With about 7,000 different species, caddis flies are easily recognized thanks to their translucent wings, covered with very thin hair.

Larvae are surrounded by a case built out of a variety of materials such as wood, grains of sand, or leaves to provide the best camouflage and protection against predators. They emerge from the water as adults.

Caddis flies exhibit different life-styles and behaviours, especially as larvae. Some have adapted to scrape the surface of rocks under water; others are filter-feeders or predators. Among adults, some species are solitary, while others tend to aggregate. Caddis fly larvae require well-oxygenated water and are very sensitive to water pollution, which makes them ideal as indicators of water quality.

Alder-flies (Megaloptera)

There are about 300 species worldwide of alder-flies and the related dobsonflies. They have membranous wings with many veins and with their anterior pair slightly longer than the posterior. They are typically dark. The larva goes through ten larval stages, which last several years in total, until pupation, which takes place on land in moist soil beneath stones or wood. The pupa is alert and can defend itself against predators.

Snake flies (Raphidioptera)

The snake fly order contains 200 species in two families. Species from both families are characterized by an elongated head and prothorax (the foremost segment of the thorax which bears the first pair of legs) which allows the head to move rapidly for catching prey. Females have a long and slender ovipositor.

Snake flies are predatory as adults and larvae, feeding on aphids and other invertebrates. The female lays up to 800 eggs under bark in living or decaying trees, where the larvae actively hunt other arthropods. As with alder-flies, the pupa is rather active and mobile.

Lacewings (Neuroptera)

There are about 5,000 species of lacewings and relatives, and the order includes mantispids, owlflies, antlions and some other groups. There are 1,300 species of lacewings known, which makes theirs the largest family of the order. They are of moderate size with the largest species having a forewing length of 34mm (1.34in). As the name suggests, their wings have a delicate lace-like appearance. This family of insects is characterized by an auditory organ, which allows them to detect bat ultrasounds, as well as by their ability to communicate through vibration transmitted by moving the legs against the abdomen. The female lacewing lays about 300 eggs, commonly on stalks. Most larvae are brown with darker spots and markings and some cover themselves with the dried remains of their prey for camouflage. Their mouthparts are a pair of strong and long curved mandibles which they sink into the victim's body to suck out the body fluids. All larvae and adults feed on aphids and other soft-bodied insects, although some adults will also feed on honeydew.

Owlflies are large, often brightly coloured day-flying insects that hunt on the wing. Antlions are famous for their voracious larvae which trap ants and other insects in sandpits. Adult antlions look rather like dragonflies but are more delicate and have broader, blotched wings.

Below: Snake flies are most common in Europe, central Asia and North America. They are named for their elongated prothorax, which gives them a vaguely snake-like appearance.

Left: Caddis flies hold their hairy wings angled roof-like when at rest. They have the appearance of moths.

Above: Lacewings have chewing mouthparts and undergo complete metamorphosis.

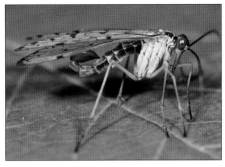

Above: Scorpion fly males are striking insects, with their scorpion-like tails and long, beak-like mouthparts.

Above: Female alder-flies lay thousands of eggs. The larvae are aquatic and have long filaments extending from the abdomen.

Scorpion flies (Mecoptera)

About 400 scorpion fly species are found around the world. They have a distinctive beak-like head with slender and serrated mandibles adapted for biting. Their abdomen is also elongated and in some males it ends in a scorpion-like tail, notably in members of the common scorpion fly family (Panorpidae). Unlike true scorpions, scorpion flies are harmless. They have membranous wings, which are often spotted, although some species are wingless. Most of the species are small, averaging 3mm (0.12in) but some can reach 30mm (1.18in). Scorpion flies are terrestrial either predatory (feeding on flies, aphids, caterpillars and moths), or scavengers or herbivores; some species prey on insects caught in spider webs.

Some scorpion flies display elaborate courtship behaviour. For instance, males offer a gift of prey before copulation to induce the female to mate or produce a column of saliva as a similar pre-copulatory gift.

Some species lay smooth and delicate eggs in crevices in small or large numbers, while others scatter their tougher eggs on the soil. Regardless of the species, the eggs generally require a moist environment from which they absorb water until they hatch. The larvae generally live in moist litter, feeding on dead insects or plants, especially mosses; some live in water and feed on live insects. Hanging scorpion flies also belong to this order. They look like craneflies and hang from a twig using their long forelegs.

Habitats of caddis flies, alder-flies, snake flies, lacewings and scorpion flies

Caddis flies have a largely aquatic lifestyle, spending most of their life in water as larvae and alongside streams or still waters as adults. The eggs, sometimes surrounded by a jelly-like substance, are laid underwater on the undersides of rocks. The larvae undergo up to seven different stages before the pupa, which is aquatic.

Snake flies are often found in the uppermost foliage of trees so are not very easy to spot. Adult females lay their eggs in cracks in the bark.

Alder-flies are found mainly in or around cool and well-oxygenated streams, but they also occur in standing waters. Females lay several thousand eggs on grass stems or other objects above water. After one to four weeks these hatch and fall into the water. The larvae are aquatic, living under stones or vegetation. They have strong, large mandibles adapted to catch other invertebrates, especially caddis flies. The adults do not feed.

Lacewings tend to live in forests but can be found in most of the habitats where aphids and their other prey exist.

Scorpion flies tend to live in habitats which provide them with shade and moisture, such as woodlands. They are often found living in mosses. Some species occur in arid regions but they are only active after rains. Antlions are found mainly in dry regions where their carnivorous larvae create sandpits to trap prey. Neuropterans generally are most diverse in the tropics.

BEETLES

The Coleoptera, or beetles, with more than 300,000 species and 166 families, constitute the largest order of insects. It has been estimated that one out of every five species of living organisms on the planet is a beetle. This order also has the largest families, with five containing more than 20,000 species.

Common features

Coleoptera means 'sheathed wing' (from the Greek: 'koleos' – sheath; and 'pteron' – wing). Beetles have two pairs of wings, but the first pair has been enlarged and thickened into a pair of hard sheaths, or elytra, which cover and protect the more delicate hindwings, as well as the dorsal surface of the abdomen. This distinctive feature has allowed beetles to exploit habitats such as leaf litter and the spaces under tree bark, and offers good protection to the wings. Although they have not adopted a truly aerial lifestyle, most beetles fly.

The mouthparts or mandibles are always of a biting type and resemble those of grasshoppers. They appear as large pincers on the front of some beetles. They are used to grasp, cut or crush prey or plant food.

Both the smallest and the largest of all insects can be found among this order, from the minute featherwing beetles (Ptiliidae), adults of which are just 0.25mm (0.01in) long, to the African Goliath and Hercules beetles (Scarabaeidae), measuring up to 15cm

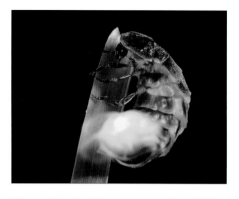

Above: Female glow-worms produce light to attract the winged males. The females retain a larva-like body form as adults.

Above: Click beetles have a mechanism which propels them into the air to escape predators, creating an audible click as they do so.

(6in) and weighing about 100g (3.5oz) true giants that are sometimes kept as unusual pets.

Beetles are exceedingly diverse ecologically and biologically. The majority of adult beetles are terrestrial herbivores, but several entire families and portions of others are predatory, fungivores or parasites, frequently with highly specialized life cycles and

Below: All beetles have claws to help them grip. They are visible on the feet of this stag beetle.

specific hosts. Though some beetles feed on nectar and pollen, this source has not been widely exploited by members of this order.

The typical life cycle of a beetle involves complete metamorphosis, from egg, through three to five larval stages, to pupa and adult. The length of the life cycle varies between species from a few weeks to many years. Some are known to spend decades as larvae.

Eggs can be laid either in clumps in substrates such as flour, or individually attached to leaves, or buried inside plant tissues. The typical larva is very voracious and usually the principal feeding stage of the beetle life cycle. Some feed on plants, but most feed on other insects. Although beetle larvae can be very diverse between species they are most often characterized by a hardened, dark head, the presence of chewing mouthparts and some openings along the sides of the body used for respiration.

Beetles provide some sort of parental care. Dung-beetles, for example, provide their young with food and shelter in the shape of a ball of dung.

Many beetle species are injurious to cultivated plants. Examples are the plum weevil or the grain weevils,

Above: Weevils are the most numerous of all beetles, accounting for around 20 per cent of beetle species.

Above: Beetles share the common feature of hardened wing cases, which protect the body of the insect.

Above: Some beetles meet and pair at flower-heads.

which live in nuts, fruits and grains and eat out the interior. The rice weevil (*Sitophilus oryzae*) and the granary weevil (*S. granarius*) are considered to be the most harmful weevils of all as they will attack all sorts of grains including rice, wheat and corn. Other beetles attack living conifer trees, such as the bark beetles of the family Scolytidae, causing damage to plantations.

On the other hand many beetle species play a beneficial role. Ladybirds are widely used for biological control, while glow-worm larvae eat snails and slugs. A few beetles are useful pollinators, while others, such as the dung-beetles, recycle animal wastes.

The following families include some of the most familiar and most numerous beetles in terms of species.

Weevils (Curculionidae)

Snout beetles or weevils are considered the most highly evolved family of Coleoptera. Containing more than 60,000 species worldwide, they comprise the largest single family in the animal kingdom. Weevils include some of the worst agricultural pests among insects.

Weevils have a distinctive long snout and clubbed antennae. Their form and size can vary greatly, with adult lengths ranging from 1–40mm (0.04–1.57in). Weevils are found on foliage or flowers as adults and are often spotted in gardens, but some species are ground dwellers or burrow in sand-dunes, and a few are aquatic or marine.

Adults are almost entirely herbivorous, feeding on seeds, fruits and other parts of plants, whereas the larvae are mostly internal plant feeders or subterranean. They are usually specialized to one species of plant.

Rove beetles (Staphylinidae)

Rove beetles are a large, varied family with very diverse modes of life; the number of species is estimated between 26,800 and 47,000. They often lurk around decaying matter.

These beetles are unusual in shape, with an elongated body and shortened elytra. Most groups have functional wings even though they are short in comparison to their body size. Some species typically run with their abdomen curled up, like scorpions. They are usually brown or black with a few species brightly coloured. Their size averages 1–10mm (0.04–0.39in) long.

Few species are herbivores. Most adults and larvae feed on invertebrates. Some feed on decaying vegetation.

Life cycle of a stag beetle

The stag beetle, *Lucanus cervus*, is typical for members of this order. The complete metamorphosis involves major changes in body shape from egg to adult. The larvae are soft-bodied grubs with hard, biting mouthparts.

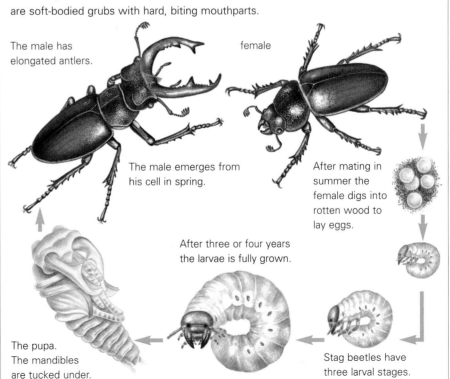

The male has elongated antlers.

female

The male emerges from his cell in spring.

After mating in summer the female digs into rotten wood to lay eggs.

After three or four years the larvae is fully grown.

The pupa. The mandibles are tucked under.

Stag beetles have three larval stages.

Above: Ladybirds, or ladybugs as they are known in North America, are among the best-loved beetles.

Above: Ladybird larvae, like the adults, are voracious predators of aphids, making them popular with gardeners.

A few species are specialized as predators or cohabit in ant or termite nests. Scarabs range in size from burrowing species less than 20mm (0.79in) up to 10cm (4in) in length.

Longhorn beetles (Cerambycidae)

Longhorn beetles include more than 20,000 species, most of which have very long antennae. Some species resemble ants, bees or wasps and can be brightly coloured. The largest beetles from this family are found in the tropics, such as the 16cm- (6in-) long giant long-horned beetle (*Titanus giganteus*) from South America.

Most are associated with decaying woody plants but some can attack living trees. These beetles can cause serious damage to trees, as well as buildings made out of wood.

Ground beetles (Carabidae)

Some of the world's best studied insects are ground beetles. They form a large and successful family with more than 40,000 named species.

They range from 1–60mm (0.04–2.36in) long. Most are flightless and if they have any wings, there are rather short and rarely used. Although around 30 per cent of species live in trees, these are predominantly ground-dwelling beetles. They are swift-running and active as adults.

Ground beetles are typically predatory and the majority of larvae have specific requirements as to the type of prey they will feed on. Some specialize in catching springtails, others in eating snails. A small percentage is herbivorous and some species feed exclusively on seeds.

Below: Many beetles are pests, destroying agricultural crops, while others are beneficial in a garden environment.

They are two types – the autumn breeders, which usually hibernate as larvae, and the spring breeders, which hibernate as adults. They usually lay eggs singly and deposit them on the ground or dig a small hollow. A few species construct mud or clay cells above the ground. Some species guard their eggs or store food for larvae. Most have three larval stages, a pupa and an adult stage.

Ground beetles can play a beneficial role in agriculture as they consume eggs of a range of pest insects, for example cabbage root-flies, cereal aphids and midges.

Scarab beetles (Scarabaeidae)

There are 27,800 species known worldwide in the scarab beetle family.

Most of the males are characterized by spectacular horns on the head used to fight over females. Scarab beetles have very diverse diets and feeding behaviour. Some larvae feed on rotting wood, others on mammal dung, carrion and roots of living vegetation. Certain beetles will choose a specific type of dung which they will provide to their larvae. Some will even tend the dung to prevent the growth of mould.

This family also contains the bulkiest of all beetles. Adults of some species such as the Japanese beetle *Popillia japonica* are significant pests as they attack living foliage, especially members of the rose family.

Darkling beetles (Tenebrionidae)

This is another large family with about 19,000 species. They are well represented in dry country, including deserts. They are highly variable in shape and size, ranging from 1–80mm (0.04–3.15in) long. They are often dark, but sometimes brightly coloured or metallic. They are often found on the ground under rocks and logs.

Jewel beetles (Buprestidae)

Characterized by their often dazzlingly bright colours, the jewel beetle family contains more than 14,000 species. The larvae are mostly wood-borers or leaf miners, with a few being considered as pests.

Below: Beetles have a wide and varied diet. A few species are reliant on a specific crop.

Above: Colouring can be a means of defence for a beetle.

Above: Beetles use their antennae to test and smell the environment around them.

Click beetles (Elateridae)

A family of around 10,000 species distributed worldwide. The click beetle name refers to a mechanism that creates a click and which allows the insects to fling themselves into the air if they are placed on their backs. The long cylindrical larvae are called 'wireworms' and are found in rotting wood. Many species feed on roots and can cause serious damage to crops. A few tropical species have bioluminescent spots.

Ladybirds (Coccinellidae)

This is one the most important families, as a majority of its species, known as ladybirds, feed on pest insects such as aphids and are widely used in biological control. However, some are also known to eat plants and crops and can be very destructive. There are around 5,000 species known.

These are small insects ranging from 1–10mm (0.04–0.39in). The most common species are yellow, orange or red with black spots. Most adults and larvae are predators on very small insects. A few species are plant eaters.

Most species breed in spring or summer and the female lays her eggs (from a few to hundreds) near aphid colonies. She sometimes lays infertile eggs as a provision for the young when food is scarce. The life cycle of a ladybird from egg to adulthood lasts four to seven weeks. The larva emerges from the egg within one to two weeks, and it then goes through four different larval stages and reaches maturity within two weeks. Pupation then takes place and the adults emerge after one or two weeks. Some species form aggregations as adults.

Water beetles (Dytiscidae)

This is the largest family of aquatic beetle, with about 4,000 species. The legs are modified as paddles for swimming. Body length ranges from 3–40mm (0.12–1.57in). These beetles typically hold air between the elytra and the abdomen when diving but they must periodically surface to renew their air supply, which restricts them to shallow water.

Most live in or near freshwater habitats. There are a few marine species, which tend to live on the shore.

Both adults and larvae are predatory; the larger species are able to kill small fish, although their main prey is tadpoles and invertebrates. The larvae in particular are voracious hunters, often known as 'water tigers'. The larvae breathe at the water surface through their tails and use their sharp jaws to catch their prey.

Habitats of beetles

Beetles exploit a variety of habitats throughout the world (except polar regions) from rainforest canopies and lakes to mountains and the driest deserts. As with most insects, they are most diverse in the tropics. They can also be found in deep caves and underground watercourses.

Rove beetles are found around decaying organic matter with a large number of species being associated with fungi, including moulds and rusts. Some have adapted to living in nests of social insects – mainly ants and termites – while others are found near water.

Ground beetles occupy every type of habitat from tropical rainforests, where they are found in the greatest numbers, to arid semi-deserts.

Scarab beetles inhabit burrows beneath their food source and come to the surface at night to feed.

Water beetles, as their name suggests, occupy many freshwater habitats.

FLIES

We see flies perhaps more than any other insects. Outdoors they are common and they are among the most visible and noisy creatures to enter our homes. As well as being a nuisance, some fly species carry diseases. A few, however, are beneficial to humans.

Common features

The order Diptera contains the insects commonly known as true flies. This is a large order with an estimated 85,000–100,000 species. Among the most common and familiar is the housefly, which has colonized every continent except Antarctica and is one of the world's most widely distributed animals.

They are mostly fairly small, with a single pair of functional wings, hence their scientific name Diptera (from the Greek 'di' – two, 'pteron' – wing). The second pair of wings is reduced. This feature distinguishes them from other insects also called flies. Although flies are predominantly aerial, about 20 families include members that are wingless or which have short wings. Some species belonging to the fungus gnat family are bioluminescent, producing their own light.

This is a diverse order with a wide range of ecological roles from herbivores and predators including the robber flies (Asilidae), to internal

parasites whose larvae develop inside a living host. Many, such as mosquitoes, are external parasites, feeding on blood, although the majority of the species feed on decaying organic matter, pollen or nectar. Some flies do not feed at all as adults.

Fly larvae also have an enormous variety of feeding habits. Many consume decaying organic matter, as seen in some mosquitoes and black flies, or are predacious; a large proportion are parasitic on other insects or organisms.

Left: Flies have large compound eyes and good eyesight, which they use to find food and avoid predators.

Life cycle

Diptera exhibit complete metamorphosis in their life cycle. Courtship and mating typically occurs in the air. Some courtship behaviours can be elaborate, such as in the dance flies, Empididae, where the male offers prey to the female. Eggs are usually small and deposited singly or in masses on or near the larval food. The larvae predominantly occur in moist to sub-aquatic habitats, and are often called maggots. They are commonly small, pale and soft-bodied. They lack true legs and have a reduced head. Adults have a short lifespan, from a few days to a few weeks.

Harmful species

Species such as the mosquitoes (Culicidae) are important as vectors of diseases including malaria, encephalitis and yellow fever, and can seriously affect humans and other animals. Black flies, deer flies and tsetse flies can also transmit diseases and house-flies can carry micro-organisms that cause dysentery. A few dipterans are pests of plants, such as root maggots, fruit flies, and leaf-mining flies, and these can seriously reduce crop yields.

Beneficial species

Perhaps surprisingly, some species can be useful. For instance, the vinegar fly *Drosophila melanogaster* has been widely used as an experimental subject in research into animal genetics and development. Other species are important pollinators, such as hover-flies. Some biting midges and sandflies are also important pollinators of tropical crops such as cacao. Overall, however, their contribution to pollination is rather small.

Habitats of flies

Dipterans have colonized most habitats and are widely distributed throughout the world. Very few species are restricted to one type of habitat or ecosystem. The majority of mosquito species live in the tropics. Midges are abundant in cooler climates. Many flies are attracted to rubbish and other waste, from which they pick up and carry disease. They are abundant in urban areas.

The larvae of some dipteran species help to provide clues in forensic science or are used in medicine to clean out wounds. Maggots are also bred commercially as bait for fishing, as well as food for captive reptiles or birds.

Some of the most important families of flies include the following.

Mosquitoes (Culicidae)

With about 1,600 species known, this family includes some important and familiar insects. They are typically slender flies, with piercing and sucking mouthparts.

Leatherjackets (Tipulidae)

Craneflies, or leatherjackets, are characterized by their large size and long legs. They are long and rather shapeless, with a sunken head. Some females can be wingless. Leatherjackets are plant root and foliage feeders.

Adults can occasionally feed on nectar or other fluids but generally do not feed. Larvae are usually found in damp soil where they feed on vegetation, mainly on plant roots, though some are scavengers and some are found in water.

A few craneflies, especially the larvae, are damaging to plants but they are otherwise harmless.

Deer or horseflies (Tabanidae)

There are about 3,500 species in this family. They are found in most of the world's ecosystems except in extreme northern and southern latitudes.

Adults feed on nectar and pollen, although females need to feed on blood for reproduction and their bites can be painful. The larvae are predatory and usually found in moist environments.

These are often considered as pests because of their bites. They can also transmit disease as well as parasites. Horseflies can cause severe damage to domesticated animals when abundant, causing some animals to lose up to 300ml (0.5 pint) of blood in one day. Some species are beneficial, especially in South Africa where they are important pollinators.

Large fruit flies (Tephritidae)

There are about 5,000 species of large fruit flies. These flies are among the most attractive and biologically interesting of the true flies, with patterned wings and often brightly coloured and patterned bodies. They also have elaborate behaviours both as adults and larvae.

Adults of many species may spend most of their life on one plant or adjacent plants of the same species. For example, the olive fruit fly feeds only on the fruit of cultivated olive

Below: All flies have just a single pair of wings. These are almost invariably transparent and point backward when the fly is at rest. Flies can land on almost any surface, including ceilings. Their feet secrete tiny droplets of slightly sticky fluid, which helps them to cling on.

Above: Craneflies have long, narrow wings and delicate dangling legs. Some craneflies can attain a length of 23cm (9in).

trees, and can ruin an olive crop. Tephritid larvae live in and feed on various plant tissues, depending on the species.

Most fruit flies lay their eggs in plant tissues, where the larvae will feed after emerging. The adults are particularly short-lived, some surviving less than a week.

Fruit flies are the most agriculturally significant family of flies. Some species are pests, causing major economic losses annually. Other species are beneficial biological control agents of weeds. Several species have been effective in destroying noxious weeds such as knapweeds.

Robber flies (Asilidae)

This family contains the robber-flies, with an estimated 4,000–7,000 species. They are moderate to large, and can be densely hairy. The family is characterized by the way they kill their prey, which are usually other insects caught in flight. They stab and inject their prey with saliva containing enzymes that paralyse the victim and enable it to be digested. These are active and powerful insects, often making rapid hunting forays from a perch.

Black flies (Simulidae)

This family of 1,800 species is widely distributed worldwide. They are also known as buffalo gnats.

These flies are characterized by a stout body with short legs and an unusually pronounced curved thorax. The average size of most species is 2–5mm (0.08–0.19in) and wing length can vary from 1.5–6.5mm (0.06–0.26in). They can be black, grey or yellow.

The aquatic larvae are found in clean, fast-moving water of streams and rivers as well as in the shallows of large lakes.

Like mosquitoes, most species are blood feeders, although the males feed mainly on nectar.

Below: Gall midges and other insects cause unsightly marks on foliage.

Above: Hoverflies are important pollinators of many flowers.

Females deposit from 100 to 600 small, shiny, creamy-white eggs on water plants, rocks, twigs or leaves, in streams, or simply scatter the eggs over the water surface where they gradually sink. The eggs hatch in four to five days. Eggs deposited in the autumn do not hatch until the following spring when the water warms.

Young larvae attach themselves to submerged objects. Most species have six to nine larval stages. The larvae elongate with the hind part of their bodies swollen. Winter may be passed in the larval form.

Pupation occurs in a cocoon, under water. Adults emerge in a bubble of air, after two to three days, when the water is warm. They are capable of immediate flight and mating. The entire life cycle lasts about four to six weeks, depending on species, as well as on the water temperature and food availability. There can be four generations per year.

Females of certain species are blood feeders and can be a problem for people and animals. They are particularly difficult to control because they often occur in large swarms and can get into hair, eyes and nostrils.

Some species can cause economic damage by killing poultry, and some transmit worms causing diseases, such as *Onchocera volvulus,* which causes blindness in humans. They can also transmit encephalitis.

Flower- or hoverflies (Syrphidae)

This family has more than 5,000 species of colourful, moderate- to large-sized flies. The black-and-yellow coloration of many species often causes them to be confused with bees or wasps. However, their flight behaviour is quite distinctive and unlike that of wasps. Hoverflies dart quickly, then hover in one spot.

Agile in flight and often seen hovering, the adults are associated with flowers, of which they are important pollinators. The larvae, on the other hand, are variable in their feeding habits; some feed on plants, others on detritus or in aquatic habitats and many are important as predators of other insects. Some hover-flies are beneficial, feeding on aphids, scale insects and other pests. Others feed within the nests of social insects.

Gall midges (Cecidomyiidae)

These are very small flies, usually no longer than 2–3mm (0.08–0.12in), with many species being less than 1mm (0.04in) long. Their name refers to the larvae of most species, which feed within plant tissues, creating abnormal plant growths called galls, which provide shelter for the larvae. The number of species is estimated at 4,000.

Gall midges can reproduce before they reach the adult form, a rare phenomenon known as paedogenesis. Sometimes the larva gets eaten by its own progeny.

They are usually considered as pests and are especially injurious to wheat. However, a number of species, mainly at the larval stage, are beneficial to humans, as they naturally predate other crop pests such as aphids or spider mites. One species, *Aphidoletes aphidimyza,* is often used for biological control and is widely sold in the USA.

FLEAS

Fleas belong to the order Siphonaptera. The order is entirely parasitic, with both adults and larvae
spending their lives on the surface of vertebrate hosts. Adult fleas are laterally flattened and their back
legs are very powerful for jumping on to their host.

Common features

This order contains the fleas, comprising 2,500 species distributed worldwide. The order name refers to their tube-like mouthparts 'siphon-' and the absence of wings '-aptera'. Fleas are highly specialized and unusual insects. External parasites, they feed on the blood of mammals and birds.

All flea species are small, measuring 1–5mm (0.04–0.19in) in length. They are dark coloured, with tube-like mouthparts adapted for piercing and sucking, and are covered with hairs and spines which project backward; these allow them to move more efficiently within hair or feathers. They are flattened laterally and wingless.

Habitats of fleas

Many fleas are mostly associated with a particular host, usually a mammal, but in some cases a bird. The adult fleas live in the fur or feathers, while the eggs and larvae develop in nests or bedding (or even in carpets).

Above: Some fleas are associated with a particular host. This is a hedgehog flea.

Fleas are agile insects with powerful hind legs adapted for jumping. Their body is tough and leathery, capable of withstanding great pressure so they cannot be killed easily. Flea larvae are worm-like and also covered with bristles. They lack eyes and have mouthparts adapted for chewing, and they have a pair of hook-like appendages on the last abdominal segment to attach more easily to their hosts.

Parasites

Adults feed on the blood of a wide variety of vertebrate animals but will usually parasitize animals that regularly return to their nests, bedding or burrows, such as rodents, bats or rabbits. They have not been observed on many animals that tend to have more nomadic behaviours, such as ungulates. More than 95 per cent of species are external parasites on mammals, with 74 per cent of fleas feeding on rodents; the remaining five per cent are parasites of birds.

Flea larvae feed on organic matter such as skin flakes and debris. As adults, they usually feed daily or every other day, but they can survive two

months to a year without a blood meal. Females require more blood than the males, for the development of eggs.

Fleas are rarely species-specific and can have up to 35 different hosts, and a host can bear up to 22 different species of flea.

Fleas lay tiny eggs in batches of up to 20, usually on the host or nearby. The eggs take around two days to two weeks to hatch; the larvae are left in dark places to avoid sunlight and pupate within one or two weeks in a silken cocoon. After one or two weeks the adult is fully developed and ready to emerge. Adults usually live four to 25 days, although some have been known to live for a year. Female fleas can lay up to 1,000 eggs in a lifetime.

Fleas are a nuisance to their hosts and can provoke allergies. They also act as vectors of bacteria, protozoans and viruses. Rat-fleas can carry the organisms that cause diseases, such as plague and murine typhus, which have killed large numbers of people.

Below: Cat fleas are highly irritating to cats when they breed in large numbers. They can also bite humans.

BUTTERFLIES AND MOTHS

With their large, opaque wings, butterflies and moths are easy to spot in flight. The two are differentiated by their antennae and the way they hold their wings at rest. In addition to this, most moths are nocturnal, whereas butterflies are active by day. They both belong to the order Lepidoptera.

Common features

The order Lepidoptera, (meaning 'scale wings') is the second largest order of insects, with more than 200,000 species comprising moths and butterflies, which have distinctive life-styles and features. The moths include about 180,000 species and more than 120 families, compared with butterflies with some 20,000 species and five families.

This is not a very diverse order in terms of biology and morphology. Members of the group have overlapping rows of scales on the wings. The body and legs are also covered by similar scales or by long hair-like scales or bristles.

Members of certain species are flightless. Most, however, can fly. The majority of members of the order are moderate-sized, averaging about 30mm (1.18in) in wingspan. However, their sizes range from 2.5mm (0.09in) wingspan in pygmy moths up to 30cm (12in) among the emperor moths and some members of the swallowtails, such as Queen Alexandra's birdwing (*Ornithoptera alexandrae*).

The shape of the antennae differs between butterflies and moths. Butterflies have thread-like antennae with the tips thickened into knobs.

Moths have very diverse types of antennae, varying from filamentous and feathered to toothed or comb-like, depending on the species.

Feeding

Lepidoptera are mainly herbivorous insects. There are exceptions. One species of owlet moth is known to suck blood using a modified proboscis to penetrate the skin, and some other species do not feed at all and live on food reserves accumulated during their larval stage, their mouthparts having atrophied. Most of the adults, however, feed on nectar, honeydew or exudates from fermenting fruit or sap. Mouthparts typically form a tubular proboscis, used to suck up liquid food.

The larvae of most species are also herbivores, feeding on a wide range of plants, from roots to leaves. The larvae have chewing mandibles that allow them to feed on plants. Many feed on a single species of plant and most of these specialize on one part of the

Below: Male moths often have more elaborate antennae than females, using them to pick up pheromones and lead them to mates. The huge 'eyes' on the lower wings act as a deterrent to potential predators.

Above: Some caterpillars live in groups and construct silk tents for protection from predators.

plant such as the leaves, buds or wood. However, larval biology can be highly variable, including aquatic forms, borers in stems, seeds or fruit, gall-inhabiting forms, scavengers, and a few predators that feed on the eggs of spiders or other moths and butterflies.

Defence

To protect themselves from predators, butterflies and moths have developed various strategies. As a result the group displays great diversity of coloration and markings, which can serve for defence as well as for courtship. For example the wing patterns which resemble eyes of vertebrates, found in many species, are believed to have the effect of startling predators. Some butterflies, especially among the hairstreaks, use their hindwing tails as 'false heads' to confuse predators.

Some butterflies and moths have developed another kind of defence – mimicry. They mimic a poisonous or distasteful species which has a striking display of warning colours. For instance, some diurnal moths have

identical colour patterns to those of certain bees or wasps in order to dissuade predators from attack. Other species display bright colours to advertise the fact that they are poisonous themselves.

Just as adult butterflies and moths have warning colours, so do many larvae. Larvae and pupae also display cryptic patterns similar to the substrate on which they rest. It is particularly important for the chrysalis (pupa), which is quite vulnerable, to remain hidden. Caterpillars may also change colour in the course of their development.

Life cycle

All lepidopterans have a complete, holometabolous (complete metamorphosis) life cycle, with egg, larva, pupa and adult stages.

Eggs are deposited on the host plant, or into the plant tissues, either singly or in groups. The total number laid can vary from a dozen to more than 18,000, with an average of 100–200 eggs; as many as 50,000 eggs have been recorded among some ghost moth females.

Some species, mainly moths, pupate on the host plant in a silken cocoon, within their leaf mine or in leaf rolls. Others pupate in the soil or leaf litter such as hawk moth larvae, which make an underground pupal cell.

Below: Butterflies at rest hold their wings closed together. The undersides are often less brightly coloured than the upper sides.

Life cycle of a red admiral butterfly

Known in many parts of the world, the red admiral butterfly, *Vanessa atalanta,* is one of the largest and most beautiful of predominantly European butterflies. It is brighter in coloration in the summer months and can vary in size depending on location. This butterfly is 'people-friendly' and will perch on a human.

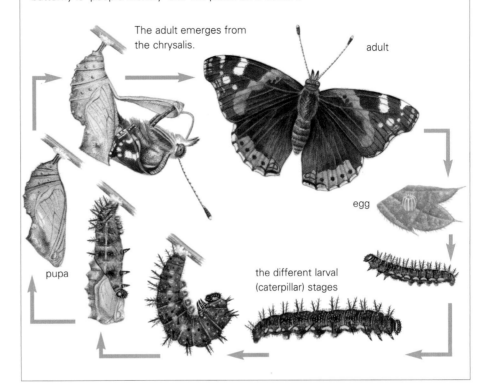

The adult emerges from the chrysalis.

adult

egg

the different larval (caterpillar) stages

pupa

Many butterflies hang head-down as a chrysalis (pupa) in a membranous covering. Depending on the species, the pupal stage may last from a few days to several years. Overall, most species have a short lifespan.

Some species are well known for their migratory behaviour, such as the well-studied monarch butterfly. The species migrates and aggregates in autumn to overwintering sites before dispersing and reproducing the following spring. Aggregation can also occur at night in some species, which is believed to be defensive behaviour against predation, for example in *Heliconius* butterflies. Other species hibernate in winter as adults.

Harmful species

Some species of moths and butterflies are known for the severe damage they cause to crops as caterpillars. The rice and grain moths (Pyralidae) cause damage to stored grain. Some also attack horticultural plants. Clothes moths, Tineidae, whose larvae can

make holes in blankets, clothes and carpets, can be a real nuisance in the home. A few species are beneficial, including *Cactoblastis cactorum,* which has been used in Australia to control the invasive prickly pear cactus, *Opuntia.*

Below: A monarch butterfly feeds on a flower. Both butterflies and moths have long, tubular mouthparts for drinking nectar.

Butterflies

The largest families of butterflies, the Nymphalidae and Lycaenidae each have about 6,000 species. One of the best known is the swallowtail family (Papilionidae) with about 600 species. The skippers (Hesperiidae) number about 3,500 species, and the whites and relatives (Pieridae) about 1,000.

Brush-footed butterflies (Nymphalidae)

This family includes well-known species such as the fritillaries, admirals, emperors, monarchs and tortoiseshells. They vary considerably in their appearance but are generally characterized by reduced and hairy front legs. Their caterpillars also vary, but are often hairy or spiny.

Many species are brightly coloured, but the undersides of the wings are generally duller, which helps the butterfly to escape detection by predators.

Lengthy migrations, territoriality and the ability to overwinter as adults have been observed in many members.

Blues, coppers and relatives (Lycaenidae)

This family contains about 40 per cent of all butterfly species. Adults usually have wingspans less than 50mm (2in).

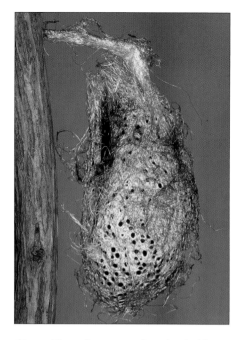

Above: These insects are hanging inside a hardened case, or chrysalis, as they metamorphose from caterpillar to adult.

They are brightly coloured, sometimes metallic, with blue and copper being the predominant colours. Larvae are more flattened than cylindrical.

Most species are herbivores but some species feed on aphids or ants.

About 75 per cent of lycaenid species have developed a strong relationship with ants, which can be parasitic, predatory or mutualistic.

In the latter, both species benefit. The larvae are tended and protected by the ants, which receive in return a sugar-rich honeydew produced by the larvae. Some species can complete their life cycle only in the presence of ants.

Swallowtail butterflies (Papilionidae)

The swallowtails and their relatives are among the most specialized lepidopterans in terms of behaviour and ecology. The name swallowtail refers to the extensions commonly found at the tips of the hindwings, although this feature is not present in all species. Many of them are large and colourful. Because of this, many swallowtail species have been collected for displays and their wings used in jewellery. Some have therefore become rare.

Powerful in flight, using alternating periods of flapping and gliding, swallowtails can cover large distances as they search for food or breeding sites. The caterpillars can easily be distinguished from those of other families thanks to the presence of a fork-like organ behind their head which can emit smelly secretions as a defence against predators. The pupae hang upside down using a hook called a 'cremaster', located on the rear of the chrysalis, another typical feature of the Papilionidae. This family includes large, magnificent birdwings.

Moths

The largest and best-known families of moths are the Noctuidae with about 26,000 species, the Arctiidae with over 11,000 species and the Tortricidae with about 9,000 species.

Owlet moths (Noctuidae)

Species belonging to the Noctuidae family occur worldwide. They have a robust appearance, a medium size and dull colours, although some species have brightly coloured hindwings. Some species that are highly predated by bats have developed a tiny ear-like

Left: Most moths have opaque wings but in a few species they are clear. Almost all moths, like butterflies, feed on nectar.

Above: Butterflies and moths lay their eggs directly on the species of plants on which their larvae feed. Many are confined to just one plant type.

organ which can detect the ultrasounds emitted by their hunters.

Noctuid caterpillars frequently hide under debris or leaf litter, emerging at night to feed. They often chew through stems, hence their common name 'cutworms'. Some species are serious horticultural and agricultural pests.

Woolly bears (Arctiidae)

The family name Arctiidae comes from the Greek 'arctos', meaning bear, referring to the hairy appearance of these caterpillars, which are popularly known as woolly bears or woolly worms. A large and diverse family, arctiids are found virtually everywhere in the world, but are most common in the tropics. The family includes the brightly coloured tiger moths, but also footmen, which are much duller, lichen moths and wasp moths. Like owlet moths, arctiids are sensitive to bat sonar. Uniquely they can produce ultrasonic sounds, which are used in mating and defence against predators; these can interfere with the ultrasounds emitted by bats, which then have difficulties locating their prey. Many species retain distasteful or poisonous chemicals obtained from the plants they fed on as caterpillars. Others produce their own chemicals. These defences are generally advertised with bright coloration, unusual postures or odours. Some arctiid caterpillars have adapted to cold temperatures by producing a protectant chemical – a sort of natural antifreeze. Many adults and larvae are active during the day.

Leaf-roller moths (Tortricidae)

This family contains the leaf-roller moths. Most of the adults are small, from 8–30mm (0.31–1.18in) in wing-span, and the caterpillars are generally smooth-skinned. Some species have particular distinguishing features, such as the bell moths whose name refers to the shape of the adult when at rest. Others have a striking resemblance to bird droppings, which helps to camouflage them from predators.

The caterpillars typically roll the leaves in which they pupate, hence the common name. Some species drop on a silken thread when disturbed. The famous Mexican jumping bean is caused by the movement of the larva of the tortricid moth *Cydia deshaisiana* which lives inside seeds.

Some species feed destructively on plants and can be of considerable economic significance, such as the

Above: Many moths are beautifully patterned and coloured, but are difficult to spot on some surfaces.

spruce budworm, *Choristoneura fumiferana*, and the fruit-tree roller, *Archips argyropsilus*. Others are leaf miners or feed on dead leaves on the forest floor. Many larvae bore into fruit, nuts or seeds and can cause severe damage. The larvae are a favourite food of many birds.

Habitats of butterflies and moths

Butterflies and moths have colonized most habitats, from the Arctic to the tropical jungles. Specially adapted to feed from flowers, butterflies are associated with rich meadows, woodland and forest clearings, and other sites where there are vegetation and flowers. They are also very welcome colourful visitors to parks and gardens. They can be found on every continent except Antarctica. They are most diverse in the tropics, but are also common in temperate regions.

The largest number of species is found in east and South-east Asia.

BEES, WASPS, PARASITIC WASPS, ANTS, SAWFLIES AND WOOD WASPS

This large order, known as Hymenoptera, contains many insect species that live highly social lives, often in large colonies. Their wings are thin and transparent and many species can deliver poisonous stings. Many species are important pollinators of flowers and crops.

Common features

Hymenoptera is the third largest order of insects, with more than 280,000 species. It includes the familiar wasps, bees and ants, as well as the rather less familiar sawflies, wood wasps and parasitic wasps (including ichneumons). It is a diverse group well known for the advanced social behaviour of some of its members, often involving separate reproductive and worker castes, as well as the complicated forms of communication displayed by some species.

Hymenopterans are characterized by their membranous wings, which has given the name to the group, derived from the Greek 'hymen' – membrane, and 'pteron' – wing. They usually have two pairs of wings. Members of

Right: Bees ingest nectar through a straw-like proboscis. Flowers produce nectar specifically to attract bees and other pollinating insects.

certain species, notably ants, only have wings at specific stages of their life cycle.

Hymenopterans have a very mobile head with well-developed mandibles adapted for chewing, as well as for attacking or defending themselves. In most species the mandibles are more often used for cutting the insect's way out of the pupal case or for nest-building rather than for feeding. However, a worker ant may dig, transport food or soil particles, manipulate prey, defend the colony, or tend grubs, all using the mandibles.

The morphological adaptations most responsible for the success of bees and wasps involve the abdomen, with the presence of a long tube called the 'ovipositor' (egg-laying tube). This has developed as a piercing organ, which can be inserted deeply into plant or animal tissues to deposit the eggs. In some it is modified into a sting; this adaptation provides efficient protection against predators. The larvae of hymenopterans are mostly grub-like. The pupae may grow in cocoons or in special cells, or develop inside the host in parasitic forms.

Bees (superfamily Apoidea)

There are more than 20,000 species of bee, found on every continent except Antarctica, and in every habitat that contains flowers. Bees have a long tube-like organ or proboscis which allows them to obtain nectar from flowers. Adults range in size from about 2–40mm (0.08–1.57in).

Bees are divided into two types: solitary bees and social bees. In solitary species, adults construct individual nests and provide their young with plant materials, usually nectar or pollen. They rear their young in the nest but do not produce honey or wax. Social bees by contrast build

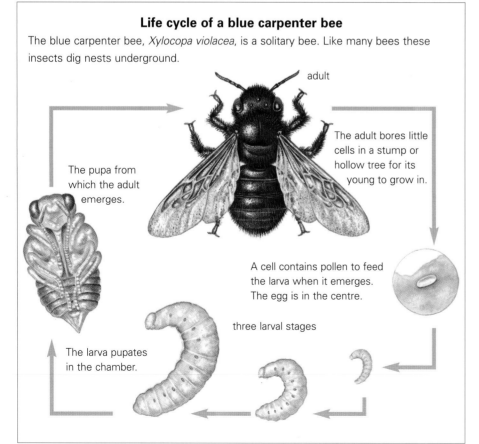

Life cycle of a blue carpenter bee

The blue carpenter bee, *Xylocopa violacea*, is a solitary bee. Like many bees these insects dig nests underground.

adult

The adult bores little cells in a stump or hollow tree for its young to grow in.

The pupa from which the adult emerges.

A cell contains pollen to feed the larva when it emerges. The egg is in the centre.

three larval stages

The larva pupates in the chamber.

communal nests in the soil (typical of bumblebees) or in cavities (honey bees). They belong to one large family (Apidae) and they are among the most familiar and well studied insects. There are two levels of sociality among social bees. Some are 'semi-social' with groups of sisters cohabiting and dividing labour between them. Others show a higher level of sociality and are called 'eusocial'. Eusocial species are characterized by the presence of a queen which gives birth to sterile females, commonly called workers. These carry out specialized tasks among the colony. Eusociality can itself be divided into a primitive form, where queens and workers are very similar and the colony relatively small, and a more evolved form where the different castes differ morphologically and the colony can contain up to 40,000 individuals.

Their strict diet of pollen and nectar makes bees very important in agriculture and horticulture worldwide. Honey bees and bumblebees will fly from flower to flower to collect nectar and pollen, using their long proboscis, and these

Below: Wasps guard the entrance to their nest. The nest appears to have a paper-like quality.

Above: As well as drinking nectar, bees sometimes collect pollen to feed to their young, carrying it in bundles on their back legs.

foods are then converted to honey to feed the larvae. During the process, the pollen remaining on the insect is transported from one flower to another and pollination may occur if a subsequent flower the bee lands on is of the same species.

Wasps (superfamilies Vespoidea and Sphecoidea)
Classification of wasps is quite complex, and includes solitary as well as social species. The following are some of the important groups.

Above: Bumblebees have more rounded, hairy bodies than honey bees, wasps or hornets.

Parasitic wasps (division Parasitica)
This group contains fig wasps and gall wasps and related hymenopterans, including ichneumons and braconids. There are 200,000 species. Many are black or yellow, with transparent wings and are rather small, averaging 1.5mm (0.06in) in length. The larvae are parasitic on other insects. The adults mainly feed on plant nectar.

Hunting wasps (Vespidae)
These are either solitary or social species that mainly catch their prey by stinging and paralysing it. This group includes common wasps ('yellow jackets'), hornets and paper wasps, as well as some solitary wasps.

Most social wasps are fairly large and there are about 5,000 species found throughout the world. A typical colony includes a queen and a number of sterile female workers. In temperate climates, colonies typically last one year, dying when winter comes. New queens and males are produced at the end of the summer, and after mating the queen hibernates over winter in sheltered locations, starting a new colony the following spring. The nests are made out of plant fibres which are chewed to make a kind of paper, hence the name 'paper wasps'. Prey are chewed before being fed to larvae, and the larvae in return produce a liquid consumed by adults.

Ants (family Formicidae)

The Formicidae are one large family of ants containing 14,000 species. The family name refers to the formic acid that ants naturally produce to defend themselves. Most ants are relatively small 2–25mm (0.08–1in), but the queens can be sizeable insects. The queens of one species, *Dorylus helvolus*, from southern Africa, grows up to 50mm (2in) long, making them the world's biggest ants. Most ant larvae are grub-like: their heads are extremely reduced and they lack legs. As such, they are completely helpless and rely on the worker ants entirely to feed them and to move them around.

Ants are highly social insects. Their colonies can consist of millions of individuals divided into castes or social groups, including sterile female workers, fertile males and fertile female queens. Each colony has at least one queen. The queens are the only individuals to lay eggs, which are then looked after by the workers. The different castes show a more or less wide range of morphological differences, depending on the species.

The lifespan of ants varies among the castes, with sterile females living from one to three years, whereas

Below: Worker ants carry partly developed larvae to safety, after their nest has been disturbed.

queens can live up to 30 years. Males usually only survive a few weeks. Colonies are large, and those of some tropical army ants may number several million individuals.

Ants are characterized by their organized permanent colonial life and their advanced social behaviour. They co-operate to an amazing extent, gathering food and hunting in large masses, even using their own bodies as bridges for other ants of the colony. Ants communicate through chemicals, which produce behavioural changes in other individuals.

Most ant species have also developed specific behaviours, some of which are strongly associated with plants. The leaf-cutter ants, for example, behave rather like farmers harvesting pieces of leaves and carrying them back to the nest (they can lift up to 20 times their own body weight). The leaves are then chewed to allow a fungus, on which the ants feed, to grow on them. Ants can also be pollinators, or protectors of plants. Some ant species in Africa live on acacia trees (in galls produced by the tree's tissues) and their presence on the twigs and branches helps to defend those trees against attack from giraffes and other herbivores.

Slave-maker ants raid the nests of other ants to steal their pupae, which are then used as slaves in the colony after they hatch.

Above: This ants' nest, built in Kenya, has become a large structure. These structures are common and litter the landscape in certain areas.

Various types of nests can be found among the different species, from the 'bivouac' type formed by the ants' own bodies observed in nomadic species such as army or driver ants, to the sophisticated leaf nests built in trees by weaver ants, or large colonies in mounds or under the soil.

Sawflies and wood wasps (suborder Symphyta)

These are considered the most primitive of all Hymenoptera, partly as they do not show the complex social behaviour observed in many other groups. They are widespread, with about 10,000 species. Their common name refers to the ovipositor, which has the appearance of a saw blade, being toothed along one edge.

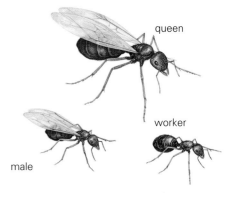

Above: Only male ants and queens have wings and these are just for the nuptial flight, when they meet and mate.

Above: Adult sawflies can fly well, but often prefer to clamber about on leaves or flowers.

A number of sawflies are considered to be pests, boring into wood or defoliating plants. Wood wasps have sturdy ovipositors and can use these to drill into timber.

Adults mostly have a dark body and legs, and look rather like wasps or flying ants. However, they lack the narrow 'waist' of most hymenopterans. The larvae look like caterpillars in general appearance and are often camouflaged in shades of green.

These hymenopterans are characteristically external foliage feeders and affect a number of different plants, depending on the species. Some feed on herbs or ferns, but most feed on trees and shrubs. Among the most common are the apple sawfly, the common gooseberry sawfly and the pear and cherry slugworm. Adults of some species are carnivorous or feed on nectar. Some of the larvae are internal feeders including leaf miners and gall formers, and are generally legless.

Adult sawflies generally do not live for more than two weeks. In many species, flower pollen forms a major part of their diet. Eggs are deposited in slits made in leaves or pine needles, and the mature larvae usually leave the host plant to pupate in a cocoon or in a cell in soil or leaf litter.

Right: An ants' nest has been carefully constructed of leaves.

Habitats of ants

Many ants nest underground, so that even where they are common they remain largely hidden. An ants' nest is made up of many different chambers. In forests they recycle huge amounts of material and in tropical rainforests in particular they play a major role in maintaining the stability of the immediate ecosystem in which they live.

Ants are common in both temperate and tropical parts of the world. Wood ants build large domed nests from pine needles. Often, as here, their nests are sited over a decaying tree stump, which provides heat as well as structure. Inside they have many different chambers used by different members of the colony.

MILLIPEDES AND CENTIPEDES

Millipedes (about 11,000 species) and centipedes (about 3,000 species) are the main classes in the superclass Myriapoda – many-legged arthropods – which has about 15,000 species in total. The other two myriapod classes are the symphylans (about 200 species) and the pauropods (about 700 species).

Millipedes (Diplopoda)

The name millipede is misleading as these creatures never have more than 350 pairs of legs. In fact, most millipedes have fewer than 50 pairs of legs. Millipedes vary in size and shape, some being long and flat-bodied, others are short with bristle-like hairs or prominent lateral projections. Some, such as the pill millipede resemble woodlice and can protect themselves by rolling into a tight ball. Millipedes have rather hardened heads, which helps them move through rotting wood or compacted soil. Adults vary from 2–30cm (0.08–12in) in length, and in colour from whitish to brown or black. They can secrete toxic chemicals through glandular openings on each side of each body segment, and this helps protect them from predators such as spiders and ants.

With elongated, segmented bodies like their close relatives the centipedes, millipedes differ from centipedes in having two pairs of legs on most body segments instead of one. In comparison they are slower-moving creatures.

Millipedes feed mainly on dead plant material, usually leaf litter, but they occasionally eat seeds as well as

Below: Millipedes and centipedes are long and slender multi-legged insects that are immediately identifiable.

roots and shoots of seedlings; some species are scavengers. They tend to wait until leaves are partially degraded before they feed on them.

The millipede life cycle is rather long compared with that of other arthropods. It can take from one to four years before millipedes complete their metamorphosis. Sperm is transferred from the male to the female by means of specially adapted legs. The eggs are deposited in spring or summer, in a cluster in the soil, sometimes in a chamber or cell. Females lay between 50 and 300 eggs, which hatch within nine or ten days. The larva looks like the adult except it only has a few segments and three pairs of legs. Body segments are added at each moult. Millipedes usually go through ten different larval stages or moults. Parthenogenesis occurs in some species.

Some millipedes can be a pest in the garden or greenhouse, such as the garden millipede *Oxidus gracilis,* introduced to temperate areas from the tropics. The spotted millipede is another pest, feeding on bulbs and roots of vegetables.

Millipedes, however, are of ecological importance, as decomposers recycling organic matter. In some habitats they ingest more than ten per cent of the annual leaf litter.

Above: Many millipedes coil up if threatened, protecting their legs and leaving just their smooth, rounded sides exposed.

Centipedes (Chilopoda)

Most centipedes move more quickly, being active hunting carnivores. They have colonized diverse habitats in many parts of the world.

Centipedes measure from 1–10cm (0.5–4in) in length on average, but some species such as the Amazonian giant centipede *Scolopendra gigantea* can reach 30cm (12in). They have long bodies but only a single pair of legs per body segment instead of the two pairs characteristic of millipedes. Many species have 15 pairs of legs, but some, such as *Gonibregmatus plurimipes*, a species found in the Pacific islands, have as many as 191 pairs. Centipedes are adapted for running and are fast. Some of the largest species have longer legs toward the end of the body, which enhances their speed. Some soil-dwelling species move by coiling their body in a worm-like movement, as well as using their legs. Centipedes have a unique adaptation in which the legs of the first body segment have been modified into claws containing poison glands; this allows them to kill their prey efficiently, as well as to defend themselves.

Right: Millipedes have long legs which stick out from the sides of their bodies.

Centipedes are predatory, feeding on insects, snails and earthworms. The largest species can feed on rodents, toads or even snakes. They use their antennae and legs to find their prey, which is then killed using their poison-bearing jaw-like claws.

Centipedes reproduce externally, the male depositing its spermatozoa in a small web, which is then taken up by the female. Reproduction is sometimes preceded by a courtship dance, depending on the species, and generally occurs in spring and summer in temperate regions. Some species lay a single egg in a hole, others lay between 15 and 60 eggs in a nest located in the soil or in rotten wood. The latter tend to show parental care as they guard and lick their eggs to prevent fungi growing on them and protect them against predators. In some species the female guards the young until they leave the nest. Many species are known to have five or six different larval stages and reach the adult stage within one to three years. The young have seven pairs of legs on hatching and develop the rest of the segments and legs with successive moults. Most species live for three to six years.

Some species can be venomous and although they are not lethal to humans they can cause severe irritation and damage to the skin. Large centipedes are best treated with respect.

Below: Leaf litter provides a hunting ground for centipedes and food for millipedes.

Habitats of millipedes and centipedes

Woodland vegetation provides an ideal millipede habitat, with plenty of decomposing plant matter for them to burrow under.

Millipedes are usually found under leaf litter or stones, or below the surface in moist environments such as deciduous forests, as many species are sensitive to desiccation. Some, such as the black millipedes, are known to climb trees. A few occur in very dry environments such as deserts, or at high altitude. Some species have developed strong relationships with ants, where the ants provide protection and the millipedes help to clean up detritus from the nests. Other millipede species form associations with termites.

Centipedes live in terrestrial habitats, either in soil, under leaf litter or beneath stones and bark, and are active mainly at night. Unlike millipedes, they tend to live in dry and arid environments including deserts, though some species require a humid environment. A few are found near the sea, among seaweeds.

SPIDERS, MITES AND TICKS

Arachnids are one of the largest groups of predatory arthropods. The most familiar are the spiders; these and the mites and ticks each have about 35,000 species. The other main arachnid groups are scorpions (about 1,500 species), pseudoscorpions (about 2,000 species) and harvestmen (about 4,500 species).

Common features

Arachnids have distinctive features that differentiate them from insects. These include a two-segmented body, four pairs of walking legs (although some mites can have fewer), and no antennae or wings. They also have simple rather than compound eyes. The number of eyes varies from none to as many as 12 in some scorpions.

Apart from some mites and the water spider (*Argyroneta aquatica*), most arachnids are terrestrial. They are mainly carnivorous, feeding on insects and other small animals.

The reproductive behaviour is unusual in many species, with prolonged and complex courtship. Parental care is common among spiders. Arachnids usually lay eggs, which hatch into offspring that look like miniature adults.

The class Arachnida consists of 11 subclasses, of which the five included here are the largest and by far the best known.

Spiders (Aranae)

Spiders produce silk, which emerges from the tip of the abdomen. Silk is very strong and flexible and is used by spiders for many purposes, including spinning webs. Most spiders have

Below: A spider clutches her egg cocoon close to her body.

Above: This spider, like many, is very well camouflaged.

Above: Spiders spin webs in which to trap their prey.

poison glands, but they are not all dangerous. The majority of species have eight eyes, arranged in two rows.

There are 35,000 species of spider in more than 100 families. Major groups include the tiny money spiders; orb-weaving spiders; wolf spiders; crab spiders; jumping spiders; trap-door spiders; and the large tarantulas.

Although a few include pollen in their diet, most spiders are carnivores. Most spiders are generalist predators. Prey are usually smaller or similar in size, but many spiders can subdue prey which are several times their own weight. Some spiders will occasionally feed on vertebrates. A few scavenge dead insects. Some spiders stalk their prey, others ambush it, while a few species steal prey captured by other spiders. Most trap their prey in some kind of web. Spiders paralyse their victims by injecting poison secreted by a pair of poison glands in the front jaws.

Female spiders lay their eggs within a few weeks of mating, usually placing them in a silken sac for protection.

Most female spiders can produce multiple egg sacs, each containing from a few to more than 1,000 eggs. In many species, the female guards the egg sacs against predators until hatching, which usually takes a few weeks. In a few species she provides food for her offspring. After one or two weeks the young are ready to disperse. In temperate regions, most species live for one or two years. Some of the longest-lived spiders are female tarantulas, which may live for up to 25 years, (the males rarely live for more than a couple of years).

Mites and ticks (Acari)

Mites and ticks are a large and diverse group of mostly very small arachnids. They greatly outnumber all other arachnids and are distributed worldwide. There are more than 34,000 named species of mites and about 850 species of tick. They vary in length from 1–7mm (0.04–0.28in) for mites and from 2–30mm (0.08–1.18in) for ticks. Their two-part body is fused into one piece. Although adults of most species have four pairs of legs some have fewer, and the larval stages of many species have only three pairs. Most larvae have six legs, as opposed to eight at the nymphal and adult stages. Some mites can produce silk in their palps. Mites and ticks are the most common external parasites of humans and other vertebrates.

Mites and ticks pass through at least four life stages – egg, larva, nymph and adult – but there is a great deal of variation, often including extra stages. A moult occurs between each stage.

Mites

Free-living and parasitic mite species exist, with the free-living being the most common. Parasitic species attack both vertebrates and invertebrates, causing damage by burrowing under skin. Other species live in or near the hair follicles of mammals including humans. The best-known mite is the house dust mite, associated with asthma. Bees and other insects are commonly attacked by mites.

Many mites cause damage to crops and stored products, and can also be vectors of important diseases of humans and domestic animals. Some species are of ecological importance and are a key element within eco-systems as decomposers, in breaking down leaf litter to produce humus.

Ticks

Soft and hard ticks exist. Both types are exclusively blood-feeding, on vertebrates. Soft ticks typically live in

Life cycle of a Mexican red-kneed tarantula

The bird-eating Mexican red-kneed tarantula, *Brachypelma smithii,* is a common pet. The young spider emerges from the egg and is a miniature version of the adult it will become. The female spider can live for 25 years.

It can take up to five years to reach full maturity.

The adult female lays 100–400 eggs.

The female stays close to the egg sac.

Young spiders moult several times a year.

Young spiders remain in the nest after hatching.

crevices and emerge briefly to feed, whereas hard ticks attach themselves to the skin of a host for a long period.

Harvestmen (Opiliones)

Like spiders, harvestmen (or harvest spiders) have jaws and pedipalps at the front, and four pairs of walking legs. The two parts of the body are broadly joined so that they appear to be just one piece. They only have two eyes and do not produce silk, neither do they have poison glands, although they do have defensive glands on each front leg that produce an odorous and distasteful fluid. They are all moderately large with a body size up to 10mm (0.39in), and are more or less brownish in colour.

Most harvestmen are nocturnal and feed on a variety of other arthropods and invertebrates, including other arachnids, insects, woodlice, small snails and earthworms. Dead and faecal material may also be eaten. They also feed on plant matter,

Below: Red velvet mites are large by mite standards.

Below: Ticks are small and have rather hard, flattened bodies.

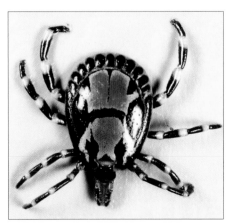

Below: Beetles are sometimes infested with parasitic mites.

but mainly for its water content. They drink from dewdrops on vegetation.

Harvestmen typically live for one year, usually passing the winter as eggs and maturing in late summer. The female harvestman uses her ovipositor to lay up to 100 eggs in damp soil.

Scorpions (Scorpiones)

Scorpions are rather large arachnids, ranging from 4–17cm (1.5–7in) long. They do not possess silk glands, but typically have a sting at the end of a long, upcurved tail, used both to subdue large prey and in defence.

While many are harmless, some are mildly toxic, and a handful are dangerous, able to inflict a lethal sting. There are several deadly species which possess highly potent neurotoxins. These include the fat-tailed scorpion, *Androctonus australis*, of Africa and Asia, which can kill an adult human within four hours. Dangerous scorpions are also found in South America, the Middle East and Africa.

Scorpions have poor eyesight and are primarily nocturnal. Like most arachnids, scorpions are predators. They use claw-like

Above: These relatives of spiders are less numerous and less often seen. Like spiders they are all predators and most count insects among their prey.

pedipalps to capture and hold their prey while stinging it with their barb-tipped abdomen.

Courtship in scorpions is complex and can last for hours or even days. The pair grasp each other's pincers and circle in a kind of dance. This is to move the female over the spermatophore (packet of sperm) that the male has deposited. The fertilized eggs develop within the female, which gives birth to live young. The baby scorpions climb on to the female's abdomen and are carried about until their first moult, after which they typically disperse. Scorpions may live for several years, even after reaching adulthood. Some species have been recorded as living for up to 20 years – nearly as long as tarantulas.

Pseudoscorpions (Pseudoscorpiones)

These creatures resemble miniature scorpions, but lack the arching, stinging tail. Few are larger than 5mm (0.19in) and they generally average 2–4mm (0.08–0.16in) in length. They are characterized by a flat, segmented

---

Habitats of spiders, mites and ticks

Spiders inhabit nearly every part of the world, from polar to tropical regions. They are particularly abundant in areas of rich vegetation.

Harvestmen are commonly found in hedgerows, parks and gardens, on tree trunks, walls and fences.

Mites and ticks have colonized nearly every ecosystem, including marine and freshwater. Unlike other arachnids, many are highly specialized, including a number of species which are parasitic on plants and animals, and a few which are exclusively aquatic. Some live in moss and leaf litter, feeding on fungus and mould. Ticks are often found in tall grass,

shrubs or trees where they will wait to attach or drop on to passing hosts.

Scorpions typically live in hot, dry climates, but they can also be found in tropical and subtropical forests, on high mountains, in grassland, caves, intertidal zones and other habitats. In deserts they survive by spending the day under rocks or in deep burrows, and only come out at night to feed and mate.

Pseudoscorpions are commonly found in moss, leaf litter, and the top layers of the soil, as well as under stones. Some live in very specialized habitats, such as under the wing cases of beetles, where they prey on mites.

abdomen, four pairs of short, pale legs and long pedipalps, with terminal pincers held out in front. Some species have poison glands in their pincers, which they use to capture small insects.

Most pseudoscorpions have one or two pairs of eyes on the side of the carapace, although some have none. The body colour ranges from yellowish to dark brown, with the pincers sometimes black. The body and appendages are sparsely covered with tactile hairs used to detect prey and predators. Pseudoscorpions produce silk from the pincers of the pedipalps, which is used for making chambers for moulting, overwintering, and brooding the young.

Pseudoscorpions are carnivorous, actively hunting for their prey, which they catch using their pincer-like pedipalps. Pseudoscorpions are beneficial as they commonly feed on pest species such as clothes moth larvae, ants, mites and small flies.

Courtship is similar to that of the scorpions, but unlike scorpions, they produce eggs which remain in a small sac held below the abdomen of the female. The larvae stay in the sac to be nourished by a milk-like secretion from the mother's ovaries. When they leave

Below: Like spiders, harvestmen have eight long legs. They are known by the common name 'daddy-long-legs'.

Life cycle of the imperial scorpion

The imperial scorpion, *Pandinus imperator*, is one of the world's largest, ireaching 20cm (8in). found in Africa, it includes termites in its diet. Species can live for up to 15 years, but more commonly for three to five years.

immature scorpion

adult

Mating ritual. The courtship dance can last for several hours.

The young, white scorpion is a small version of its parents.

Fertilized eggs develop inside the female for a year before live young are born.

---

the sac, the young attach themselves to the sides of the mother's abdomen. Usually there are fewer than two dozen young, but there can be more than one brood per year. The young moult three times, taking one to several years to reach adulthood. Some species disperse by clinging to the legs of passing flies or other animals. Adults can live up to three years and typically overwinter in a silken cocoon.

Below: Pseudoscorpions are well named, for they resemble true scorpions in overall shape. However, they are much smaller and lack the long, stinging tail.

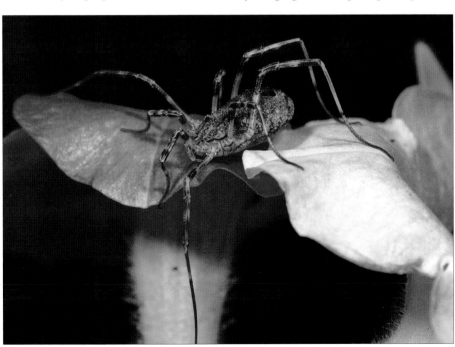

INSECTS OF THE WORLD

Insects are an extremely numerous and successful group, and are adapted to a wide range of habitats. Almost a million species have been described and at least as many probably await discovery or scientific description and classification.

Insects are found all over the world, from the complex habitats of the tropical forests, through temperate latitudes to the polar regions – some are able to survive even on snow and ice. They reach their highest diversity and numbers in the tropics, especially in Central and South America where their numbers and variety are amazing.

Of the 30 orders that make up the class Insecta, the largest by some margin is the beetles (order Coleoptera), with about 300,000 known species, closely followed by the bees, wasps, ants and relatives (Hymenoptera) and the moths and butterflies (Lepidoptera), each with over 200,000 species. Next come the flies (Diptera), with 120,000 species, and the true bugs (Hemiptera) with 82,000. Spiders and their relatives in the class Arachnida are arthropods – like insects, having jointed legs and a comparable exoskeleton. There are about 36,000 spider species and a similar number of their relatives, the mites and ticks. Compare these huge numbers with the relatively small number of vertebrate species: there are about 5,400 species of mammal and about 9,850 bird species. In terms of numbers alone then, the insects can claim to be the dominant and arguably the most successful group of animals on Earth.

This part of the book takes the form of a directory of insects from all over the world. The insects are grouped into chapters by the area in which they are found – Europe and Africa, the Americas, and Australia and Asia. Within each chapter a selection of insects are included, arranged by family.

Left: Three wasps pause at the entrance to their communal nest. Wasp colonies have a clearly defined caste system, where jobs are carefully delineated between members.

HOW TO USE THE DIRECTORY

The directory of insects includes a diverse selection of some of the most common, unusual and well-known insects that roam the planet. Each insect order, or group of orders, has a concise introduction, providing some key features of the groups described.

The insects in the following directory are organized first according to the broad geographical region in which they thrive. The three regions included here are Europe and Africa, the Americas, and Australia and Asia. Although each region encompasses a large area of landmass with vastly differing terrain, climate and biodiversity, many species are found across the whole region. In fact, some species of insect thrive all over the world.

Within each broad region, the insects are arranged according to insect order. There are 30 insect orders in total, each of which contains many

Insect order or suborder
Each insect species belongs to a larger family of insects that have common physical characteristics. Each family is part of a larger order or suborder.

families of insects that are grouped together according to a set of common features. Included too are the spiders, which belong in the class Arachnida.

The insects of each order are hugely varied in terms of appearance, size, habitat, method of locomotion, preferred food source, predatory behaviour and life cycle. Those that have been chosen for inclusion in the directory represent a very small selection of the entire insect world, but are each representative of the families to which they belong. The insects that are featured have been chosen for a variety of reasons. Many are well known throughout the region; some are endangered or rare. Others may have unusual characteristics that make them worthy of note. Some may be highly visible and welcome additions

to our immediate environment, pollinating flowers, adding colour to the garden, and collecting nectar for honey. Others may be less welcome. They may be pests of agricultural crops, vectors of disease, or may bite and sting.

Together they give an overview of the huge range and diversity of insects that populate the globe. Included too are many that are not visible to the naked eye but which may live within our vicinity without us ever being aware of their presence.

Learning to identify insects and distinguish between species by studying their appearance or behaviour can be fascinating. It's easy to find out more about the insects around you, then marvel at the range of shapes and patterns they display.

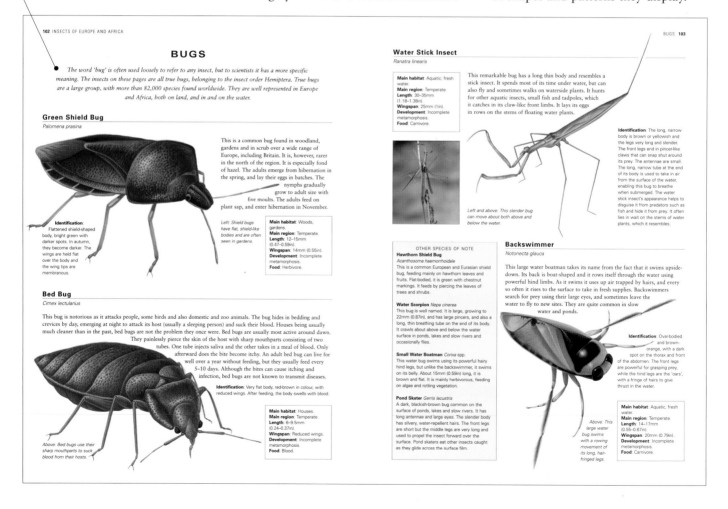

BUGS

The word 'bug' is often used loosely to refer to any insect, but to scientists it has a more specific meaning. The insects on these pages are all true bugs, belonging to the insect order Hemiptera. True bugs are a large group, with more than 82,000 species found worldwide. They are well represented in Europe and Africa, both on land, and in and on the water.

Green Shield Bug
Palomena prasina

This is a common bug found in woodland, gardens and in scrub over a wide range of Europe, including Britain. It is, however, rarer in the north of the region. It is especially fond of hazel. The adults emerge from hibernation in the spring, and lay their eggs in batches. The nymphs gradually grow to adult size with five moults. The adults feed on plant sap, and enter hibernation in November.

Identification: Flattened shield-shaped body, bright green with darker spots. In autumn, they become darker. The wings are held flat over the body and the wing tips are membranous.

Left: Shield bugs have flat, shield-like bodies and are often seen in gardens.

Main habitat: Woods, gardens.
Main region: Temperate.
Length: 12–15mm (0.47–0.59in).
Wingspan: 14mm (0.55in).
Development: Incomplete metamorphosis.
Food: Herbivore.

Bed Bug
Cimex lectularius

This bug is notorious as it attacks people, some birds and also domestic and zoo animals. The bug hides in bedding and crevices by day, emerging at night to attack its host (usually a sleeping person) and suck their blood. Houses being usually much cleaner than in the past, bed bugs are not the problem they once were. Bed bugs are usually most active around dawn. They painlessly pierce the skin of the host with sharp mouthparts consisting of two tubes. One tube injects saliva and the other takes in a meal of blood. Only afterward does the bite become itchy. An adult bed bug can live for well over a year without feeding, but they usually feed every 5–10 days. Although the bites can cause itching and infection, bed bugs are not known to transmit diseases.

Identification: Very flat body, red-brown in colour, with reduced wings. After feeding, the body swells with blood.

Above: Bed bugs use their sharp mouthparts to suck blood from their hosts.

Main habitat: Houses.
Main region: Temperate.
Length: 6–9.5mm (0.24–0.37in).
Wingspan: Reduced wings.
Development: Incomplete metamorphosis.
Food: Blood.

Water Stick Insect
Ranatra linearis

Main habitat: Aquatic; fresh water.
Main region: Temperate.
Length: 30–35mm (1.18–1.38in).
Wingspan: 25mm (1in).
Development: Incomplete metamorphosis.
Food: Carnivore.

This remarkable bug has a long thin body and resembles a stick insect. It spends most of its time under water, but can also fly and sometimes walks on waterside plants. It hunts for other aquatic insects, small fish and tadpoles, which it catches in its claw-like front limbs. It lays its eggs in rows on the stems of floating water plants.

Left and above: This slender bug can move about both above and below the water.

Identification: The long, narrow body is brown or yellowish and the legs very long and slender. The front legs end in pincer-like claws that can snap shut around its prey. The antennae are small. The long, narrow tube at the end of its body is used to take in air from the surface of the water, enabling this bug to breathe when submerged. The water stick insect's appearance helps to disguise it from predators such as fish and hide it from prey. It often lies in wait on the stems of water plants, which it resembles.

OTHER SPECIES OF NOTE

Hawthorn Shield Bug
Acanthosoma haemorrhoidale
This is a common European and Eurasian shield bug, feeding mainly on hawthorn leaves and fruits. Flat-bodied, it is green with chestnut markings. It feeds by piercing the leaves of trees and shrubs.

Water Scorpion *Nepa cinerea*
This bug is well named. It is large, growing to 22mm (0.87in), and has large pincers, and also a long, thin breathing tube on the end of its body. It crawls about above and below the water surface in ponds, lakes and slow rivers and occasionally flies.

Small Water Boatman *Corixa* spp.
This water bug swims using its powerful hairy hind legs, but unlike the backswimmer, it swims on its belly. About 15mm (0.59in) long, it is brown and flat. It is mainly herbivorous, feeding on algae and rotting vegetation.

Pond Skater *Gerris lacustris*
A dark, blackish-brown bug common on the surface of ponds, lakes and slow rivers. It has long antennae and large eyes. The slender body has silvery, water-repellent hairs. The front legs are short but the middle legs are very long and used to propel the insect forward over the surface. Pond skaters eat other insects caught as they glide across the surface film.

Backswimmer
Notonecta glauca

This large water boatman takes its name from the fact that it swims upside-down. Its back is boat-shaped and it rows itself through the water using powerful hind limbs. As it swims it uses up air trapped by hairs, and every so often it rises to the surface to take in fresh supplies. Backswimmers search for prey using their large eyes, and sometimes leave the water to fly to new sites. They are quite common in slow water and ponds.

Identification: Oval-bodied and brown-orange, with a dark spot on the thorax and front of the abdomen. The front legs are powerful for grasping prey, while the hind legs are the 'oars', with a fringe of hairs to give thrust in the water.

Above: This large water bug swims with a rowing movement of its long, hair-fringed legs.

Main habitat: Aquatic; fresh water.
Main region: Temperate.
Length: 14–17mm (0.55–0.67in).
Wingspan: 20mm (0.79in).
Development: Incomplete metamorphosis.
Food: Carnivore.

Common name
This is the most popular, non-scientific name for the insect entry.

Latin name
This is the internationally accepted Latin name for the insect entry.

Introduction
This provides a general introduction to the insect and may include information on life cycle, habitat, behaviour and food sources.

Bed Bug
Cimex lectularius

This bug is notorious as it attacks people, some birds and also domestic and zoo animals. The bug hides in bedding and crevices by day, emerging at night to attack its host (usually a sleeping person) and suck their blood. Houses being usually much cleaner than in the past, bed bugs are not the problem they once were. Bed bugs are usually most active around dawn. They painlessly pierce the skin of the host with sharp mouthparts consisting of two tubes. One tube injects saliva and the other takes in a meal of blood. Only afterward does the bite become itchy. An adult bed bug can live for well over a year without feeding, but they usually feed every 5–10 days. Although the bites can cause itching and infection, bed bugs are not known to transmit diseases.

Identification: Very flat body, red-brown in colour, with reduced wings. After feeding, the body swells with blood.

Main habitat: Houses.
Main region: Temperate.
Length: 6–9.5mm (0.24–0.37in).
Wingspan: Reduced wings.
Development: Incomplete metamorphosis.
Food: Blood.

Above: Bed bugs use their sharp mouthparts to suck blood from their hosts.

Profile
The profile is an illustration of the insect, usually an adult.

Identification
This description will enable the reader to properly identify the insect. It gives an overall impression of the visual appearance of the insect.

Main habitat
The insect's natural geographical location. It may live, for example, in houses, near freshwater streams, or among specific crops.

Main region
This describes the insects natural distribution throughout the world.

Length
Describes the average dimensions the insect will grow to, given optimal conditions.

Wingspan
For many insects such as butterflies and moths, the wings are the main defining feature and method of movement.

Development
Describes the life cycle of the insect.

Main habitat: Houses.
Main region: Temperate.
Length: 6–9.5mm (0.24–0.37in).
Wingspan: Reduced wings.
Development: Incomplete metamorphosis.
Food: Blood.

Food
Insects have preferences for specific foods. They may be herbivores or carnivores.

Other species of note
The insects featured in this tinted box are usually less well-known species of the order. They are included because they have some outstanding features worthy of note.

Species names
The name by which the insect is most commonly known is presented first, followed by the Latin name and any other common name by which it is known.

Entries
The information given for each entry describes the insect's main characteristics and the specific features it has that distinguish it from similar species.

● OTHER SPECIES OF NOTE

Hawthorn Shield Bug
Acanthosoma haemorrhoidale
This is a common European and Eurasian shield bug, feeding mainly on hawthorn leaves and fruits. Flat-bodied, it is green with chestnut markings. It feeds by piercing the leaves of trees and shrubs.

Water Scorpion *Nepa cinerea*
This bug is well named. It is large, growing to 22mm (0.87in), and has large pincers, and also a long, thin breathing tube on the end of its body. It crawls about above and below the water surface in ponds, lakes and slow rivers and occasionally flies.

Small Water Boatman *Corixa* spp.
This water bug swims using its powerful hairy hind legs, but unlike the backswimmer, it swims on its belly. About 15mm (0.59in) long, it is brown and flat. It is mainly herbivorous, feeding on algae and rotting vegetation.

Pond Skater *Gerris lacustris*
A dark, blackish-brown bug common on the surface of ponds, lakes and slow rivers. It has long antennae and large eyes. The slender body has silvery, water-repellent hairs. The front legs are short but the middle legs are very long and used to propel the insect forward over the surface. Pond skaters eat other insects caught as they glide across the surface film.

INSECTS OF EUROPE AND AFRICA

Europe and Africa contain a very wide range of habitats, from the Arctic tundra of the far north, through the varied forests and other temperate seasonal habitats of much of Europe, to the Mediterranean, and the deserts, grasslands and tropical forests of Africa. Insects have been very successful in adapting to all of these habitats. Some insects, such as the biting midges of northern Europe and the mosquitoes and tsetse flies of Africa, are annoying or even dangerous, but there are other beneficial insects too – especially bees that are important pollinators of crops, and honey bees for the honey they produce. Termites and locusts do enormous damage to wooden buildings and crops but they also form part of the traditional diet of people in many parts of Africa.

Above from left: A wood wasp, a click beetle and a stick insect.

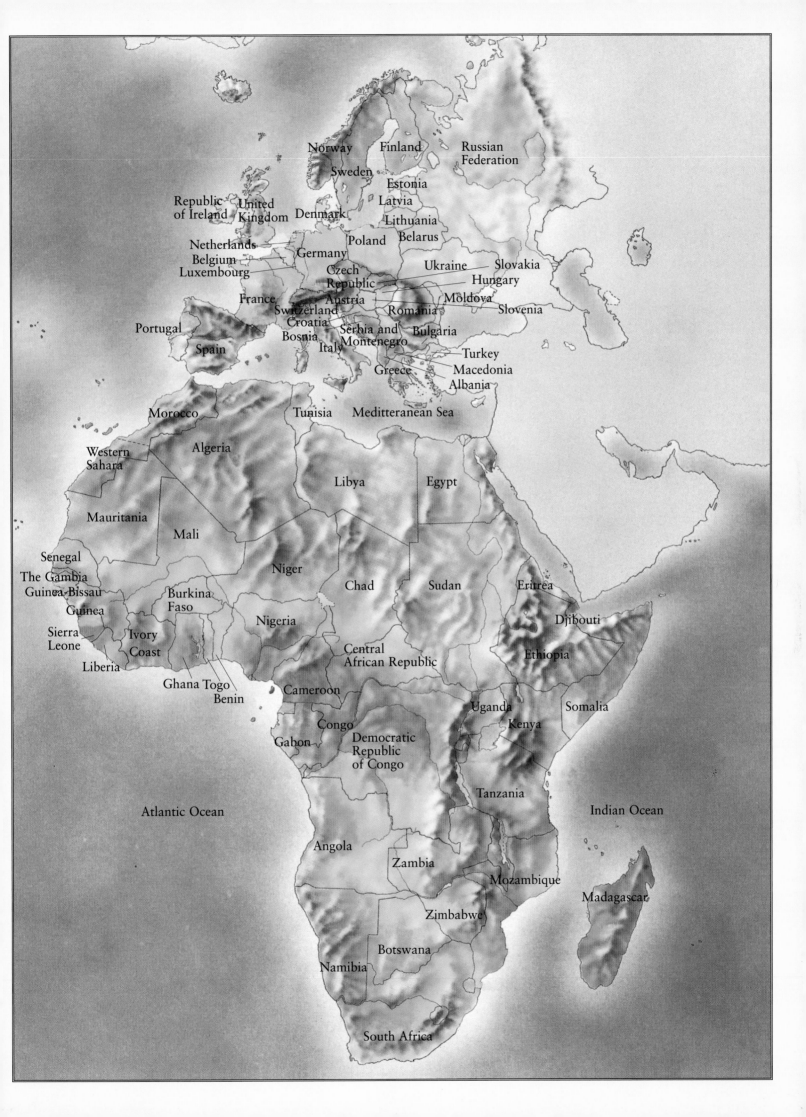

WINGLESS NEAR-INSECTS AND WINGLESS INSECTS

Proturans (Protura), diplurans (Diplura) and springtails (Collembola) are all small or tiny and wingless and are not now regarded as true insects. Bristletails (Archaeognatha) and silverfish (Thysanura) are true, wingless insects often seen in and around houses throughout this region.

Springtail

Tomocerus longicornis

This is a very common springtail found, for example, in leaf litter and humus. A sample of soil often reveals large numbers of springtails, and this species, being relatively large, is easier to spot than many – most are less than 5mm (0.19in) long. They take their name from the spring-like hinged appendage at the end of the body. This normally lies under the body, but it can be released suddenly, propelling the springtail into the air.

Main habitat: Soil litter.
Main region: Temperate.
Length: 4–5mm (0.16–0.19in).
Development: No obvious metamorphosis.
Food: Omnivore.

Above: This springtail often jumps suddenly in order to escape capture by predators.

Identification: A rather large springtail, with long legs, long 'spring' appendage and also unusually long antennae – longer than the body in this species. It has clear segmentation. In colour it is mottled brown, giving it good camouflage in the soil. Like all springtails it has a curious tube-shaped organ on its underside. This acts both as an aid to breathing and as an adhesive organ to help the springtail cling to surfaces.

Springtail

Sminthurus spp.

These tiny springtails live mainly on living plants, and some of them can become serious crop pests if they reach large numbers, feeding on the tissues of growing crops such as clover or peas. *S. viridis* is known as the lucerne flea. A related species is semi-aquatic, living on pond surfaces and feeding on duckweed.

Main habitat: Plants.
Main region: Temperate.
Length: 1–2mm (0.04–0.08in).
Development: No obvious metamorphosis.
Food: Herbivore.

Identification: Very small rotund springtails with no obvious segmentation, some as small as 1mm (0.04in). The antennae are quite long and distinctly elbowed, and the body ends in a two-pronged 'tail'. Depending on the species they are pale yellow, green or dark brown.

Right: Tiny in size, the members of this genus are hard to spot unless present in large numbers.

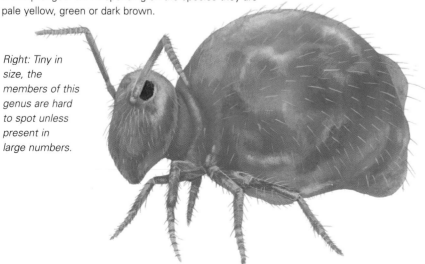

OTHER SPECIES OF NOTE

Proturans *Protura*
These are tiny, mostly less than 2mm (0.08in), usually very pale or white, mainly found in the soil where they feed on rotting humus and fungi. They have simple biting and sucking mouthparts, and no eyes, or antennae.

Diplurans *Diplura*
These look a little like tiny earwigs, with their elongated bodies ending in a two-pronged 'tail' (claw-like in some species). Like proturans, they are blind, but they do have antennae. They range in size from about 6–50mm (0.24–2in).

Water Springtail *Podura aquatica*
A tiny springtail found, sometimes in large numbers, on the surface of ponds. The colour varies from brown to a bluish-black. It feeds on small, floating organic matter. Like other springtails, it can jump well.

Rock Springtail *Anurida maritima*
This is a small species often found in large numbers on the seashore. Hundreds of these tiny, dusty-blue springtails may be seen clustered on the still surface of rock-pools on the seashore where they scavenge on dead animals or other debris. Their bodies are waterproof and they can move with ease on the surface of still water.

Common Silverfish

Lepisma saccharina

The common silverfish is usually found in damp places such as food cupboards, cellars, kitchens and bathrooms, but being strictly nocturnal it is not often seen in daytime. With today's generally higher standards of hygiene it is not as common as it used to be in houses. It lays its eggs in cracks and feeds on food scraps such as flour, damp wallpaper and damp cloth. It can go without food for months if necessary.

Identification: Silverfish have widely separated eyes and a three-pronged 'tail'. This silverfish is brownish-grey with a bright silvery scaly sheen and looks rather fish-like as it scuttles rapidly across the floor, twisting its supple body as it goes. The head has compound eyes and long sensitive antennae. It can grow to about 25mm (1in).

Main habitat: Houses.
Main region: Temperate.
Length: 8–20mm (0.31in–0.79in).
Development: No obvious metamorphosis.
Food: Omnivore.

Above: Lithe and agile, silverfish generally appear at night, seeking scraps of food.

Firebrat

Thermobia domestica

This silverfish takes its name from one of its favoured habitats, the hearth of a fire or the vicinity of an oven in a bakery or kitchen. It lives on scraps of food found on the floor or in cracks. Fast-moving, firebrats disappear rapidly into the shelter of crevices when disturbed. Firebrats tend to be found in warm sites although they also like quite high humidity as well. In the wild they lurk under rocks, logs and in leaf litter, but are most often spotted in houses or buildings.

Below: Firebrats, like silverfish, are harmless, though they may sometimes contaminate food.

Main habitat: Houses.
Main region: Temperate.
Length: 8mm (0.31in).
Development: No obvious metamorphosis.
Food: Omnivore.

Identification: The firebrat is less silvery and significantly smaller than the common silverfish and it also has much longer antennae and tail filaments. It has a flattened body and runs rapidly, close to the ground, and can easily hide in narrow crevices.

Shore Bristletail

Petrobius maritimus

This bristletail lives at the seashore and is often found in large numbers on rocks within the splash zone. It can jump and run rapidly if surprised or splashed by a wave but it cannot survive for long if immersed in the sea. Shore bristletails feed mainly on lichens and micro-organisms scraped off the rocks. They are mainly nocturnal. They lay their eggs in cracks in the rocks, up to 100 in a batch.

Main habitat: Seashore.
Main region: Temperate.
Length: 18mm (0.71in).
Development: No obvious metamorphosis.
Food: Omnivore.

Identification: Rather mottled and metallic grey in colour, with three long bristles at the end of the abdomen. It can grow up to 18mm (0.71in).

Above: These shore-dwelling arthropods are found above high water, usually under stones and on screes or in chinks of rock and wood.

MAYFLIES AND STONEFLIES

Mayflies (Ephemeroptera) are rather delicate winged insects with short-lived adults and longer-lived aquatic larvae, known as nymphs. Stoneflies (Plecoptera) are found in similar habitats and also have aquatic nymphs, but the adults live for two or three weeks. Both groups are well represented in Europe and Africa. On some rivers, especially in Europe, mass hatchings of mayflies are an impressive spectacle.

Green Drake

Ephemera danica

This is a common mayfly of temperate regions, emerging usually in May or June. As with several other mayfly species there may be mass hatchings and large swarms at certain sites. The adults live only a short time, often only a matter of days, during which they pair and mate. Most of the mayfly's life is spent under water, as a larva (nymph).

Main habitat: Slow rivers and streams, lakes.
Main region: Temperate.
Length: 10–30mm (0.39in–1.18in).
Wingspan: 40mm (1.57in).
Development: Incomplete metamorphosis.
Food: Adult does not feed.

Identification: The nymph lives for two years or more and grows to about 25mm (1in). It is yellowish with dark triangular spots and three tail filaments. It lives in a tube in the mud of a riverbed, usually in a slow-flowing stream. The adult only lives for a day or so, or sometimes for just a few hours. It has one or two pairs of thin wings and a weak, fluttery flight and is often seen around May in northern regions. At rest it holds its wings upright. The adult has delicate, brownish veined wings and long tail filaments. The hindwings are short.

Right: Note the long, net-veined forewings, short hindwings, and long tail filaments.

Sepia Dun

Leptophlebia marginata

This is a favourite of fly-fishermen and is quite common in northern Europe. It is found mainly in acidic waters in small lakes and also in slow-moving rivers and streams. The adult emerges in May or June and lives for only about a day. Technically the name sepia dun is used for the subadult when it first emerges.

Main habitat: Lakes and slow rivers.
Main region: Temperate.
Length: 12mm (0.47in).
Wingspan: 25mm (1in).
Development: Incomplete metamorphosis.
Food: Adult does not feed.

Identification: The nymph is about 12mm (0.47in) long and grey-brown with pointed, leaf-shaped gills. The tail filaments are longer than its body and are splayed outward. The adult is rather small and dark in colour with a weak, fluttery flight.

Left: This dark mayfly is a favourite food of freshwater fish and is therefore well known to anglers.

OTHER SPECIES OF NOTE

Pond Olive *Cloeon dipterum*
This mayfly is a common species of ponds and stagnant water. The nymph is grey or greenish-brown and about 20mm (0.79in) long. The adult is mainly yellow and lacks hindwings.

Mayfly *Metretopus norvegicus*
This mayfly belongs to a family found in Scandinavia, and this is one of the commonest species. Members of this family have oval-shaped hindwings.

Stonefly *Capnia bifrons*
A small, blackish stonefly common in much of Europe. It prefers small streams, often in woods. The nymph is small, about 11mm (0.43in), and slender. The adult male has tiny wings.

Blue-winged Olive Mayfly
Ephemerella ignita
A pretty mayfly that flies between June and October. The adult has reddish eyes and a reddish or yellow-brown body. The nymph is about 10mm (0.39in) long and may be abundant in underwater vegetation in fast-flowing rivers such as clear streams favoured by trout.

Large Stonefly

Perlodes microcephala

Main habitat: Chalk streams.
Main region: Temperate.
Length: 25mm (1in).
Wingspan: 30mm (1.18in)
(female).
Development: Incomplete
metamorphosis.
Food: Adult feeds little or
not at all.

Identification: A relatively large
stonefly, the nymph is about
25mm (1in) long, strong and
sturdy, and dark brown. The
adults have two long antennae
and two long tail filaments.

This is a fairly common stonefly of European streams, especially
chalk streams. The adult is a weak flier and in this species only the
female has large enough wings to fly. It spends a lot of time crawling
and resting in vegetation or on stones near to the water's edge. The
nymph lives mainly under stones and crawls on to rocks and stones
when ready to hatch into an adult from March to June. It prefers
fast streams with plenty of oxygen and spends about two years as
a nymph. Stoneflies are restricted to cooler regions, such as
northern Europe and favour mountain streams. They mostly need
oxygen-rich clean water and are very sensitive to pollution. They
have therefore been reduced in numbers and have disappeared from
many rivers and streams in heavily populated regions.

*Right: A healthy population of stoneflies is a sure sign of a clean aquatic
environment, so they are good indicators of water quality.*

February Red

Taeniopteryx nebulosa

The common name of this stonefly is given to it by anglers
and refers to its reddish colour. It is unusual in preferring
muddy streams and rivers, rather than clear water. The adult
emerges quite early in the year, in March or April. Stoneflies
are so-named for their habit of crawling over rocks,
stones and wood at the margins of ponds and rivers. Fossil
evidence suggests that they have changed little in basic body
plan since the Carboniferous Period. They are sturdier than
mayflies and their four wings are roughly equal in size.
Another difference is that stoneflies hold their wings flat
over their bodies when at rest.

Identification: The nymph grows to about 15mm (0.59in), and is dark
brown or blackish, with long antennae and two long tail filaments. It has
gills at the base of its legs and spines along its body. The long-winged
adult flies in sunny weather in spring.

Main habitat: Slow streams
and rivers.
Main region: Europe.
Length: 10mm (0.39in).
Wingspan: 25mm (1in).
Development: Incomplete
metamorphosis.
Food: Adult feeds little or
at all.

*Left: Unlike mayflies, stoneflies
have prominent antennae.*

Drake Mackerel

Ephemera vulgata

This large mayfly is related to the green drake, which it resembles quite
closely. It prefers still or slow, rather muddy rivers or ponds and
often emerges in large numbers from May to July. The nymph lives
in burrows on the riverbed.

Main habitat: Slow streams
and rivers.
Main region: Europe.
Length: 20mm (0.79in).
Wingspan: 35mm (1.38in).
Development: Incomplete
metamorphosis.
Food: Adult feeds little or
not at all.

Identification: The nymph grows to about 15mm
(0.59in), and is dark brown or blackish, with long
antennae and two long tail filaments. It has gills at
the base of its legs and spines along its body. The
adult has short hindwings and long, broad forewings,
marked with an obvious brownish mark roughly
halfway down the wings. Its body is
yellowish with dark
triangular markings above.

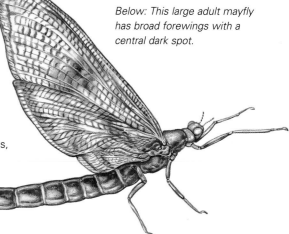

*Below: This large adult mayfly
has broad forewings with a
central dark spot.*

DRAGONFLIES

Dragonflies and damselflies (Odonata) are large, active insects with large compound eyes and a flexible neck. The order contains about 6,000 species, of which about 125 are found in Europe. Dragonflies are some of Europe's most fascinating insects and several species are expanding their range northward as the Earth's climate warms up.

Emperor Dragonfly

Anax imperator

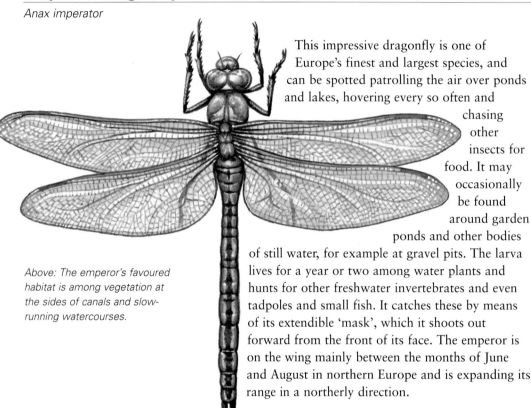

This impressive dragonfly is one of Europe's finest and largest species, and can be spotted patrolling the air over ponds and lakes, hovering every so often and chasing other insects for food. It may occasionally be found around garden ponds and other bodies of still water, for example at gravel pits. The larva lives for a year or two among water plants and hunts for other freshwater invertebrates and even tadpoles and small fish. It catches these by means of its extendible 'mask', which it shoots out forward from the front of its face. The emperor is on the wing mainly between the months of June and August in northern Europe and is expanding its range in a northerly direction.

Above: The emperor's favoured habitat is among vegetation at the sides of canals and slow-running watercourses.

Main habitat: Large ponds, lakes.
Main region: Temperate Europe.
Length: 82mm (3.23in).
Wingspan: 11cm (4.5in).
Development: Incomplete metamorphosis.
Food: Carnivore.

Identification: Large dragonfly, about 82mm (3.23in) long and with a wingspan of 11cm (4.5in). The male has a green thorax and blue abdomen (green in the female). The larva is black and white at first, turning green or greenish-brown later. Fully grown larvae are quite capable of catching other aquatic insects and other animals and they feed voraciously, stalking their prey and then striking fast. The adult is strong and powerful in flight.

Broad-bodied Chaser

Libellula depressa

A medium-sized dragonfly, widespread in Europe. It lives near shallow ponds and lakes, and like most dragonflies is mainly active in sunny weather, mainly in June and July. It quickly colonizes suitable new ponds, including quite small garden ponds. The larva lives partly buried in the mud of shallow water. Unlike many dragonflies, this species spends quite long periods resting on plants, watching for interlopers, which it then chases out of its territory. Both male and female are yellowish-brown when newly emerged, attaining a pretty powder-blue in the mature male and brown in the mature female, in both cases with a line of yellow spots along the sides.

Right: Fast and aggressive, the broad-bodied chaser is energetic in flight.

Main habitat: Shallow ponds.
Main region: Temperate.
Length: 45mm (1.77in).
Wingspan: 75mm (2.95in).
Development: Incomplete metamorphosis.
Food: Carnivore.

Identification: Chasers have broad, rather flat bodies giving them a bulky appearance in flight. The male of this species has a bright blue abdomen (yellow in the young male and brownish in the female). Males occupy territories and chase off intruders, often returning to the same perch. The flight period is roughly from the middle of May until August.

Common Hawker

Aeschna juncea

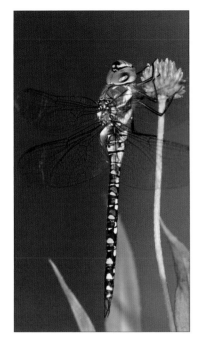

Hawkers are fast-flying dragonflies, spending lots of their time in flight, cruising after other flying insects, often far from water, for example in woodland clearings. It breeds in still water, from large lakes to small ponds. Its larvae takes 3–4 years to develop and have a stripy appearance. The brown hawker (*A. grandis*) is similar, but its body is mainly brown and it has a yellow tint on the wings.

Identification: It has broad yellow stripes on the side of its thorax. The adult male is dark, with blue spots and smaller yellow marks, while the female is normally brown with yellow spots.

Below and far left: Hawkers fly strongly and are often seen well away from water.

Main habitat: Pools and woodland edges.
Main region: Temperate.
Length: 75mm (2.95in).
Wingspan: 90mm (3.54in).
Development: Incomplete metamorphosis.
Food: Carnivore.

OTHER SPECIES OF NOTE

Orange Emperor *Anax speratus*
This mainly orange dragonfly is one of Africa's largest and most impressive species, found in west, east, central and southern Africa. It has a wingspan of more than 12cm (4.5in).

Golden-ringed Dragonfly *Cordulegaster boltonii*
This large dragonfly has bold wasp-like markings – black with golden-yellow bands – and green eyes. It breeds in acidic ponds and hunts over moorland or at woodland edges.

Four-spotted Chaser *Libellula quadrimaculata*
(Above) An unmistakable dragonfly, with a flat, brown abdomen and two brown spots at the front of each wing. It often perches on waterside plants. It occurs throughout most of Europe.

Common Darter *Sympetrum striolatum*
Small and active, this species is widespread in Europe, except the far north. It has an orange abdomen. Common darters breed in lakes, ponds, ditches and rivers. They often perch high up in a bush or tree.

Scarlet Dragonfly

Crocothemis erythraea

Common in southern Europe and Africa, this is a very attractive and distinctive species. It is on the wing from April to November and often hovers. At rest the wings are held slightly forward. Its bright coloration makes this species one of the easiest of dragonflies to spot, although it is actually quite shy and difficult to approach closely. This is another species that is gradually moving northward in regions where the climate is warming.

Main habitat: Pools.
Main region: Temperate.
Length: 34mm (1.34in).
Wingspan: 50mm (2in).
Development: Incomplete metamorphosis.
Food: Carnivore.

Identification: Large and active, the male is bright red on its head and on its rather broad abdomen. The female is mainly yellow. Both sexes have a yellow patch at the base of the wings.

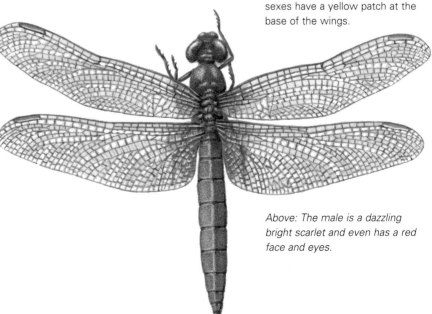

Above: The male is a dazzling bright scarlet and even has a red face and eyes.

DAMSELFLIES

Damselflies are rather daintier than dragonflies, and have a more fluttering, less direct and slower flight. They hold their wings upright over their backs when at rest, a feature which can be used to differentiate them from dragonflies. Their larvae (nymphs), like those of dragonflies, are aquatic. In Europe, the demoiselles are arguably the most beautiful damselflies.

Beautiful Demoiselle

Calopteryx virgo

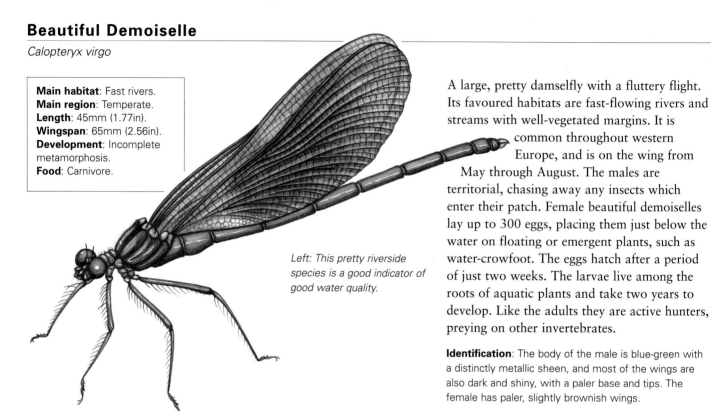

Main habitat: Fast rivers.
Main region: Temperate.
Length: 45mm (1.77in).
Wingspan: 65mm (2.56in).
Development: Incomplete metamorphosis.
Food: Carnivore.

Left: This pretty riverside species is a good indicator of good water quality.

A large, pretty damselfly with a fluttery flight. Its favoured habitats are fast-flowing rivers and streams with well-vegetated margins. It is common throughout western Europe, and is on the wing from May through August. The males are territorial, chasing away any insects which enter their patch. Female beautiful demoiselles lay up to 300 eggs, placing them just below the water on floating or emergent plants, such as water-crowfoot. The eggs hatch after a period of just two weeks. The larvae live among the roots of aquatic plants and take two years to develop. Like the adults they are active hunters, preying on other invertebrates.

Identification: The body of the male is blue-green with a distinctly metallic sheen, and most of the wings are also dark and shiny, with a paler base and tips. The female has paler, slightly brownish wings.

Banded Demoiselle

Calopteryx splendens

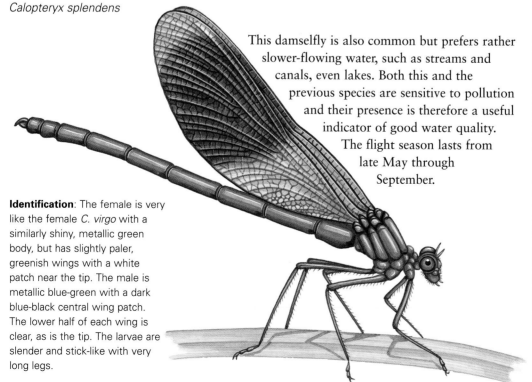

This damselfly is also common but prefers rather slower-flowing water, such as streams and canals, even lakes. Both this and the previous species are sensitive to pollution and their presence is therefore a useful indicator of good water quality. The flight season lasts from late May through September.

Main habitat: Slow rivers.
Main region: Temperate.
Length: 45mm (1.77in).
Wingspan: 65mm (2.56in).
Development: Incomplete metamorphosis.
Food: Carnivore.

Identification: The female is very like the female *C. virgo* with a similarly shiny, metallic green body, but has slightly paler, greenish wings with a white patch near the tip. The male is metallic blue-green with a dark blue-black central wing patch. The lower half of each wing is clear, as is the tip. The larvae are slender and stick-like with very long legs.

Left and above: This long-legged species often perches close to slow-flowing water.

Emerald Damselfly

Lestes sponsa

Main habitat: Ponds, lakes, ditches.
Main region: Temperate.
Length: 35mm (1.38in).
Wingspan: 42mm (1.65in).
Development: Incomplete metamorphosis.
Food: Carnivore.

Identification: The wings are held half open when at rest. Both males and females are green, the female rather duller, and the male has a pale powdery blue thorax. The larvae are long and slender, and their three gills at the tip of the abdomen are striped.

This is one of the commonest damselflies in central and northern Europe and it has a wide distribution farther east as well. It is usually seen near to pools, ditches, canals and lakes, and often rests on waterside plants. Its larvae prefer slightly acidic water and it is often particularly common around acid ponds and lakes, although it avoids moors and other exposed upland areas. In flight it is rather weak, like most damselflies. At rest, it holds its wings slightly open, which is unusual for a damselfly. This habit has earned it another common name – the common spreadwing. The larvae feed on small crustaceans and insect larvae. The adults emerge in late June and fly until late September.

Right: Both sexes of this damselfly are green, the male with a definite metallic sheen.

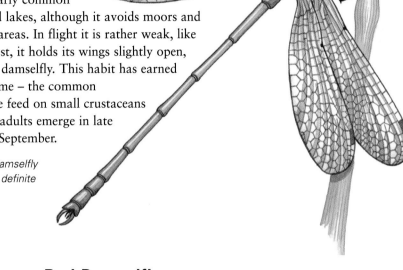

OTHER SPECIES OF NOTE

Swamp bluet *Africallagma glaucum*
This very pretty damselfly has a wide range in Africa from the west and central regions into South Africa. The male is a bright sky blue with a dark back; the female is duller and brownish.

Dancing Jewel *Platycypha caligata*
This must be a candidate for one of the most beautiful of all damselflies. The orange-red thorax of the male contrasts with its bright blue abdomen, and the lower legs are bright red outside and pure white inside. It lives near wooded streams in southern and central Africa.

Azure Damselfly (Left)
Coenagrion puella
A dainty species found commonly in Europe and also south to northern Africa. The male is mainly blue with black rings while the female is darker, and often green, especially below. It can often be spotted perched close to sheltered ponds.

Goldtail *Allocnemis leucosticta*
This attractive damselfly is endemic to South Africa where it occurs mainly near shady streams. Mostly dark, it has a bright blue underside to the thorax and a vivid gold tip to the abdomen that seems almost to glow as it flutters slowly in and out of the sun.

Large Red Damselfly

Pyrrhosoma nymphula

In Europe this is mainly a northern species, found on acid bogs and also around canals, ditches and ponds. However, it can often be seen away from the water, feeding at the edges of woodlands and by hedges. The main flight season is from May through July and this is often the first damselfly to be spotted in the spring. Its distribution in northern Europe reaches farther than many others and it is common in Britain, even as far as the Orkney Islands. Studies of the larvae have shown them to be territorial, defending a patch from neighbours.

Main habitat: Bogs, ponds, canals.
Main region: Temperate.
Length: 34mm (1.34in).
Wingspan: 45mm (1.77in).
Development: Incomplete metamorphosis.
Food: Carnivore.

Identification: The larva is rather dark and has pointed gills, each with an x-shaped mark. Midge larvae feature in the larval diet. The adult is bright red, with black markings toward the rear, and bright red eyes. The legs are black.

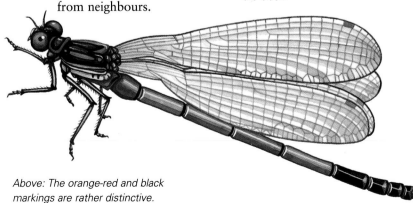

Above: The orange-red and black markings are rather distinctive.

COCKROACHES AND TERMITES

Cockroaches (Blattodea) are rather low-slung insects with flattened, oval bodies and rather a leathery texture, some being quite shiny. There are about 3,500 species. In Europe they are mainly found in buildings. Termites (Isoptera) are highly social insects living in colonies, some containing more than a million. There are about 2,500 species, mainly found in tropical regions, notably in Africa.

Oriental Cockroach

Blatta orientalis

Main habitat: Buildings.
Main region: Widespread.
Length: 20mm (0.79in).
Wingspan: 30mm (1.18in).
Development: Incomplete metamorphosis.
Food: Omnivore.

Identification: Dark brown and shiny with a flattened body. The females produce their eggs in a cylindrical egg case which they carry for a while, protruding from their rear.

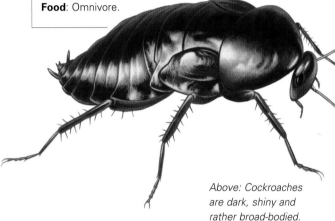

Above: Cockroaches are dark, shiny and rather broad-bodied.

This is one of the commonest of all cockroaches and has become a widespread pest. It is now often found in association with people and buildings, especially where food scraps are left on floors such as in larders, bakeries, warehouses and the like. Oriental cockroaches emerge at night to forage for anything edible. Their numbers can build up quickly in the right conditions and they can cause problems by fouling food. Cockroaches have long legs and can move with remarkable speed, scuttling across the ground and disappearing quickly into even quite narrow crevices. Some males are capable of short flights but the wings of the female are very short and useless. In nature they tend to inhabit damp places, such as leaf litter, but they are most likely to be seen inside, as they cannot withstand cold weather, being tropical in origin.

Hissing Cockroach

Gromphadorhina portentosa

This large cockroach is particularly fascinating as it produces an audible hissing sound when disturbed. This it does by quickly pushing air out of its spiracles. This often startles a predator (such as a rodent), giving the cockroach time to make good its escape. Hissing is also used for display and to dispute territory. This cockroach is sometimes kept as a pet.

Left: Native to Madagascar, the hissing cockroach needs a warm and dark environment in which to thrive. It is the only creature known to produce a hissing sound by pushing air through spiracles in its body.

Identification: Large and dark brown, with an orange-and-brown abdomen. The latter is very tough and leathery and the cuticle is tough and armour-like. The thorax is strong, with hard bumps.

Left: This species makes an unusual and fascinating pet, feeding on fruit and vegetables.

Main habitat: Forest floor.
Main region: Tropical.
Length: 70mm (2.76in).
Wingspan: Mainly wingless.
Development: Incomplete metamorphosis.
Food: Omnivore.

Drywood Termite

Kalotermes flavicollis

Main habitat: Trees and timber.
Main region: Temperate.
Length: 10mm (0.39in).
Wingspan: Mainly wingless.
Development: Incomplete metamorphosis.
Food: Wood.

Above and right: Although their colonies are small, these termites can cause damage to timber.

This termite is rather common in parts of southern Europe. It lives in simple colonies numbering only a few hundred. The king and queen are attended by young workers and a smaller number of soldiers. They live mainly in dead trees and fallen timber, but they do sometimes attack wooden buildings and can then be a nuisance.

Identification: The body is soft and rather pale, but the thorax and head are slightly harder. At the tip of the abdomen there are two short projections (cerci). The antennae are relatively short and simple. The queen has an enlarged abdomen full of developing eggs. The soldier caste has mouthparts with large crossed mandibles.

OTHER SPECIES OF NOTE

German Cockroach *Blatella germanica*
This is a widespread species and is sometimes a pest. It is a yellowish colour with darker markings on the thorax. Originally tropical, it is not native to Germany.

Dusky Cockroach *Ectobius lapponicus*
This small cockroach is about 10mm (0.39in) long. It is found in many parts of Europe, mainly on the ground in leaf litter and sometimes in bushes. Both sexes are winged, but the female flies more readily.

Harvester Termite *Hodotermes* spp.
This genus, found in Africa in dry grassland, can damage grass yield and quality. The colonies feed in grasses and therefore compete for resources with grazing domestic livestock such as sheep, cattle and goats.

Mediterranean Termite
Reticulitermes lucifugus (Above)
This termite is quite common in southern Europe and the Middle East. It causes considerable damage to wooden structures. Like most termites, Mediterranean termites live mainly below the ground.

Mound-building Termite

Macrotermes subhyalinus and *M. bellicosus*

This termite is responsible for the often very tall and massive mounds that are a feature of many areas of African savannas. These mounds can be as tall as 13m (43ft) and are usually constructed when the colony is a couple of years old. Inside the mound is a complex system of tunnels, shafts and chambers, some of which act as air-conditioning, others as brood chambers, fungus gardens and food storage.

Main habitat: Savanna.
Main region: Tropical.
Length: 4–14mm (0.16–0.55in); 60mm (2.36in) for the queen.
Wingspan: Most are wingless.
Development: Incomplete metamorphosis.
Food: Fungi and plants.

Identification: The workers are blind, wingless and infertile. They do most of the jobs in the colony. The members of the soldier caste have large sharp jaws to help them defend the colony. They are fed by the workers. The fertile members are the king and queen, which are larger. The single, enormous queen, up to 60mm (2.36in) long, is an egg-laying machine with a hugely distended abdomen. Her sole function is reproductive.

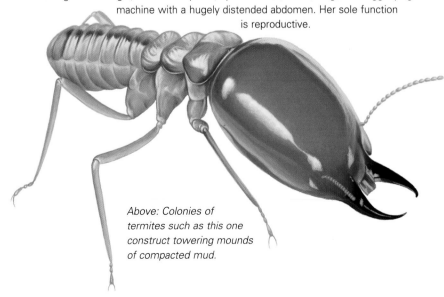

Above: Colonies of termites such as this one construct towering mounds of compacted mud.

MANTIDS AND EARWIGS

Mantids (Mantodea) are a mainly tropical group containing mostly quite large carnivorous species with claw-like grasping front legs used for catching their prey. There are about 2,000 species found throughout the warmer regions. A few can be found in the warmer countries of Europe. Earwigs (Dermaptera) number 1,900 species. They are often seen in gardens in Europe.

Praying Mantis

Mantis religiosa

This well-known insect is found in southern Europe. It sits and waits for its prey, mainly in bushes. When suitable prey strays close it quickly grabs it with its spiny claws. The name comes from the fact that it sits in a posture akin to someone at prayer. It flies fairly readily. Like most mantids, this species is very effectively camouflaged as it sits, sometimes swaying gently, among the twigs and leaves. The large eyes and flexible neck enable it to track its prey without moving the rest of its body, before striking.

Main habitat: Bushes and dry scrub.
Main region: Warm temperate.
Length: 80mm (3.15in).
Wingspan: 80mm (3.15in).
Development: Incomplete metamorphosis.
Food: Carnivore.

Identification: The praying mantis has long legs, a long neck, and a triangular head with large compound eyes. The front legs are spiny and clawed. It is green or brown-green in colour, and has well-developed wings.

Above: The praying mantis catches its victims in its spiny front claws.

Flower Mantis

Harpagomantis discolor

Main habitat: Flowering shrubs.
Main region: Tropical.
Length: 10cm (4in).
Wingspan: 10cm (4in).
Development: Incomplete metamorphosis.
Food: Carnivore.

Identification: This large mantis is beautifully patterned in pink and green, and also has petal-like projections on its body. The eyes are bright yellow. The wings have bright colours and these can be flashed to frighten a predator.

Far right: This predator has a gaudy, almost clown-like appearance.

This mantis from South Africa is one of a group known as flower mantises, as they tend to sit and wait for their prey on a flower, using their colours to blend into the background, sometimes swaying gently, as in a breeze. They can even alter the intensity of their colours to match the background. Insects visiting the flower may be unaware of the mantis until too late. A large proportion of its prey consists of other insects that visit the flowers to feed on the nectar, where they are quickly dispatched. Most flower mantises lie in wait among the flowers of a particular plant species. As well as being coloured to match these flowers, their bodies often have flattened petal-like projections to help them blend in.

European Earwig

Forficula auricularia

Main habitat: Soil.
Main region: Temperate.
Length: 12mm (0.47in).
Wingspan: 10mm (0.39in).
Development: Incomplete metamorphosis.
Food: Omnivore.

Right and below: Earwigs are familiar insects, commonly found in gardens.

This earwig is common in Europe and has spread to many other regions, where it is sometimes a pest of vegetables and flowers. In most gardens, however, earwigs do little damage and help to recycle organic matter. They live in the soil and underneath stones and leaves, emerging at night to feed. Female earwigs lay their eggs in a chamber and look after them, guarding the newly hatched young. Newly moulted earwigs are very pale, but soon become dark. Despite their pincers, earwigs are harmless insects.

Identification: A medium-sized, rather flat-bodied insect with fairly long antennae and characteristic pincers at the end of the abdomen. These rather sharp pincers are longer and curved in the male. Mainly red-brown, with yellowish wings (seldom used).

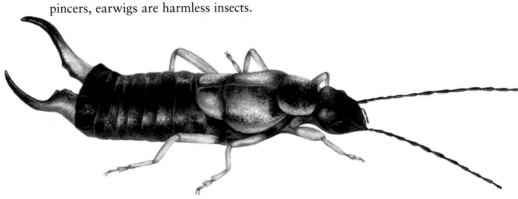

OTHER SPECIES OF NOTE

Mediterranean Mantis *Iris oratoria*
This widespread species is found in southern Europe and has also been introduced elsewhere, including the USA. It is smaller than the praying mantis, reaching about 65mm (2.56in). Its hindwings are brightly coloured, mainly yellow with a red leading edge and a dark central spot.

African Mantis *Sphodromantis lineola*
This fine species is widespread in Africa and is also often kept as a pet, being quite easy to feed and breed. Bright green or brown, it resembles a leaf when at rest. In many ways the African mantis resembles the praying mantis and it grows to a similar size. Like the praying mantis it has powerful mandibles for crushing and cutting up prey.

Small Earwig *Labia minor*
Another common European species. This earwig is often found in compost heaps or around dung. It takes to the wing more readily than most other earwigs and occasionally flies to lighted windows. This is the smallest earwig in Europe and is sometimes mistaken for an ant especially as it runs more rapidly than the European earwig and is often active by day as well as at night.

Woodland Earwig *Chelidurella acanthopygia*
This is smaller than the European earwig and has smaller pincers. It is also a more even brown colour. It lacks wings and therefore cannot fly. It is mainly found in woods, where it hides in leaf litter or in crevices on tree bark.

Tawny Earwig

Labidura riparia

Sometimes called the giant earwig, this is certainly one of the largest species. It prefers sandy habitats where it lurks under stones. It can also be found on sandy coasts and occasionally at riverbanks. In some regions it has become endangered due partly to habitat loss. Unlike many other earwigs, this species is entirely carnivorous, feeding on a range of other invertebrates. Males use their pincers in jousting matches.

Identification: Similar in shape to the European earwig, but larger and mainly pale yellow in colour, with brown stripes on the thorax and brownish patches along the abdomen. The pincers are long and rather straight, and darker than the body.

Right: The tawny earwig boasts an impressive pair of pincers.

Main habitat: Coastal and sandy sites.
Main region: Temperate.
Length: 25–35mm (1–1.38in).
Wingspan: 20mm (0.79in).
Development: Incomplete metamorphosis.
Food: Carnivore.

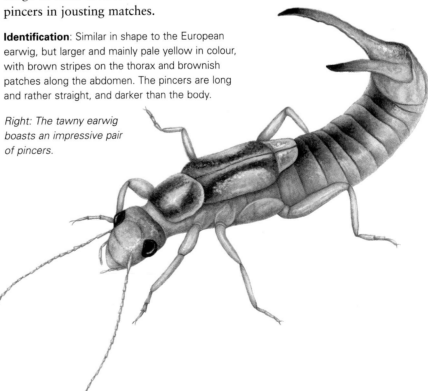

GRASSHOPPERS

Grasshoppers (part of Orthoptera) are active insects with powerful hind legs for jumping. Many species produce loud 'songs' by rubbing movements of their legs or wings. In Europe grasshoppers are most often found in grassland and dry habitats. All grasshoppers are herbivores. The locusts form swarms, which in Africa can cause massive damage to crops.

Meadow Grasshopper

Chorthippus parallelus

A common European grasshopper found in long grass, on which it feeds. It lays its eggs in the soil in summer, and these hatch the following spring, becoming adult in late June. Adults are active into October. The song is a series of 10–15 short chirps, each bout repeated at irregular intervals.

Identification: Small and usually greenish-brown, this grasshopper usually has only very short wings and so cannot fly. The underside of the abdomen is grey-green or yellow-green. Females may be greenish, brown or even purplish in colour. They normally lay their eggs in bare soil or in ant hills.

Below: The winged grasshopper is illustrated here. However, wing length varies considerably and the species often has much shorter wings.

Main habitat: Grassland.
Main region: Temperate.
Length: 15–20mm (0.59–0.79in).
Wingspan: Usually flightless.
Development: Incomplete metamorphosis.
Food: Herbivore.

Red-winged Grasshopper

Oedipoda germanica

Although mainly cryptically coloured, this grasshopper is able to fly a short distance when disturbed; it then reveals its brightly coloured hindwings. It will soon land nearby, the hindwings being obscured once more, and become difficult to find again. The sudden flash of colour often succeeds in confusing a predator such as a bird. This attractive species is quite common in southern Europe, but rare and endangered in central Europe.

Main habitat: Grassland.
Main region: Temperate.
Length: 25mm (1in).
Wingspan: 30mm (1.18in).
Development: Incomplete metamorphosis.
Food: Herbivore.

Identification: A medium-sized grasshopper with well-developed wings. The hindwings are mainly covered by a bright red patch, with a dark brown margin and translucent edge. It jumps well but often flies for a short distance as well.

Left: This species flashes bright colours when it takes off and flutters low over the ground.

OTHER SPECIES OF NOTE

Field Grasshopper *Chorthippus brunneus*
This is a common grasshopper of Europe. Small, about 20mm (0.79in), its colour varies and can be green, yellow, reddish, brown or even grey or almost black. Its wings are well developed, and it often flies. The antennae are short. This grasshopper sometimes turns up in large gardens and wasteland. Its chirping consists of short calls.

Desert Locust *Schistocerca gregaria*
This locust of Africa and the Middle East is even more of a pest than the migratory species. When local food is plentiful, for example after the rains, desert locusts are solitary. But in drier times they can become gregarious, forming swarms that may number billions of insects. These can travel hundreds of miles, stripping landscapes of vegetation.

Large Marsh Grasshopper
Stethophyma grossum
A large greenish-yellow species with black, yellow and red markings on its legs. Its favoured habitat is wet, boggy country, especially with some open water. In hot weather it often flies for a short distance. Its chirp is distinctive, being a series of ticking sounds.

Common Green Grasshopper
Omocestus viridulus
A grasshopper of long grasses, also found in clearings in woods and at roadsides. Its loud ticking song is often heard, but the insect can be hard to spot.

Migratory Locust

Locusta migratoria

This is the famous locust that causes serious problems to crops and grassland, mainly in Africa, especially when its population undergoes a periodical explosion. At these times clouds of locusts, numbering millions, may ravage the landscape for miles around, destroying crops and stripping the vegetation. It is found from southern Europe to Africa and the Middle East, and also in Asia and Australia. Migratory locusts can fly over 100km (60 miles) a day and each can eat its own weight in food each day. The species can exist in either a solitary or gregarious (flocking) phase.

Main habitat: Grassland.
Main region: Temperate and tropical.
Length: 60mm (2.36in).
Wingspan: 80mm (3.15in).
Development: Incomplete metamorphosis.
Food: Herbivore.

Identification: The colour varies from brown and green to greenish-yellow. Colour varies with age and whether the locust is in the solitary or gregarious phase.

Below: This locust is a large powerful species capable of sustained flight.

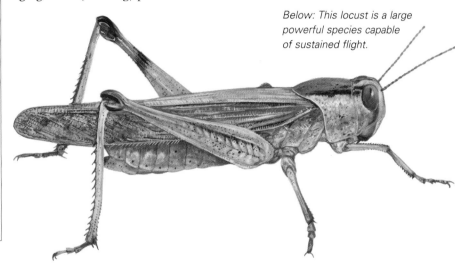

Variegated Grasshopper

Zonocerus variegatus

Found mainly in Africa south of the Sahara, this attractive insect is often a pest of crops such as cassava and maize. It is particularly common in west and central Africa. It stores toxic chemicals from some plants in its body so that it tastes bad to predators, and its bright colours warn them to stay clear. It is one of the most brightly coloured grasshoppers and is usually easy to spot against the greens and browns of its bushland habitats.

Main habitat: Savanna, fields.
Main region: Tropical.
Length: 35–55mm (1.38–2.17in).
Wingspan: Normally short-winged.
Development: Incomplete metamorphosis.
Food: Herbivore.

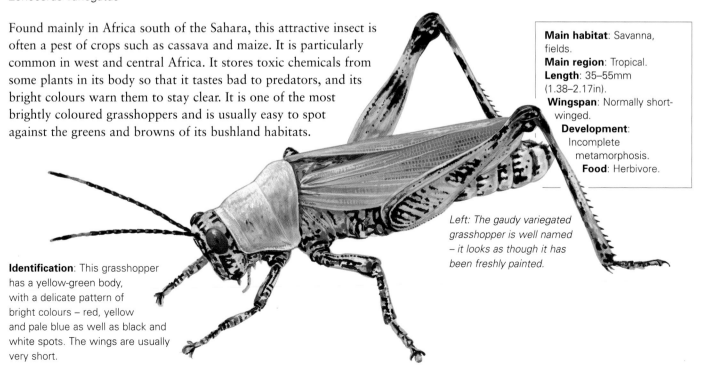

Left: The gaudy variegated grasshopper is well named – it looks as though it has been freshly painted.

Identification: This grasshopper has a yellow-green body, with a delicate pattern of bright colours – red, yellow and pale blue as well as black and white spots. The wings are usually very short.

CRICKETS

Most members of this order (Orthoptera) are known as crickets. Crickets are mostly bulkier than grasshoppers and most species do not jump as far or as readily. They make distinctive sounds by rubbing their forewings and have long antennae (very long in bush crickets). They are found across Europe, especially in the south, and in Africa.

Great Green Bush Cricket

Tettigonia viridissima

This is one of Europe's largest crickets and is found over much of Europe, mainly in dry grassland and scrub, between July and November. It eats mainly other insects using its powerful jaws, and is capable of inflicting a nasty bite. It jumps and flies well. Its loud song can be heard from afternoon well into the night during warm weather. It lays its eggs, up to 300 at a time, in loose soil.

Main habitat: Dry grassland.
Main region: Temperate.
Length: 36–42mm (1.42–1.65in).
Wingspan: 60mm (2.36in).
Development: Incomplete metamorphosis.
Food: Carnivore.

Identification: Mainly green with a pale brown stripe down its back, and long antennae. The female has a long ovipositor that reaches to the end of the wings. The ovipositor is used for laying eggs.

Above: This bush cricket has long wings and is able to fly well.

Field Cricket

Gryllus campestris

This cricket is widespread over much of Europe where it lives in tunnels on warm grassy banks. Its chirping, almost bird-like song is often heard between April and high summer. Field crickets feed on a wide range of material, including grasses and decaying matter such as dead insects. In Britain it is very rare.

Identification: Relatively small, yet bulky and large-headed, this cricket is dark brown or black. The males call by rubbing their forewings together making a chirping almost bird-like sound.

Main habitat: Dry grassland.
Main region: Temperate.
Length: 23mm (0.91in).
Wingspan: Wings reduced.
Development: Incomplete metamorphosis.
Food: Omnivore.

Left and below: Field crickets are found in warm, dry grassland from Europe to North Africa.

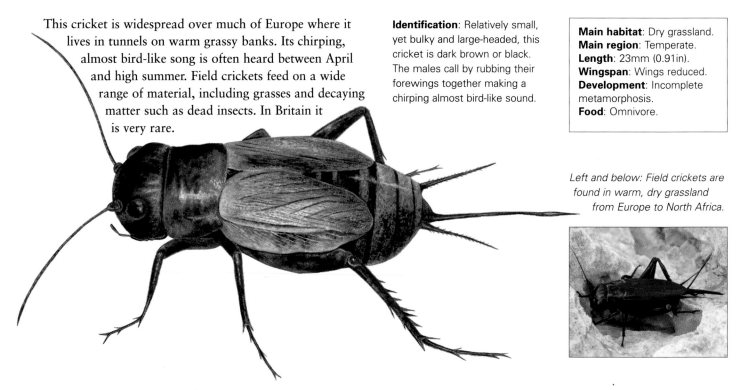

Predatory Bush Cricket

Saga pedo

This remarkable cricket is one of Europe's largest insects, reaching well over 10cm (4in) in length. It is found mainly in southern Europe where it inhabits dry scrub and grassland. It is rather rare in many places and has a scattered distribution. It is very unlike most bush crickets, having a long thin body and being a fierce predator. Males are virtually unknown, the females usually reproducing without sex (parthenogenesis). It eats mainly other insects, which it grabs with its claw-like front legs.

Below: This large cricket is unusually long and thin.

Identification: Large size, and long thin body with strong claw-like front limbs make it unmistakable. The body is bright green and the antennae very long.

Main habitat: Dry scrub and grassland.
Main region: Warm temperate.
Length: 12cm (4.5in).
Wingspan: Wingless.
Development: Incomplete metamorphosis.
Food: Carnivore.

OTHER SPECIES OF NOTE

Hedgehog Katydid *Cosmoderus erinaceus*
This bulky bush cricket from the forests of central Africa is well named. Its body is covered with sharp protective spines, especially on its thorax and legs.

Cave Cricket *Speleiacris tabulae*
One of a group of crickets adapted to life in caves. This species is found in the caverns of Table Mountain and some other parts of the Cape in South Africa. It has reduced eyes, long legs and antennae and is wingless. It feeds partly on bat droppings.

Tobacco Cricket *Brachytrupes* spp.
These noctural insects are pests of tobacco, maize and other food crops. They are found mainly in the savannas of Africa. In some places the local people regard them as a delicacy and collect them as food.

Oak Bush Cricket *Meconema thalassinum*
This rather delicate bush cricket is common in Europe and often turns up in gardens where it can be found crawling on twigs and leaves. It is pale yellow-green, with very long, narrow antennae. The female has a long, upcurved ovipositor.

Speckled Bush Cricket
Leptophyes punctatissima
Resembles the oak bush cricket but is flightless, and with a hunched appearance and broader ovipositor in the female. It is also often seen in gardens.

Mole Cricket

Gryllotalpa gryllotalpa

Mole crickets are some of the strangest of all the crickets. Like moles they are adapted to a largely underground existence and their front claws are massive and built like shovels for excavating burrows in muddy soil. Their bodies are also covered in short, velvety hairs, helping them slither through the soil. However, they do have functional wings and sometimes fly at night. They can be attracted to lights. They feed on roots and also grubs. Males sit at the entrance to their burrows on warm evenings or at night producing long bursts of song – a churring, bird-like sound not unlike that of the European nightjar, swaying from side to side if a female approaches. They feed mainly on plant material and insect grubs.

Main habitat: Damp grassland and muddy sites.
Main region: Temperate.
Length: 35mm (1.38in).
Wingspan: 30mm (1.18in).
Development: Incomplete metamorphosis.
Food: Omnivore.

Identification: Greatly enlarged front legs, broad and toothed. Powerful hind legs are used to push out the soil.

Below: The mole cricket has spade-like front legs and subterranean habits.

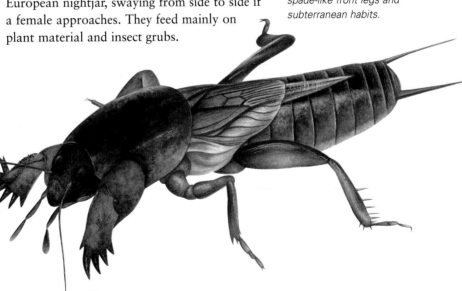

STICK INSECTS

Stick insects (Phasmida) have amazing camouflage and can be hard to spot. Most of the 2,500 or so species live in warm or tropical regions including Africa, but a few stick insects are found in Europe, mainly in the warmer south. Stick insects have long, thin bodies and resemble twigs, often escaping detection by birds and other predators by the use of mimicry.

Mediterranean Stick Insect

Bacillus rossius

Main habitat: Bushes.
Main region: Temperate.
Length: 80–90mm
(3.15–3.54in).
Wingspan: Usually wingless.
Development: Incomplete
metamorphosis.
Food: Herbivore.

This medium-sized stick insect is quite common in southern Europe. It feeds on a variety of leaves, including bramble, and can be found in bushes, grasses and forest edges, mainly in warm, dry sites. In most regions males are unknown or rare and the females reproduce without fertilization (parthenogenesis). In winter the adults tend to die and the eggs are left to overwinter.

Identification: Slim, long, rather smooth green or brownish body, with very long narrow legs. The antennae are rather short. The abdomen has a series of large, obvious segments.

Left: This is one of only a few stick insects found in Europe.

OTHER SPECIES OF NOTE
French Stick Insect *Clonopsis gallica*
This is a very slender stick insect found in southern Europe, as far north as Brittany. It is pale green and can be hard to spot when resting on a blade of grass. It has long legs and very short antennae.

Thunberg's Stick Insect *Macynia labiata*
This is a common species in the Cape region of South Africa where it feeds on many plants including heathers. Males are about 50mm (2in) and light brown-green. The female is about 55mm (2.17in) and rather plump and green with cream stripes.

Bactrododema hippotaurum
This giant stick insect is one of the largest found in Africa. It grows to 26cm (10in), although its body is not bulky and is actually very light. It lives mainly in trees and shrubs in southern and central Africa.

Spanish Stick Insect *Leptynia hispanica*
This is a small stick insect found in Spain and southern France. The antennae are short and reddish, contrasting with the green body. It is found on bushes and shrubs in open sunny country.

Giant Stick Insect

Bactrododema krugeri

This huge stick insect is found in southern Africa, mainly in Botswana, Namibia, South Africa, Swaziland and Zimbabwe. It moves rather slowly around twigs and branches, chewing leaves. Occasional winged forms occur and these can glide and flutter short distances, although their wings are small compared with the body size.

Identification: One of the world's longest insects, reaching nearly 30cm (12in) with legs outstretched. The body is long and thin, with slight bumps, and mainly brown. Males are smaller than females and winged forms are found.

Below: This very large stick insect looks rather like a thorny twig.

Main habitat: Trees and bushes.
Main region: Tropical.
Length: 16cm (6in) (male); 19cm (7.5in) (female).
Wingspan: 80mm (3in).
Development: Incomplete metamorphosis.
Food: Herbivore.

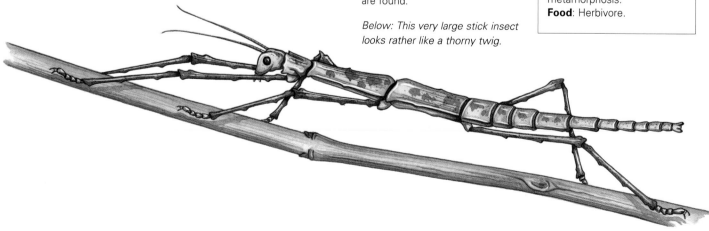

Pink-winged Stick Insect

Sipyloidea sipylus

| Main habitat: Trees and bushes.
Main region: Tropical.
Length: 10cm (4in).
Wingspan: 10cm (4in).
Development: Incomplete metamorphosis.
Food: Herbivore. |
| --- |

Like many stick insects, males are not always produced and the females can also lay fertile eggs without mating. When disturbed, this species can spray an irritant chemical at its attacker, but it is not dangerous to people. In fact, it makes a good pet and can be fed on leaves such as bramble, rose and hawthorn.

Above and below: Stick insects use their long, thin legs to clamber slowly among foliage.

Identification: The name comes from the relatively large wings, which are an attractive pale pink colour. The body is a pale, pinkish brown and rather slim and delicate with quite long, narrow antennae. This stick insect has very slender legs and is somewhat unusual in having wings that are fully functional. It is native to Madagascar. It is widely kept, and breeds readily in captivity. The adult females glue their small grey eggs to plant stems.

Above: Though often kept as a pet, this stick insect can emit a chemical if it is not used to being handled.

Cape Stick Insect

Phalces brevis

This is a common stick insect of South Africa, especially the Cape Province and also in Natal. The adults live for about six months and each female lays a few hundred eggs. As with many stick insects, the eggs are simply dropped on to the ground. Ants sometimes carry away the eggs, eating an edible part of the covering without harming the eggs and thus helping to disperse the insect. The young nymphs take about six months to reach maturity. Both males and females can be found. This stick insect feeds on a range of plants including members of the heather family, pea family and rose family, as well as the tea tree (*Leptospermum*).

| Main habitat: Scrub.
Main region: Subtropical.
Length: 50–80mm (1.97–3.15in).
Wingspan: Wingless.
Development: Incomplete metamorphosis.
Food: Herbivore. |
| --- |

Identification: Very thin and twig-like with short antennae; wingless. The male is green-brown with white bands and blue-green markings on the thorax. Females are brown or grey, sometimes green. The slender legs are green, with a brown base. As with many stick insects, this species can be kept easily and eats a range of leaves.

Above: The Cape stick insect is an excellent twig mimic.

BOOKLICE, THRIPS, LICE AND FLEAS

Booklice and bark-lice are small, ranging from 1–10mm (0.04–0.39 in). Web spinners are mostly small, usually wingless, with soft bodies, living in damp places in silken tunnels. Thrips are tiny insects with hair-like wings and sucking mouthparts. Lice and fleas are wingless parasites of other animals (mammals and birds). They feed on fragments of skin, fur or feathers and also on blood.

Web Spinner

Embia ramburi

This is a common European species of web spinner. It lives under stones, bark and logs and the like, where it spends most of the time inside an untidy web spun of fine silk. The web is spun from its front legs. Both sexes are wingless, except some males. Web spinners are usually found in colonies of many individuals.

Identification: The front legs are swollen with special glands that produce the silk from which the webs are created. The body is brownish, elongated and soft. The legs are rather short; the hind legs more powerful.

> **Main habitat**: Damp sites.
> **Main region**: Temperate.
> **Length**: 8mm (0.31in).
> **Wingspan**: Wingless.
> **Development**: Incomplete metamorphosis.
> **Food**: Omnivore.

Left: Being small and secretive, web spinners are seldom spotted.

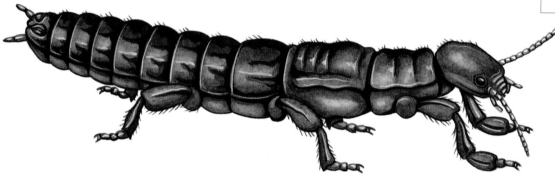

Grain Thrip

Limothrips cerealium

This is a widespread species of thrip, found notably in the ears of wheat and other grain cereals, as well as wild grasses. Both the nymphs and young adults feed on these and when present in large numbers they can cause shrivelling of the grain and ears. But they also help by pollinating crops. This is one of several species that are known commonly as 'thunderflies' as they tend to emerge in warm, humid weather.

Identification: A tiny, almost speck-like, insect, 1–2mm (0.04–0.08in) long, and dark brown or black, with paler legs and antennae. The male is slightly smaller and paler and lacks wings.

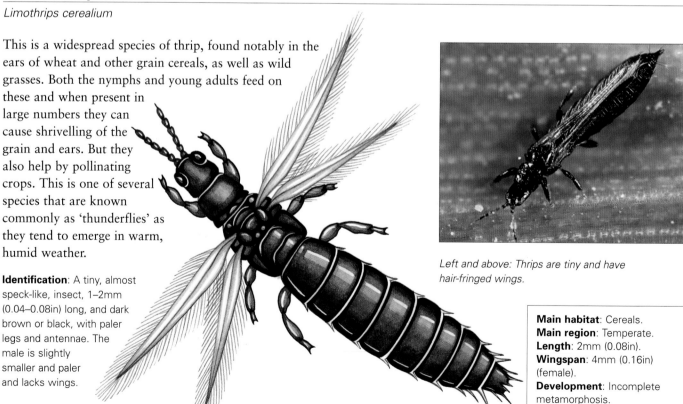

Left and above: Thrips are tiny and have hair-fringed wings.

> **Main habitat**: Cereals.
> **Main region**: Temperate.
> **Length**: 2mm (0.08in).
> **Wingspan**: 4mm (0.16in) (female).
> **Development**: Incomplete metamorphosis.
> **Food**: Grain.

Head Louse

Pediculus humanus capitis

This is one of the best known of all lice, not least because it is found on people, most notably in the hair of children. Even clean hair is no deterrent and infection from head to head is often very rapid in places where children gather, particularly in schools. The adults and nymphs both feed on blood and the eggs (known as nits) are glued to the base of hairs. Each adult female head louse can lay about ten eggs a day, so their numbers can build up rapidly. Infected hair should be inspected regularly and the nits removed by using a special comb. Various chemicals are used (for example in special shampoos) to kill head lice, but physical removal of the nits is also necessary to prevent re-infection. Clean hair can easily be infected by brushing against someone already carrying head lice, so regular inspection is recommended. The body louse (*P. humanus humanus*) sometimes carries and spreads the disease typhus, but the head louse is usually merely an irritant.

Main habitat: Human hair.
Main region: Worldwide.
Length: 2–5mm (0.08–0.19in).
Wingspan: Wingless.
Development: Complete metamorphosis.
Food: Blood.

Identification: Small, pale, wingless insects with a pear-shaped flat body and hairy legs with sharp, curved claws. The head is narrow with small eyes and short antennae.

Below: A head louse uses its sharp, curved claws to cling to the hair of its host.

Cat Flea

Ctenocephalides felis

Familiar to most cat owners, this is the most common domestic flea. These irritating insects can also attack people, resulting in itchy spots. An adult flea can live for about 25 days. The eggs fall off the host into its bedding or carpets and the larvae that hatch live there, and turn into pupae in cocoons. Hatching is stimulated by movement, ideally by the host mammal and the new adult fleas jump and (if lucky) attach themselves to their host.

Right: Fleas use their muscular hind legs to jump and disperse on to a new host.

Identification: The larvae are pale, worm-like and about 2mm (0.08in) long. They turn into pupae, which inhabit silk cocoons and hatch as adult fleas. The adult fleas are red-brown and compressed laterally. They have powerful hind legs, and can jump and scuttle through fur.

Main habitat: Fur and bedding of cats.
Main region: Temperate.
Length: 1–3mm (0.04–0.12in).
Wingspan: Wingless.
Development: Complete metamorphosis.
Food: Blood.

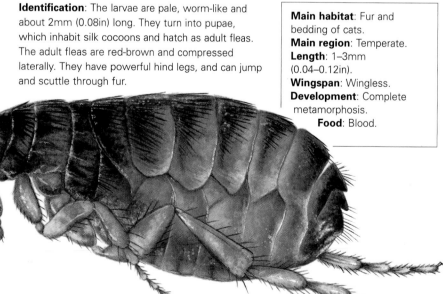

BUGS

The word 'bug' is often used loosely to refer to any insect, but to scientists it has a more specific meaning. The insects on these pages are all true bugs, belonging to the insect order Hemiptera. True bugs are a large group, with more than 82,000 species found worldwide. They are well represented in Europe and Africa, both on land, and in and on the water.

Green Shield Bug

Palomena prasina

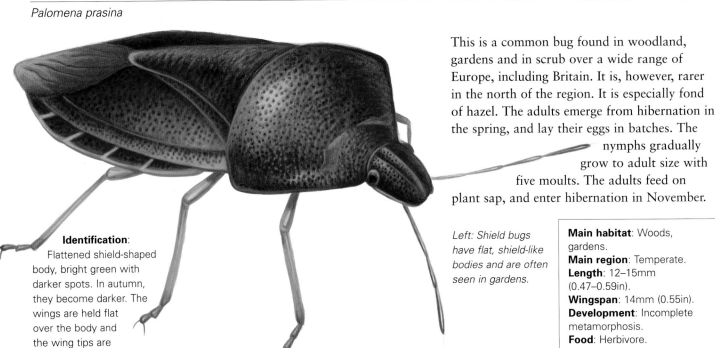

This is a common bug found in woodland, gardens and in scrub over a wide range of Europe, including Britain. It is, however, rarer in the north of the region. It is especially fond of hazel. The adults emerge from hibernation in the spring, and lay their eggs in batches. The nymphs gradually grow to adult size with five moults. The adults feed on plant sap, and enter hibernation in November.

Identification: Flattened shield-shaped body, bright green with darker spots. In autumn, they become darker. The wings are held flat over the body and the wing tips are membranous.

Left: Shield bugs have flat, shield-like bodies and are often seen in gardens.

Main habitat: Woods, gardens.
Main region: Temperate.
Length: 12–15mm (0.47–0.59in).
Wingspan: 14mm (0.55in).
Development: Incomplete metamorphosis.
Food: Herbivore.

Bed Bug

Cimex lectularius

This bug is notorious as it attacks people, some birds and also domestic and zoo animals. The bug hides in bedding and crevices by day, emerging at night to attack its host (usually a sleeping person) and suck their blood. Houses being usually much cleaner than in the past, bed bugs are not the problem they once were. Bed bugs are usually most active around dawn. They painlessly pierce the skin of the host with sharp mouthparts consisting of two tubes. One tube injects saliva and the other takes in a meal of blood. Only afterward does the bite become itchy. An adult bed bug can live for well over a year without feeding, but they usually feed every 5–10 days. Although the bites can cause itching and infection, bed bugs are not known to transmit diseases.

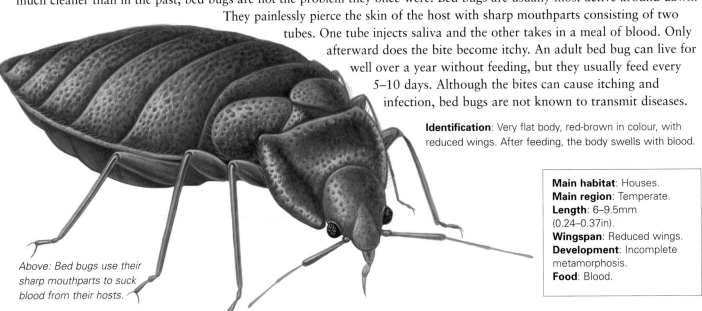

Identification: Very flat body, red-brown in colour, with reduced wings. After feeding, the body swells with blood.

Above: Bed bugs use their sharp mouthparts to suck blood from their hosts.

Main habitat: Houses.
Main region: Temperate.
Length: 6–9.5mm (0.24–0.37in).
Wingspan: Reduced wings.
Development: Incomplete metamorphosis.
Food: Blood.

Water Stick Insect

Ranatra linearis

Main habitat: Aquatic; fresh water.
Main region: Temperate.
Length: 30–35mm (1.18–1.38in).
Wingspan: 25mm (1in).
Development: Incomplete metamorphosis.
Food: Carnivore.

This remarkable bug has a long thin body and resembles a stick insect. It spends most of its time under water, but can also fly and sometimes walks on waterside plants. It hunts for other aquatic insects, small fish and tadpoles, which it catches in its claw-like front limbs. It lays its eggs in rows on the stems of floating water plants.

Left and above: This slender bug can move about both above and below the water.

Identification: The long, narrow body is brown or yellowish and the legs very long and slender. The front legs end in pincer-like claws that can snap shut around its prey. The antennae are small. The long, narrow tube at the end of its body is used to take in air from the surface of the water, enabling this bug to breathe when submerged. The water stick insect's appearance helps to disguise it from predators such as fish and hide it from prey. It often lies in wait on the stems of water plants, which it resembles.

OTHER SPECIES OF NOTE

Hawthorn Shield Bug
Acanthosoma haemorrhoidale
This is a common European and Eurasian shield bug, feeding mainly on hawthorn leaves and fruits. Flat-bodied, it is green with chestnut markings. It feeds by piercing the leaves of trees and shrubs.

Water Scorpion *Nepa cinerea*
This bug is well named. It is large, growing to 22mm (0.87in), and has large pincers, and also a long, thin breathing tube on the end of its body. It crawls about above and below the water surface in ponds, lakes and slow rivers and occasionally flies.

Small Water Boatman *Corixa* spp.
This water bug swims using its powerful hairy hind legs, but unlike the backswimmer, it swims on its belly. About 15mm (0.59in) long, it is brown and flat. It is mainly herbivorous, feeding on algae and rotting vegetation.

Pond Skater *Gerris lacustris*
A dark, blackish-brown bug common on the surface of ponds, lakes and slow rivers. It has long antennae and large eyes. The slender body has silvery, water-repellent hairs. The front legs are short but the middle legs are very long and used to propel the insect forward over the surface. Pond skaters eat other insects caught as they glide across the surface film.

Backswimmer

Notonecta glauca

This large water boatman takes its name from the fact that it swims upside-down. Its back is boat-shaped and it rows itself through the water using powerful hind limbs. As it swims it uses up air trapped by hairs, and every so often it rises to the surface to take in fresh supplies. Backswimmers search for prey using their large eyes, and sometimes leave the water to fly to new sites. They are quite common in slow water and ponds.

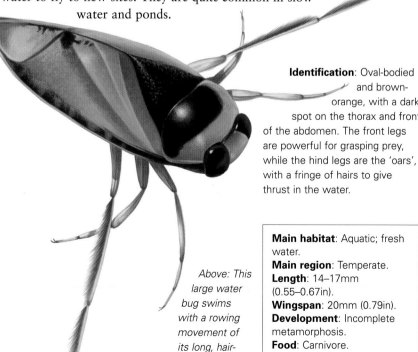

Identification: Oval-bodied and brown-orange, with a dark spot on the thorax and front of the abdomen. The front legs are powerful for grasping prey, while the hind legs are the 'oars', with a fringe of hairs to give thrust in the water.

Above: This large water bug swims with a rowing movement of its long, hair-fringed legs.

Main habitat: Aquatic; fresh water.
Main region: Temperate.
Length: 14–17mm (0.55–0.67in).
Wingspan: 20mm (0.79in).
Development: Incomplete metamorphosis.
Food: Carnivore.

Common Green Capsid

Lygocoris pabulinus

Main habitat: Trees and bushes.
Main region: Temperate.
Length: 5mm (0.19in).
Wingspan: 8mm (0.31in).
Development: Incomplete metamorphosis.
Food: Herbivore.

This bug sucks sap from shoot tips and buds, causing damage to plant tissues. It sometimes attacks garden plants such as chrysanthemums, fuchsias, clematis and roses, as well as fruit trees, cane fruits and strawberries. The eggs overwinter and the nymphs hatch in the spring, feeding on trees, shrubs and other plants. Each female can lay as many as 100 eggs and there are normally two generations each year.

Identification: Bright green and winged as adults, about 5mm (0.19in) long, the nymphs are slightly smaller and wingless.

Above: Often seen in gardens, this bug can damage plants, especially while it is in the nymph stage.

Striped Mirid

Miris striatus

This is a common bug found on a range of trees and bushes, especially on oak, elm and hazel, but also on birch, alder and willow. The adults are most active in June and July. As well as feeding on plant sap, this bug also eats some caterpillars. The female lays her eggs in the bark of deciduous trees, where they overwinter, hatching the following spring.

Identification: Black head and body, with yellow and orange lines and patches on the wings. The thorax is also black, and has a large yellow patch. The legs are a bright orange-red, darker toward the tips, and with black claws. The antennae are rather long, with angled joints and a pale mark about halfway along.

Main habitat: Trees and bushes.
Main region: Temperate.
Length: 10–12mm (0.39–0.47in).
Wingspan: 15mm (0.59in).
Development: Incomplete metamorphosis.
Food: Omnivore.

Above and below: The striped mirid is found throughout most of Europe.

Cicada

Cicada orni

Main habitat: Trees and bushes.
Main region: Warm temperate.
Length: 35–40mm (1.38–1.57in).
Wingspan: 80mm (3.15in).
Development: Incomplete metamorphosis.
Food: Herbivore.

Right: Cicadas are usually heard rather than seen.

This medium-sized cicada is quite common in southern Europe. The adults, which live for only a few weeks, make their presence felt (or rather heard) by high-pitched fizzing 'singing' in hot sunny weather, audible over a long distance. It is the males that sing, using their special drum-like membranes. The nymphs live underground, taking some years to complete their development.

Identification: The adult is a mottled greyish-brown colour, blending well with the bark of the trees in which it lives. The front wings are clear with black spots and dark veins.

Frog-hopper

Cercopis spp.

Main habitat: Grasses.
Main region: Temperate.
Length: 10mm (0.39in).
Wingspan: 15mm (0.59in).
Development: Incomplete metamorphosis.
Food: Herbivore.

Identification: Squat and dumpy, with round eyes, this bug is easily recognized by the bold red-and-black chequerboard markings on its forewings. The hind legs are sturdy and bear short spines. They are powerful and used to jump suddenly when disturbed.

These colourful bugs are found in grassy habitats over much of Europe. As the name suggests, frog-hoppers are adept at jumping and often escape predation or disturbance this way. This genus advertises its distastefulness through its bold markings. The nymphs live underground, feeding on roots. Frog-hopper nymphs surround themselves with froth to prevent drying out. This is sometimes called 'cuckoo-spit', and the bugs 'spittlebugs'.

Above: The bright colours of this frog-hopper warn birds of its unpleasant taste.

Masked Assassin Bug

Reduvius personatus

This strange, rather sinister bug, also known as the fly bug, belongs to a group known as assassin bugs. It pierces small invertebrates and feeds on their body fluids; both adults and nymphs feed in this manner. It is fairly common in Europe and lives in hollow trees, but is also commonly found in buildings, including houses, hiding in cracks. It can deliver a painful bite so should be handled with care. These bugs emerge at night and feed on bed bugs and other small invertebrates.

Main habitat: Trees, houses.
Main region: Temperate.
Length: 15mm (0.59in).
Wingspan: 20mm (0.79in).
Development: Incomplete metamorphosis.
Food: Carnivore.

Identification: The body is elongated, with long legs and long antennae. The adult is dark blackish-brown and rather hairy. The nymph is unusual in that it covers (masks, hence the name) itself with particles of dust, sand and other debris and can be quite hard to spot.

Right: The masked assassin bug is a stealthy predator with a nasty bite.

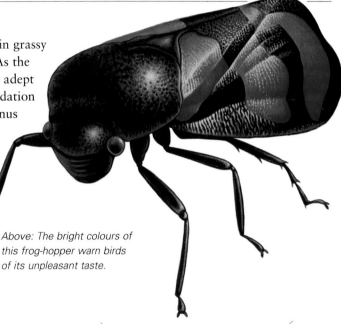

CADDIS FLIES, LACEWINGS, ANTLIONS AND SCORPION FLIES

Caddis flies (Trichoptera) look rather like moths, but they have hairy bodies. Lacewings (Neuroptera) have net-veined wings. As well as the lacewings, this order contains antlions, mantis flies and ascalaphids. Scorpion flies (Mecoptera) are named for the curved abdomen tip of some of the males.

Caddis Fly

Phryganea grandis

Identification: It is light grey in colour, with some darker stripes on the hairy wings. The antennae are nearly as long as the forewings and are usually held pointing forward.

Below: Adult caddis flies are very moth-like in general appearance.

Main habitat: Still or slow water (larvae).
Main region: Temperate.
Length: 12–26mm (0.47–1in).
Wingspan: 50mm (2in).
Development: Complete metamorphosis.
Food: Omnivore (larva).

This large caddis fly is one of the bigger European species. Caddis fly adults are rather moth-like, but they have hairy rather than scaly bodies and they hold their wings over their bodies in a roof-like shape. Like moths they are mainly active at dusk or at night. They feed rather sparingly as adults, simply licking up nectar or juices from plants. The aquatic larvae of caddis flies are as well known as the adults and most build protective tubes out of bits of plant tissue or stones. The larvae of this caddis fly live in slow-moving rivers and streams and build their cases from a spiral of plant tissues. Anglers often refer to caddis flies as 'sedges'.

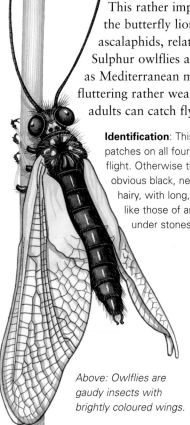

Sulphur Owlfly

Libelloides coccajus

This rather impressive insect, sometimes called the butterfly lion, belongs to a group known as ascalaphids, related to lacewings and antlions. Sulphur owlflies are typical of warm sites, such as Mediterranean maquis, and may be spotted fluttering rather weakly in sunny weather. The adults can catch flying insects.

Identification: This species has bright sulphur-yellow patches on all four wings, very visible when it takes flight. Otherwise the wings are transparent with obvious black, net-like veins. The body is dark and hairy, with long, clubbed antennae. The larvae are like those of antlions and live in soft soil or under stones.

Main habitat: Warm sunny sites.
Main region: Temperate.
Length: 30mm (1.18in).
Wingspan: 45mm (1.77in).
Development: Complete metamorphosis.
Food: Carnivore.

Above: Owlflies are gaudy insects with brightly coloured wings.

OTHER SPECIES OF NOTE

Antlion *Myrmeleon formicarius*
This antlion is common over much of (mainly) southern Europe. The fierce larva, about 20mm (0.79in) long, sits at the bottom of a pit of loose sand. When an ant trundles near, the larva flicks sand at it to make it slide down into its waiting jaws. The adult is mostly brown with a wingspan of about 60mm (2.36in) and flies mainly at dusk.

Antlion *Tomatares citrinus*
This large antlion is found in Africa. Its slender abdomen is brown and its head and thorax have red and black markings. The wings are beautifully marked with brown blotches and stripes on a yellow background.

Scorpionfly
Panorpa spp. (e.g. *P. communis*; *P. cognata*)
This insect is quite common in Europe, on bushes and scrub at forest edges, for instance. Its wings have dark spots and the head has a long downward-projecting point. The male has the characteristic swollen upcurving abdomen tip. The larvae live in tunnels in the soil.

Caddis Fly *Rhyacophila obliterata*
This is a medium-sized caddis fly found around fast-flowing streams. The larvae, which are bright green and brown, hunt actively around the stony stream bed, searching for scraps of edible material. The wings are rather narrow and slightly pointed; the forewings a pale yellow colour, the hindwings clear.

Green Lacewing

Chrysopa septempunctata

Main habitat: Shrubs, gardens.
Main region: Temperate.
Length: 10mm (0.39n).
Wingspan: 30mm (1.18in).
Development: Complete metamorphosis.
Food: Carnivore.

This common European lacewing is often seen on windows at night as it is attracted to light. This type of lacewing has a clever method of protecting its eggs – each egg sits on top of a thin thread, which gives it some protection from ants. The larvae feed on small insects such as aphids.

Above and below: The pretty green lacewing is a very common visitor to gardens.

Identification: Both the body and the delicate veins in the wings are a pale shade of green and the large eyes are bright and shiny, with a golden tint. The antennae are long and narrow. This species has seven tiny black spots on its head.

Large Antlion

Palpares libelluloides

Main habitat: Scrub.
Main region: Warm temperate.
Length: 55mm (2.17in).
Wingspan: 10cm (4in).
Development: Complete metamorphosis.
Food: Carnivore.

Identification: The larvae have large pincer-like jaws. The adult looks rather like a dragonfly, with a long body and large, mottled clear wings. Its antennae are curved and end in a club.

This splendid insect is the largest European antlion and the species mainly inhabits southern regions. Antlions take their name from their voracious predatory larvae which live partly submerged in loose sand and soil, awaiting insect prey (often ants). The larvae is an oval shaped grub with short legs on which it shuffles. It constructs a sandy pit in which to capture prey.

Left: The adult antlion looks like a rather weak dragonfly when in flight. When at rest, the adult holds its wings roof-like above its body and its flight is rather weak and fluttery.

Snake Fly

Raphidia notata

Snake flies such as this have very long necks (pronoti) with a triangular head which they can raise in a snake-like manner. This insect is fairly common, especially on conifers and in mature oak woods, and feeds on smaller insects such as aphids. The larvae live beneath loose bark.

Below: The snake fly takes its name from its long, flexible 'neck'.

Main habitat: Woodland.
Main region: Temperate.
Length: 15mm (0.59in).
Wingspan: 30mm (1.18in).
Development: Complete metamorphosis.
Food: Carnivore.

Identification: Resembles a lacewing, with transparent, net-veined wings. The key feature is the curiously long neck-like thorax and snake-like head. The female has a long, narrow ovipositor.

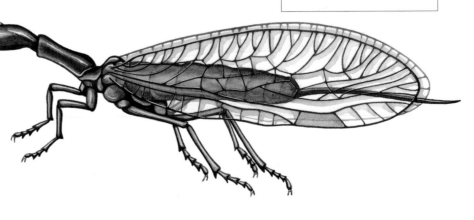

BEETLES

With more than 300,000 known species, beetles (Coleoptera) are by far the largest of the insect orders: about a third of all animal species are beetles. In Europe and Africa they are a common sight and range from small wood-boring beetles to giant longhorns and Goliath beetles. The beetles on the following pages are some of the most spectacular and commonly seen species.

Violet Ground Beetle

Carabus violaceus

Identification: Long and dark, with a slight violet tinge, especially around the edge of the body. The legs are long and powerful.

Main habitat: Soil.
Main region: Temperate.
Length: 20–30mm (0.79–1.18in).
Wingspan: Flightless.
Development: Complete metamorphosis.
Food: Carnivore.

Above: This fast-moving predator is useful in consuming garden pests.

This medium-sized and active beetle is common on dry ground, notably in woodland, and is most likely to be spotted in the summer. Both larvae and adults hibernate. Ground beetles are long-bodied and long-legged and most, like this species, hunt for other invertebrate prey on the ground. When disturbed they can exude a nasty-smelling fluid from the tip of their abdomen. They are mainly active at night, catching varied prey including worms and slugs and the larvae of other insects. In the day they hide in dark places such as underneath logs or stones.

Green Tiger Beetle

Cicindela campestris

This lively and colourful tiger beetle is common, especially on sandy soils such as in heathland and dunes in Europe. It is most active in dry, warm sunny weather. Tiger beetles are similar in shape to ground beetles and are also long-legged and fast-running. They hunt other insects, such as ants, and can also fly. The larvae live in vertical tunnels in the ground and lie in wait for passing prey which they dart out to grab in their jaws. Handsome and lively, this is one of Europe's most common tiger beetles. It uses speed and acute eyesight to track and catch live prey. Tiger beetles include some of the world's fastest running insects. This species can reach 8kmh (5mph) – faster than a man at walking pace.

Identification: Bright, metallic green with yellow patches and fairly long antennae. The underside and upper legs are a coppery colour.

Left: This predatory beetle has large eyes and fearsome jaws in which it seizes insect prey.

Main habitat: Sandy soils.
Main region: Temperate.
Length: 12–16mm (0.47–0.63in).
Wingspan: 30mm (1.18in).
Development: Complete metamorphosis.
Food: Carnivore.

Poplar Leaf Beetle

Chrysomela populi

The leaf beetles are a large family with more than 35,000 species. They mostly feed on leaves, hence the name, and many are brightly coloured. This species is common on poplar trees, and sometimes on willows, usually on the leaves or suckering shoots. The adults are most active from May through August. The larvae live on the leaves and can do considerable local damage to the leaf tissues.

Right: The poplar leaf beetle is chunky and brightly coloured.

Main habitat: Trees and bushes.
Main region: Temperate.
Length: 10–12mm (0.39–0.47in).
Wingspan: 15mm (0.59in).
Development: Complete metamorphosis.
Food: Herbivore.

Identification: This beetle has a plump, oval body that is rounded at the rear end. The wing covers are tough, shiny and brick red. The head and prothorax are blackish.

Cereal Leaf Beetle

Oulema melanopus

This leaf beetle genus is from Africa and Asia, where it has become an agricultural pest in some areas. It feeds on many plants, including cereals such as barley, rye, millet and rice. It spends the winter in grassland, entering crop fields in the spring. The larvae disguise themselves as bird droppings by covering themselves with their own droppings. Both larvae and adults feed on grasses or cereals.

Right: In parts of its range this beetle affects cereal crops.

Main habitat: Cereals.
Main region: Warm temperate.
Length: 4–5.5mm (0.16–0.22in).
Wingspan: 8mm (0.31in).
Development: Complete metamorphosis.
Food: Herbivore.

Identification: Mainly blue-black, with a bright red prothorax and orange legs. The antennae are rather sturdy and quite long.

Burying Beetle

Nicrophorus investigator

This rather brightly coloured beetle and its close relative *N. vespilloides* are also known as sexton beetles. They play a great role in recycling, feeding mainly on carrion and fungi, and thus helping to dispose of the dead bodies of other animals, hence the common names. They bury dead animals (including mammals), feed on the corpses and lay their eggs close by, so the larvae also have a convenient food supply.

Main habitat: Soil.
Main region: Temperate.
Length: 12–18mm (0.47–0.71in).
Wingspan: 16mm (0.63in).
Development: Complete metamorphosis.
Food: Carnivore.

Left and above: Burying (or sexton) beetles fly mainly at night, following the scent of corpses.

Identification: A sturdy beetle with clubbed antennae and the elytra not covering the whole abdomen. It is completely black except for orange-red bands across the back. The legs are quite powerful for digging.

Devil's Coach Horse

Staphylinus (Ocypus) olens

This is one of the best-known European beetles and it is common in gardens as well as woodland, living in the soil and underneath stones, moss and leaves. When disturbed, it curves its abdomen upward and can also inflict a bite with its large jaws. Its diet includes spiders, worms, woodlice and other insects. To deter predators it exudes a nasty liquid from glands in its abdomen. Although it has wings, it rarely flies, but scuttles rapidly away.

Main habitat: Soil and litter.
Main region: Temperate.
Length: 25mm (1in).
Wingspan: 20mm (0.79in).
Development: Complete metamorphosis.
Food: Carnivore.

Identification: Black and glossy with a flexible body and short elytra covering only the thorax. At first sight the devil's coach horse resembles an earwig and it moves in a similar fashion. The segmented abdomen is completely exposed. The head is broad and the mandibles large. If threatened, the mandibles are opened wide.

Left: This flightless beetle is commonly found under stones or wood in gardens.

Timberman

Acanthocinus aedilis

This impressive beetle belongs to the family known as longhorn beetles. These are mostly long and slender, with long legs and very long antennae. They are active by day and fly in warm sunny weather. They lay their eggs in the wood of trees, where the larvae grow. This species is widespread in much of Europe, and likes pine stumps, dead trunks and wood piles. The adults are mainly active from April through June. Infestations can kill weak trees.

Identification: Grey-brown with brown spots on the elytra and a dark tip to the abdomen. There are four yellow spots on the pronotum. The female has antennae twice as long as the body, but those of the male are up to five times as long.

Above left and above right: The enormous antennae of the male are the most obvious feature of this remarkable beetle.

Main habitat: Pine trees.
Main region: Temperate.
Length: 20mm (0.79in).
Wingspan: 30mm (1.18in).
Development: Complete metamorphosis.
Food: Herbivore.

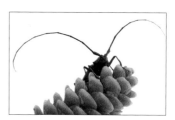

OTHER SPECIES OF NOTE

Jewel Longhorn Beetle *Sternotomis* spp.
These sturdy longhorns are found in tropical Africa and have very beautiful shiny and colourful bodies. They are therefore collected by enthusiasts and also feature on stamps in several African countries. *S. mirabilis* for example, from Gabon and Cameroon, has bright red-and-black markings.

Wasp Beetle *Clytus arietis*
This attractive longhorn beetle has bold black and yellow warning colours and long, narrow legs. It resembles a wasp and even behaves like a wasp, moving rather erratically and tapping its antennae. This gives it some protection from predators such as birds. It is quite common in Europe and often visits flowers.

Poplar Longhorn Beetle *Saperda carcharias*
This longhorn has a yellow-brown body that blends well with the leaves of poplars and willows, its staple food. The antennae are striped black and white. It lays its eggs in cracks in tree bark and the larvae tunnel into the wood.

House Longhorn Beetle *Hylotrupes bajulus*
Like so many longhorns, this species feeds on dead wood, but is also occasionally found inside houses where its larvae attack timbers. It is sometimes called the old house borer, but it is also found in new timber, especially pine.

Yellow Longhorn Beetle

Phosphorus jansoni

This striking beetle is from Africa, where it is found in forests. In some places it has become a pest as its larvae can cause considerable damage.

Identification: The body is black in background colour with bright yellow patches on the sides and toward the back of the elytra. The antennae, which taper noticeably toward the tip, are a little longer than the body. The body is rather parallel-sided.

Main habitat: Forests.
Main region: Tropical.
Length: 35mm (1.38in).
Wingspan: 55mm (2.17in).
Development: Complete metamorphosis.
Food: Herbivore.

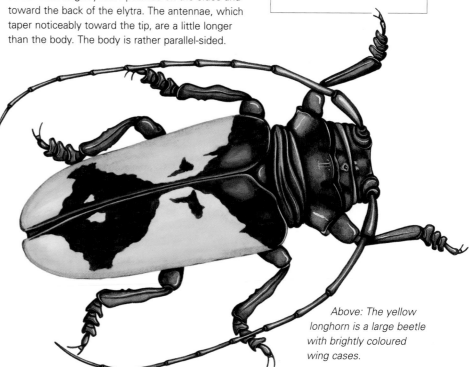

Above: The yellow longhorn is a large beetle with brightly coloured wing cases.

Stag Beetle

Lucanus cervus

This magnificent beetle is Europe's largest species, but has rather a patchy distribution and is rare in the north. The males 'antlers' are actually modified mandibles used in displays and jousting with other males. It cannot close them tightly and the beetle is quite harmless. The larvae feed on rotting wood, such as old tree stumps, especially oak and beech.

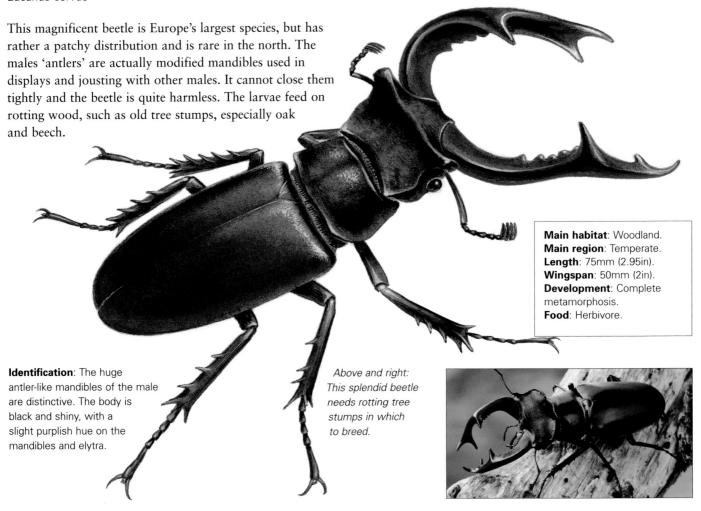

Main habitat: Woodland.
Main region: Temperate.
Length: 75mm (2.95in).
Wingspan: 50mm (2in).
Development: Complete metamorphosis.
Food: Herbivore.

Identification: The huge antler-like mandibles of the male are distinctive. The body is black and shiny, with a slight purplish hue on the mandibles and elytra.

Above and right: This splendid beetle needs rotting tree stumps in which to breed.

Lousy Watchman

Geotrupes stercorarius

This common European beetle belongs to a family known as the dor-beetles. They are very important in ecosystems as their larvae feed on dung and help to recycle animal waste. This species is often found on forest tracks or in meadows and flies on warm summer evenings. The adults feed on a range of food, including dung and fungi. The adults bury lumps of dung and then the female lays her eggs in it. This beetle is often found on farmland where livestock leave plenty of droppings. The common name refers to the fact that adults often carry yellowish mites that feed on their bodies.

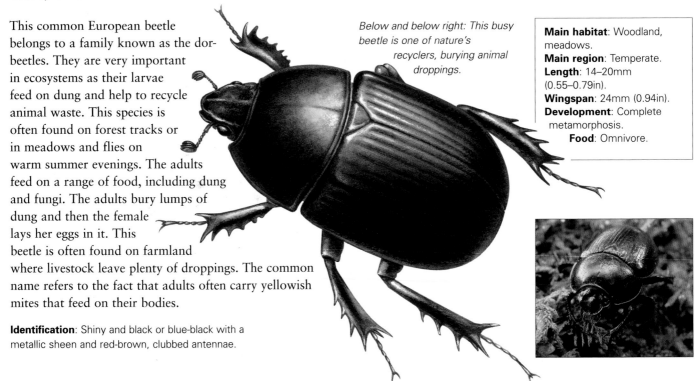

Below and below right: This busy beetle is one of nature's recyclers, burying animal droppings.

Main habitat: Woodland, meadows.
Main region: Temperate.
Length: 14–20mm (0.55–0.79in).
Wingspan: 24mm (0.94in).
Development: Complete metamorphosis.
Food: Omnivore.

Identification: Shiny and black or blue-black with a metallic sheen and red-brown, clubbed antennae.

Sacred Scarab Beetle

Kheper aegyptiorum

Main habitat: Dry grassland, savanna.
Main region: Tropical.
Length: 30mm (1.18in).
Wingspan: 60mm (2.36in).
Development: Complete metamorphosis.
Food: Omnivore.

Identification: This has a shovel-shaped head and a hard, reddish body with a sheen. The antennae are clubbed, with finger-like projections. The front legs are strong and used for digging and the hind legs are long and also powerful, designed for rolling the ball of dung. Some species have horns of varying colours.

This beetle was revered by the ancient Egyptians who compared its dung-rolling behaviour with the passage of the sun across the sky. Scarabs such as this can live for up to three years. They are able to roll balls of dung using their hind legs, and they lay their eggs inside the dung ball to provide the larvae with a supply of food. The balls are then buried to protect them from predators. This and other scarab beetles are found on the plains of Africa.

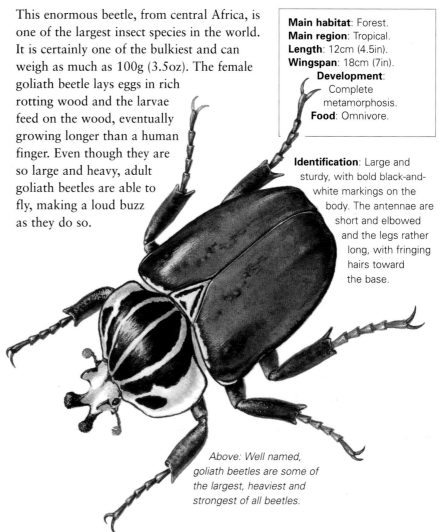

Above: Scarab beetles are large, with hard, shiny, often colourful bodies.

OTHER SPECIES OF NOTE

Common Cockchafer *Melolontha melolontha*
(Above) This well-known beetle is sometimes called the 'may bug'. The adults often emerge in numbers in May and June and have a buzzing, rather haphazard flight, frequently colliding with windows. The larvae ('rookworms') live underground, feeding on the roots of grasses and other plants.

Giant Dung-beetle *Heliocopris andersoni*
This large dung-beetle is found in the drier areas of southern and central Africa. It measures about 60mm (2.36in) long by about 33mm (1.3in) wide and has a wingspan of about 13.5cm (5.5in). It rolls balls of herbivore dung, and buries these to feed its larvae.

Royal Goliath Beetle *Goliathus regius*
This species, found mainly in the rainforests of western equatorial Africa, is one of the largest of all beetles, reaching over 10cm (4in). It has bold black-and-white markings on the thorax and abdomen.

Goliath Beetle

Goliathus druryi

This enormous beetle, from central Africa, is one of the largest insect species in the world. It is certainly one of the bulkiest and can weigh as much as 100g (3.5oz). The female goliath beetle lays eggs in rich rotting wood and the larvae feed on the wood, eventually growing longer than a human finger. Even though they are so large and heavy, adult goliath beetles are able to fly, making a loud buzz as they do so.

Main habitat: Forest.
Main region: Tropical.
Length: 12cm (4.5in).
Wingspan: 18cm (7in).
Development: Complete metamorphosis.
Food: Omnivore.

Identification: Large and sturdy, with bold black-and-white markings on the body. The antennae are short and elbowed and the legs rather long, with fringing hairs toward the base.

Above: Well named, goliath beetles are some of the largest, heaviest and strongest of all beetles.

Seven-spot Ladybird

Coccinella septempunctata

One of Europe's most familiar beetles, this bright, active insect is common and is often found in gardens. Here, like many other ladybirds, it is a useful agent of biological control as both the adults and larvae feed on aphids. The bright colours warn predators such as birds not to attack, as ladybirds are very unpalatable.

Above and right: Bright and cheerful, this gardeners' friend is boldly marked with black spots.

Identification: Small and rounded, with bright red elytra and seven black spots. The prothorax is black with a cream spot behind each eye.

Main habitat: Shrubs.
Main region: Temperate.
Length: 5.5–8mm (0.22–0.31in).
Wingspan: 15mm (0.59in).
Development: Complete metamorphosis.
Food: Carnivore.

Orange Ladybird

Halyzia 16-guttata

This species has the typical rounded ladybird shape but is less brightly coloured and somewhat smaller than most. It tends to be most common in late summer and autumn and is associated with bushes and trees, notably ash, sycamore and birch. It feeds mainly on mildew fungus scraped from the leaves, trunks and branches. Like many insects it can be attracted to lighted windows as it frequently flies at night.

Identification: The body is a pale orange colour, with 16 cream spots. The legs are pale and the head is yellow. The thorax and border of the wing cases are partly translucent.

Main habitat: Trees.
Main region: Temperate.
Length: 6mm (0.24in).
Wingspan: 15mm (0.59in).
Development: Complete metamorphosis.
Food: Herbivore.

Left: The orange ladybird is common and increasing in parts of Europe.

Glow-worm

Lampyris noctiluca

Below: The larva-like adult female glow-worm emits light from the tip of her abdomen.

This is a most famous but confusingly named insect, best known for its ability to emit light. It is mainly the grub-like, wingless female glow-worm that glows, to attract the attention of flying males nearby, although males, pupae and even eggs can also glow weakly. The light produced is greenish-yellow. The larvae feed on snails and slugs.

Main habitat: Hedges and banks.
Main region: Temperate.
Length: 12mm (0.47in).
Wingspan: 18mm (0.71in).
Development: Complete metamorphosis.
Food: Carnivore.

Identification: The male is grey-brown with a dark patch on the prothorax. The female looks quite different, being wingless and resembling the larva. The last three segments of her abdomen can emit light.

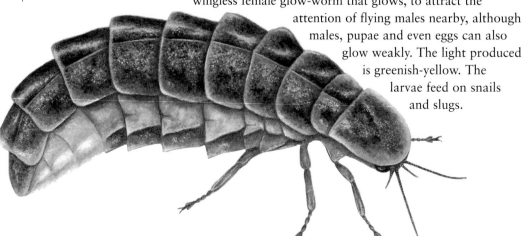

Large Pine Weevil

Hylobius abietis

Like most weevils, this common European species has a long snout (rostrum) with the mouthparts at its tip. It is usually found on spruce or pine between June and September, but it also attacks other coniferous species including larch and Douglas fir. The weevil gnaws holes in the bark of shoots, especially on young trees, which may even be killed if enough of these insects attack. The larvae live in the roots or stumps of dead or dying trees. This weevil is regarded as the most significant pest of forest plantations in northern Europe, where it causes major damage to young conifers. Attempts have been made to control it by using parasitic nematode worms and wasps, and also fungi, as well as chemical sprays.

Identification: The antennae are sharply elbowed and emerge from near to the tip of the snout. The body is black with yellow spots and the legs are rather sturdy.

Above: This weevil attacks and damages young coniferous trees.

Main habitat: Conifers.
Main region: Temperate.
Length: 8.5–13mm (0.33–0.51in).
Wingspan: 12mm (0.47in).
Development: Complete metamorphosis.
Food: Herbivore.

OTHER SPECIES OF NOTE

Firefly *Luciola lusitanica*
This beetle is a close relative of the glow-worm, but both sexes flash their lights, the males blinking in the night air as they fly over females. The females are also winged, but they do not fly. This firefly lives in southern Europe.

Engraver Beetle *Ips typographus*
A small bark-beetle found in spruce and some other conifers in Europe, this glossy brownish-black beetle is about 4.2–6mm (0.17–0.24in) long. This and other bark-beetles leave characteristic tunnelled galleries under the bark and may cause considerable damage by transmitting a fungus that attacks the bark.

Eyed Ladybird *Anatis ocellata*
This ladybird is mainly found on conifers, notably spruce and pine. About 8mm (0.31in) long, it has orange-coloured elytra with about ten black eye-spots, each with a pale ring. The eyed ladybird is the largest species of ladybird in Britain. Like most other ladybirds it is a predator, feeding mainly on aphids.

Nut Weevil *Curculio nucum*
About 6–8.5mm (0.24–0.33in) long and rather nut-like in colour and shape, this weevil has a long snout with antennae emerging from about halfway down. The larva feeds on hazelnut kernels and the adults often visit hawthorn flowers. Due to their diet, the larvae are considered pests.

Giraffe Weevil

Trachelophorus giraffa

This is one of the most remarkable of weevils and perhaps the strangest looking of all beetles. Named for the amazingly long snout of the male, it is found in Madagascar where it feeds on seeds, leaves and other plant tissues. The eggs are laid on leaves on which the newly hatched larvae feed.

Main habitat: Forests.
Main region: Tropical.
Length: 80mm (3.15in).
Wingspan: 40mm (1.57in).
Development: Complete metamorphosis.
Food: Herbivore.

Identification: Shiny black, with red elytra. The male has a huge, articulated snout and long antennae.

Below and right: Surely one of the most bizarre of insects, this weevil has an enormous 'neck' and snout.

Furniture Beetle

Anobium punctatum

This is the notorious beetle whose larvae, known as 'woodworms' bore neat holes into furniture and other exposed timber inside houses. They are also found outdoors in dead trees, but have adapted especially well to living indoors, much to the distress of many people. The tunnelling of the larvae gradually turns wood to sawdust, weakening furniture and timber.

Left: Seldom seen, this very small beetle causes much damage to timber.

Main habitat: Dead wood and timber.
Main region: Temperate.
Length: 3–5mm (0.12–0.19in).
Wingspan: 7mm (0.28in).
Development: Complete metamorphosis.
Food: Herbivore.

Identification: Small and brown, with a humped prothorax and long stripes along the elytra. The larvae bore tunnels in timber and the adults emerge through round flight holes about 2mm (0.08in) across.

Fur Beetle

Attagenus pellio

This small beetle is also known as the two-spotted carpet beetle or bug. In the wild its main habitat is in birds' nests or in the fur of dead animals, but it has also adapted to live inside houses, especially in carpets, old blankets and the upholstery of furniture, where the larvae feed and develop. The adults feed mainly on pollen from various flowers.

Identification: Small, with an oval shape. The carapace is dark, with two pale spots in the centre of the elytra and three pale marks at the rear edge of the thorax. The larvae of the fur beetle feed on dry material, including carpets.

Main habitat: Dry sites, carpets and the like.
Main region: Temperate.
Length: 5mm (0.19in).
Wingspan: 10mm (0.39in).
Development: Complete metamorphosis.
Food: Herbivore.

Left: It is the larvae, not the adult, that gives this insect its name.

Deathwatch Beetle

Xestobium rufovillosum

Main habitat: Wood and timber.
Main region: Temperate.
Length: 6–9mm (0.24–0.35in).
Wingspan: 12mm (0.47in).
Development: Complete metamorphosis.
Food: Herbivore.

Less common than the furniture beetle, this beetle can nevertheless be very dangerous to old timbers, such as roof joists in halls and churches. In the wild it lives in dead wood such as oak, preferring timber that has been weakened by fungal rot. The name comes from the eerie tapping noises produced as signals by both sexes as they knock their heads against their tunnels.

Left: This is another beetle that attacks old timbers.

Identification: Larger than the furniture beetle, this species is also brownish, but has irregular patches of yellowish hairs along the abdomen, and more powerful legs. The flight holes from which they emerge are about 3.5mm (0.14in) across.

Great Diving Beetle

Dytiscus marginalis

Main habitat: Still water.
Main region: Temperate.
Length: 30–35mm
(1.18–1.38in).
Wingspan: 60mm (2.36in).
Development: Complete
metamorphosis.
Food: Carnivore.

This is perhaps the best known of European diving
beetles and one of the largest. It is quite common
in ponds and lakes with abundant plant life, often
in or near woodland. Great diving beetles spend
most of the time submerged, coming to the
surface for fresh supplies of air. The larva is a
ferocious predator, capable of catching tadpoles
and even small fish. The adults are winged, and sometimes fly off to find
new ponds, occasionally landing by mistake on shiny surfaces. The
female lays her eggs in the soft stems of water plants in the early spring.

*Left: One of Europe's largest
water beetles, this is an active
predator, both as larva
and adult.*

Identification: A large water
beetle, and a powerful swimmer,
using its strong legs, fringed with
hairs, to row itself through the
water. Red-brown, with a
greenish sheen and a yellow
border. The larva is up to 50mm
(2in) long, with a large head and
powerful jaws.

OTHER SPECIES OF NOTE
Great Silver Beetle *Hydrophilus piceus*
A fine water beetle and, at 50mm (2in)
one of Europe's largest beetles. It is a
herbivore but can be dangerous to handle
as it has a sharp spine on its underside.
The body is glossy black. This splendid
beetle has declined in numbers through
loss and pollution of its habitats. It is
typically found in still water, often in grazed
freshwater marshes.

Crawling Water Beetle *Haliplus fulvus*
This is a small water beetle that prefers to
crawl among water plants rather than swim.
It is about 2.5mm (0.09in) long, orange with
darker patches and has an oval, slightly
pointed body. It is common in reed swamps
and shallow water.

Platambus maculatus
This small water beetle, about 8mm (0.31in)
long, is very common on sandy and stony
bottoms of lakes and rivers. The head and
prothorax are reddish and the elytra boldly
marked in black and yellow.

Scavenger Water Beetle
Helophorus brevipalpis
A small water beetle, with a slightly pointed
oval abdomen and a bright green thorax with
five furrows. Unlike many water beetles, this
species does not swim, but rather crawls
about in the semi-aquatic plants at the
water's edge, feeding on decomposing
plant tissues.

Whirligig Beetle

Gyrinus natator

These small beetles are a common sight on still
waters (ponds, lakes and slow-flowing rivers)
where they swim rapidly in tight circles on the
surface, often in large numbers, especially in
calm, warm weather. Their eyes are divided so
that they can watch above and below the water
simultaneously. Although they are usually on
the surface, they can dive down, carrying a
supply of air trapped at the tip of the abdomen.
They can fly, usually at night. Both larvae and
adults are predators, catching small crustaceans
and insect larvae. When alarmed, the beetles
will dive down quickly into the water, to rest
clinging to submerged water plants until the
danger has passed, after which they usually
return to the surface film. There are several
rather similar species in Europe.

Main habitat: Still or slow-
moving fresh water.
Main region: Temperate.
Length: 5.6–6.6mm
(0.22–0.26in).
Wingspan: 10mm (0.39in).
Development: Complete
metamorphosis.
Food: Carnivore.

Identification: Small and dark,
the female is larger than the
male. The body is mainly black
and glossy and with a rather
metallic sheen. The antennae are
small, and the short, hair-fringed
legs are adapted for swimming.
The larva has a row of gills along
the sides of its body.

*Below: Whirligig beetles are
spotted twisting and turning
in groups on pond
surfaces.*

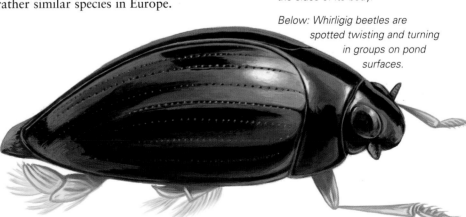

FLIES

Although not the best-loved insects, flies (Diptera) are nonetheless one of the most fascinating of the insect orders. Flies are very diverse, with about 120,000 known species, and they exist in virtually all habitats outside the oceans. They are common across the whole of Europe and Africa. A few species, such as horseflies and mosquitoes are a nuisance, some spreading disease.

Housefly

Musca domestica

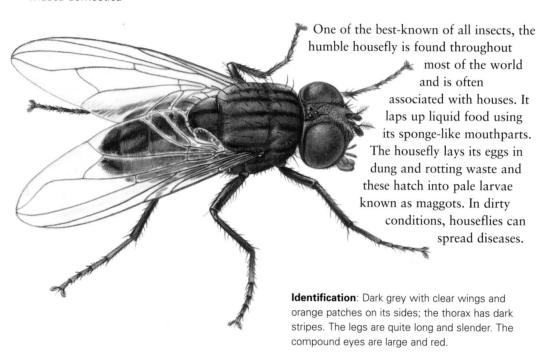

One of the best-known of all insects, the humble housefly is found throughout most of the world and is often associated with houses. It laps up liquid food using its sponge-like mouthparts. The housefly lays its eggs in dung and rotting waste and these hatch into pale larvae known as maggots. In dirty conditions, houseflies can spread diseases.

Main habitat: Houses.
Main region: Temperate, but scattered worldwide.
Length: 9mm (0.35in).
Wingspan: 20mm (0.79in).
Development: Complete metamorphosis.
Food: Omnivore.

Left and below: The housefly is one of the most familiar flies.

Identification: Dark grey with clear wings and orange patches on its sides; the thorax has dark stripes. The legs are quite long and slender. The compound eyes are large and red.

Bluebottle

Calliphora vomitoria

The bluebottle belongs to a family known as blowflies, which are large and hairy, often with bright shiny colours. They can be seen throughout the year and often overwinter inside sheds or houses. The eggs are laid preferably on meat or fish, or carrion, which the maggots consume. Bluebottles can breed rapidly where rotting food is abundant. Each female fly lays up to 200 eggs and the resulting maggots can grow and turn into adult flies in three weeks.

Main habitat: Gardens, houses.
Main region: Temperate.
Length: 9–12mm (0.35–0.47in).
Wingspan: 20mm (0.79in).
Development: Complete metamorphosis.
Food: Omnivore.

Identification: It is blue-black, with red eyes and a striped, hairy thorax. The abdomen has a definite metallic sheen.

Above and right: Bluebottles are large and shiny with a hairy body.

Drone fly

Eristalis tenax

Main habitat: Shrubs, flowers.
Main region: Temperate.
Length: 15mm (0.59in).
Wingspan: 24mm (0.94in).
Development: Complete metamorphosis.
Food: Herbivore.

This fly belongs to the family known as hoverflies from their habit of hovering in one spot. Hoverflies mostly have bright colours and can often be seen feeding at flowers. Many hoverflies mimic stinging insects such as bees and wasps, although they are themselves quite harmless. The drone fly is active in warm, sunny weather, in late summer and autumn especially. Its larva, known as a rat-tailed maggot, lives in ponds and drains and breathes through a long retractable tube.

Identification: The drone fly mimics a honey bee drone. It has a brown-and-yellow abdomen, darker thorax and large eyes. The wings are transparent and have a clear, vein-free rear margin, a feature common to all hoverflies.

Above and far left: This stingless fly gains some protection from its resemblance to a honey bee.

OTHER SPECIES OF NOTE

Lesser Housefly *Fannia canicularis*
Like a miniature housefly, about 5mm (0.19in) long and rather slender of build. Seen indoors, but less often than the housefly. The larvae live in rotting manure and other organic material and have rows of spiny tufts. These enable them to 'swim' through semisolids. The lesser housefly is a significant pest on many poultry farms, breeding prolifically in battery sheds in particular.

Greenbottle *Lucilia* spp.
This common blowfly genus is found throughout the world. The species are bright green in colour and *L. sericata* has become famous in medicine because its larvae are used to clean wounds. It even destroys dangerous bacteria such as MRSA (methicillin-resistant *Staphylococcus aureus*).

Wasp Hoverfly *Syrphus ribesii*
This hoverfly is boldly striped in black and yellow, mimicking a wasp. It also emits a wasp-like buzz as it flies. It is common in Europe. This and many other hoverflies are useful in the garden as their larvae feed on aphids.

Thistle Gallfly *Urophora cardui*
This small fly has wavy dark markings on its wings and a dark abdomen, pointed at the rear end in the female. She lays her eggs in the leaf veins or leaf base of thistles and the larvae develop inside rounded galls on the thistle stems. Gallflies of this genus have been widely used to control invasive weeds such as thistles.

Bee fly

Bombylius major

Like hoverflies, bee flies are adept fliers and often hover. This species is a mimic of bumblebees, but like hoverflies it has no sting. The bee fly is on the wing mainly in spring, darting from flower to flower and using its proboscis to suck nectar. The larvae are parasites of solitary bees.

Identification: The body is hairy and rounded, like a bumblebee, but its legs are very long and thin. It has large eyes and a long, narrow proboscis.

Main habitat: Scrub, gardens.
Main region: Temperate.
Length: 7–12mm (0.28–0.47in).
Wingspan: 14mm (0.55in).
Development: Complete metamorphosis.
Food: Herbivore.

Below: The furry bee fly likes to visit flowers in sunny, sheltered sites.

Mosquito

Culex pipiens

This species is a common European mosquito. It can be found wherever there is still, open water with no predators such as fish. Ideal breeding sites are pools, water butts, wet holes and tree hollows. The female mosquito lays rafts of waxy, floating eggs on the water surface. These then hatch into active, wriggling larvae that filter their food from the water, and eventually pupate and hatch into flying adults at the surface. The adult females suck blood from mammals, but the males feed mainly on nectar.

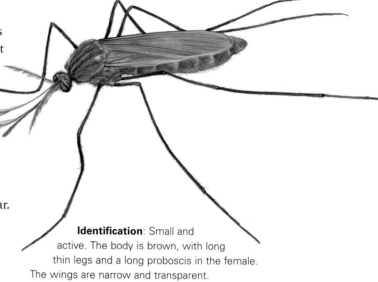

Right: Also known as gnats, these mosquitoes require still, predator-free water in which to breed.

Main habitat: Pools (larvae).
Main region: Temperate.
Length: 8mm (0.31in).
Wingspan: 10mm (0.39in).
Development: Complete metamorphosis.
Food: Carnivore/herbivore.

Identification: Small and active. The body is brown, with long thin legs and a long proboscis in the female. The wings are narrow and transparent.

Non-biting Midge

Chironomus annularis

In humid, still weather the males of this species can often be seen dancing in display flights above the ground, to attract females. The larvae are aquatic, often bright red in colour, and are known as 'bloodworms'; they are found mainly in ponds, stagnant pools and the like. The larvae are a foodsource for fish.

Left: The male of this midge has feathery antennae.

Identification: Small fly, with a blackish-brown body, thin legs and transparent wings. The species swarm over water in the dark. The thorax is distinctly humped over the head. The antennae of the male are very feathery. The adults do not feed, and are harmless.

Main habitat: Pools (larvae).
Main region: Temperate.
Length: 12mm (0.47in).
Wingspan: 15mm (0.59in).
Development: Complete metamorphosis.
Food: Omnivore (larva).

St Mark's fly

Bibio marci

This rather hairy, black fly is so-named as it tends to appear in Europe around St Mark's Day (25 April). On warm days the males swarm at woodland edges or over grassland, flying slowly, with their legs dangling down. These display flights attract females, after which the flies pair and mate. The female lays her eggs in a tunnel in the ground and the larvae feed on organic matter in the soil and leaf humus.

Main habitat: Pastures.
Main region: Temperate.
Length: 12mm (0.47in).
Wingspan: 18mm (0.71in).
Development: Complete metamorphosis.
Food: Herbivore (larva).

Left: In warm weather the male flies participate in communal display flights.

Identification: The body is black and shiny with long rather hairy legs. The males have large eyes.

Cranefly

Tipula maxima

This is a common European cranefly, sometimes called a 'daddy-long-legs'. It can often be seen over grassland and in gardens, and is attracted to lights at night. Most active in May through August. The larvae ('leatherjackets') live in the soil and can damage root crops. The adults are harmless.

Right: Craneflies are often considered a pest because the larvae feed on plant roots. This causes significant damage, especially to turf, which can develop bald patches.

Identification: Very large fly, with a slender slightly pinched brown abdomen, long, delicate legs and long, narrow wings with dark markings.

Main habitat: Woods, gardens.
Main region: Temperate.
Length: 40mm (1.57in).
Wingspan: 65mm (2.56in).
Development: Complete metamorphosis.
Food: Omnivore (larva).

OTHER SPECIES OF NOTE

Biting Midge *Culicoides* spp.
These are tiny bloodsucking midges that sometimes occur in large numbers, when they can be a nuisance. They tend to emerge in the evenings and are found near water or in boggy country. The adults are only a few millimetres long, mostly seen flying from May through August. This genus contains around 500 species.

Black Fly *Simulium* spp.
These small dark brown or black flies have short antennae and broad wings. The larvae live in running water and the adult females can inflict a painful bite. They are usually on the wing from April to June. Black flies are widespread in Europe and are particularly common during the summer in northerly regions. Their swarms often cause significant irritation to both wild animals and livestock.

Malaria Mosquito *Anopheles gambiae*
This mosquito, found mainly in Africa south of the Sahara, is notorious as the carrier of the dangerous disease malaria. The larvae live in pools and the adult females sometimes carry the dangerous malaria protozoan *Plasmodium falciparum*. They infect people with malaria when feeding on their blood, transferring the parasite in their saliva. Malaria is still a huge problem in many tropical regions, notably in Africa, especially during the wet season when many pools appear in which the mosquitoes can lay their eggs. Various methods are used to control them, including spraying with insecticides. Netting is also used to exclude them from houses and beds.

Winter Gnat

Trichocera annulata/T. hiemalis

Winter gnats are rather like craneflies, but they fold their wings over their backs when at rest. They are named for their habit of flying in late autumn and winter, after most other flies have ceased to be active. The males form dancing swarms, drifting and bobbing in the still air. The larvae live among rotting leaves and humus, in the soil. Like many flying insects, winter gnats are attracted to light and they may gather at windows on mild winter evenings.

Identification: Rather like a small cranefly, but with clear wings, longer antennae and a less pinched abdomen. The abdomen is usually banded. Due to its appearance, this species is often mistaken for a mosquito.

Main habitat: Soil (larvae).
Main region: Temperate.
Length: 10mm (0.39in).
Wingspan: 15mm (0.59in).
Development: Complete metamorphosis.
Food: Omnivore (larva).

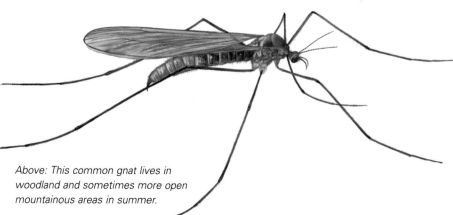

Above: This common gnat lives in woodland and sometimes more open mountainous areas in summer.

Brown Horsefly

Tabanus bromius

This large horsefly is quite widespread in Europe, where it is found typically in open pastures, especially where there is livestock, such as horses and cattle. While the males are harmless, feeding on nectar, the females can inflict really painful bites using their blade-like mouthparts and leaving a nasty swelling. They fly rather silently and can therefore go undetected – until they bite. The larvae live in damp soil.

Main habitat: Pastures.
Main region: Temperate.
Length: 20mm (0.79in).
Wingspan: 40mm (1.57in).
Development: Complete metamorphosis.
Food: Herbivore (male); carnivore (female).

Identification: A large, powerfully built fly with very large eyes. The thorax is greenish with darker stripes and the abdomen mottled brown and black. The wings are clear.

Left and right: This fly also goes by the common name of band-eyed brown horsefly due to its prominent eyes.

Tsetse Fly

Glossina spp.

The tsetse fly is a biting fly found in Africa, mainly south of the Sahara. In fact, the word 'tsetse' simply means 'fly', so this insect should more properly be called simply 'tsetse'. This dangerous insect can carry the disease sleeping sickness and a similar disease of livestock, and is therefore a major problem in some parts of its range. Both male and female tsetse flies feed by biting mammals (including people) and sucking their blood. If the fly has already fed on someone suffering from sleeping sickness (caused by a tiny trypanosome protozoan) the disease may be transferred as it feeds. Female tsetse flies produce live maggots as the eggs hatch within their body.

Identification: A large, sturdy, yellow-brown fly with an obvious proboscis.

Right: The tsetse is one of Africa's most dangerous biting flies.

Main habitat: Woodland and scrub.
Main region: Tropical.
Length: 15mm (0.59in).
Wingspan: 25mm (1in).
Development: Complete metamorphosis.
Food: Carnivore.

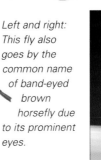

OTHER SPECIES OF NOTE

Cleg-fly *Haematopota pluvialis*
Often simply called clegs, these horseflies can be really irritating as they fly silently before landing gently and then biting hard. They are most common near water and fly from June to August.

Warble Fly *Hypoderma bovis*
This fly is a nasty parasite of cattle. The adult fly is hairy and rather bee-like. It is short-lived and does not feed. Warble flies lay their eggs on the legs of cattle and the larvae bore into the skin, eating their way up through the tissue until they end up on the host's back, where they produce swellings known as warbles. The larvae drop from these warbles and pupate on the ground before hatching as flies.

Long-legged Fly
Dolichopus ungulatus
Named for its long legs, this fly is quite common in Europe. It rests in a characteristic posture, its front raised on its long legs. The body is a bright shiny green. This fly is carnivorous, as are the larvae, which live in wet mud.

Fruit Fly

Drosophila funebris

> **Main habitat**: Gardens, orchards, houses.
> **Main region**: Temperate.
> **Length**: 3mm (0.12in).
> **Wingspan**: 6mm (0.24in).
> **Development**: Complete metamorphosis.
> **Food**: Fruit.

Identification: Tiny, brownish fly with branched antennae.

Far right: These very small flies are often seen near ripe fruit, or near sweet drinks.

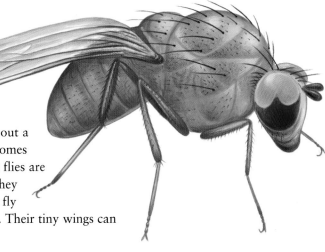

Very small and neat, fruit flies (or vinegar flies) are famous for all the genetic research that has been carried out on *D. melanogaster*. The short life cycle (which can be completed in about a week) and large salivary gland chromosomes make it suitable for such research. Fruit flies are often found on rotting fruit, on which they lay their eggs. The adult flies hover and fly around sweet drinks and fruit outdoors. Their tiny wings can beat at up to 220 times a second.

Hornet Robber Fly

Asilus crabroniformis

Robber flies are large, strong-flying, hairy flies with powerful legs. They catch other insects in midair then suck their bodies dry. This species lives in grassland, heath and chalk grassland in western Europe. The larvae feed on the larvae of beetles such as dung-beetles. They emerge in mid-June and the adults also prey on grasshoppers, bees and wasps, hunting from a perch.

> **Main habitat**: Grassland.
> **Main region**: Temperate.
> **Length**: 25mm (1in).
> **Wingspan**: 40mm (1.57in).
> **Development**: Complete metamorphosis.
> **Food**: Carnivore.

Identification: Large and hairy and golden-brown in colour, the first few segments of the abdomen are black. Note the deep groove between the eyes, which is typical of robber flies.

Above: Robber flies are fast-flying, powerful predators.

Yellow Dung-fly

Scathophaga stercoraria

This fly is commonly found on fresh dung. The adults catch and eat other insects which may be attracted to the dung, and are most active from April to October. It is mainly the ginger-coloured male flies that we see buzzing around cowpats. The greyer females lay their eggs in dung and the larvae feed on this nutritious substance, helping to recycle it.

> **Main habitat**: Pasture.
> **Main region**: Temperate.
> **Length**: 10mm (0.39in).
> **Wingspan**: 25mm (1in).
> **Development**: Complete metamorphosis.
> **Food**: Carnivore.

Identification: The male is covered in bright yellow hairs, and the widely separated eyes are bright red. The female is mainly greyish.

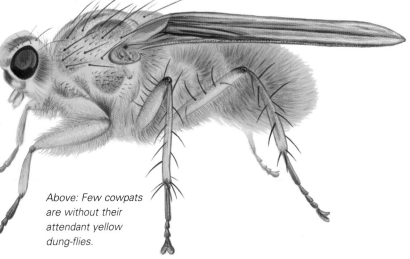

Above: Few cowpats are without their attendant yellow dung-flies.

BUTTERFLIES AND MOTHS

A truly cosmopolitan group, butterflies and moths (Lepidoptera) are found in every habitat where there are flowers. In Europe and Africa they are well represented with many large and colourful species, especially in southern Europe and tropical Africa. Containing some of the most colourful members of the insect world, these creatures are appreciated by people all over the world for their beauty.

Painted Lady

Vanessa cardui

This is one of the region's best-known and widespread species, and one of the most successful in the world. It is common over much of Europe, North Africa and the Middle East, and is also found in America. A strong migrant, it regularly moves north from its strongholds in the dry semi-deserts of North Africa and Arabia, sometimes even reaching the Arctic. Migrating swarms can number thousands or even millions of individuals. They tend to arrive in northern Europe early in the year and the males then establish territories, usually in dry habitats such as heathland. The mated females lay their eggs on thistles and sometimes on mallows or nettles, and there may be several generations in the summer quarters.

Main habitat: Varied.
Main region: Temperate, subtropical.
Length: 20mm (0.79in).
Wingspan: 55mm (2.17in).
Development: Complete metamorphosis.
Food: Herbivore.

Identification: Prettily patterned, with a pinkish-orange background with scattered black spots and patches. It has white spots at the tips and outside edges of the forewings. The hindwings are greyish beneath with small spots and pale patches.

Far left: The migratory painted lady is one of the most widespread species.

Scarce Swallowtail

Iphiclides (Papilio) podalirius

The common name of this attractive butterfly is misleading as it is actually quite common over most of its range, though it is rather rare in the north. It is found in Europe (but not the British Isles) and across the Middle East and much of Asia. It is rather common in southern Europe. A close relative, *I. feisthameli*, breeds in North Africa, Morocco, Algeria and Tunisia. The larvae feed on blackthorn, almond, cherry and related trees.

Main habitat: Mixed.
Main region: Temperate.
Length: 25mm (1in).
Wingspan: 70–90mm (2.76–3.54in).
Development Complete metamorphosis.
Food: Herbivore.

Identification: One of Europe's most beautiful butterflies. Large, with a roughly triangular shape and long tail streamers on the hindwings. The forewings have bold, long, roughly parallel black-and-white stripes, tinted yellow toward the edges. The hindwings have scalloped blue patches and two eye-spots, ringed red. Powerful in flight, with long periods of gliding.

Above and right: The scarce swallowtail is powerful and agile in flight.

Giant Peacock Moth

Saturnia pyri

> **Main habitat**: Varied.
> **Main region**: Temperate-subtropical.
> **Length**: 30mm (1.18in).
> **Wingspan**: 10–13cm (4–5in).
> **Development**: Complete metamorphosis.
> **Food**: Herbivore.

Identification: Very large. Mainly grey, shading from light grey at the leading edge to darker grey at the trailing edge. The wings are bordered in white and there is a round eye-spot at the centre of each wing.

This splendid moth is Europe's largest. It is found in southern Europe and also the Middle East and in northern Africa. It is on the wing from April to June. The larvae feed on certain trees including walnut, pear, ash and poplar. The fully grown larva is large and bright green with a yellow stripe along each side. The emperor moth (*S. pavonia*) is a close relative. The giant peacock moth is just as beautifully marked, but is rather smaller with a wingspan of about 60–80mm (2.36–3.15in). It is typically seen on heathland, as heather is one of the most common larval food plants.

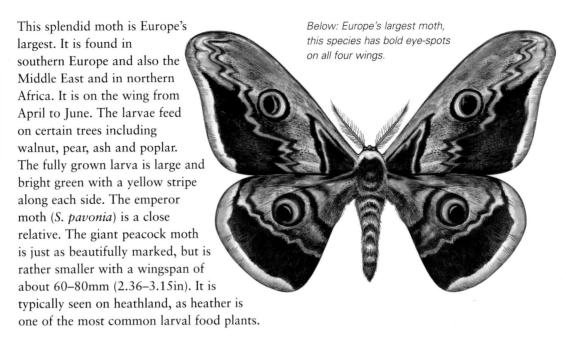

Below: Europe's largest moth, this species has bold eye-spots on all four wings.

OTHER SPECIES OF NOTE

Two-tailed Pasha *Charaxes jasius*
Europe's largest butterfly, this magnificent, almost hand-sized insect has a wingspan of more than 10cm (4in). It is typical of the Mediterranean region and also North Africa. Its larval plant is the strawberry tree (*Arbutus*). The wings are intricately patterned in chocolate, orange, black and cream and the hindwings have tails. It is highly territorial and will defend its area energetically, almost like a bird.

Peacock *Inachis io*
This is one of Europe's most familiar butterflies and is common in parks and gardens. Its wings are reddish above, with a bright blue, yellow and black eye-spot on each wing. Underneath, the wings are very dark and look very like a dead leaf. When disturbed, the peacock suddenly flashes its wings open to scare the attacker, while making a snake-like hissing sound. Stinging nettles are the usual larval food plant.

Silver-washed Fritillary *Argynnis paphia*
One of Europe's largest fritillaries, bright orange above with black spots and scribbles. The undersides of the hindwings have a pretty greenish sheen with silver streaks, hence the common name. It is mainly a woodland species and the adults often fly fast over the tops of the trees, pausing to sip aphid honeydew, or diving down to clearings to feed from flowers. The larvae feed on violets.

Oleander Hawk Moth *Daphnis nerii*
This is a large hawk moth with a greenish body and beautifully patterned wings – olive green with pink and white markings. Its wingspan may reach 12cm (4.5in). Found from North Africa and southern Europe to Asia, its larvae feed mainly on oleander.

Sunset Moth

Chrysiridia croesus

> **Main habitat**: Forest and nearby open country.
> **Main region**: Tropical.
> **Length**: 25mm (1in).
> **Wingspan**: 75–90mm (2.95–3.54in).
> **Development**: Complete metamorphosis.
> **Food**: Herbivore.

Identification: The amazing patterned wings are black, overlaid with a medley of iridescent colours that glint in differing tones according to the angle – patches of green, yellow, pink and orange. The hindwings each have three prominent thin tails and scalloped edges. The undersides of the wings are equally colourful.

This beautiful day-flying moth lives in eastern Africa, notably in Tanzania. A similar species, *C. rhipheus*, is endemic to Madagascar. The bright colours serve to warn predators of its nasty taste. The caterpillars feed on poisonous plants of the euphorbia family (genus *Omphalea*) and build up the plant toxins in their own bodies. These insects are some of the most attractive in the world and are therefore sought after by collectors and sold as ornaments and jewellery.

Below: The colours in the wings of this splendid moth explain its common name.

WASPS, BEES, ANTS AND SAWFLIES

These highly specialized insects, with chewing mouthparts and four membranous wings, are very familiar in Europe and Africa. Some of them are important as pollinators of garden plants and crops, and many help to control other insect species, including many pests. A large proportion of wasps, bees, ants and sawflies are social insects, living in large groups.

Giant Wood Wasp

Urocerus gigas

Wood wasps are sometimes known as 'horntails' because of the long ovipositor of the female. The female looks especially frightening, with wasp-like warning colours and the sting-like (but harmless) ovipositor. She uses this to bore into timber, usually unhealthy pines or other conifers, and lay her eggs. The larvae tunnel in the wood, probably eating fungi.

Main habitat: Conifer woods.
Main region: Temperate.
Length: 15–40mm (0.59–1.57in).
Wingspan: 50mm (2in).
Development: Complete metamorphosis.
Food: Herbivore.

Identification: The female is large and banded bright yellow and black, with an obvious ovipositor. The wings are clear and the antennae quite long. The male is smaller, with a black thorax and more orange abdomen.

Above and right: Adult giant wood wasps are active from May until October.

Sabre Wasp

Rhyssa persuasoria

Main habitat: Woodland.
Main region: Temperate, tropical.
Length: 20–35mm (0.79–1.38in) (plus ovipositor of 40mm (1.57in)).
Wingspan: 40mm (1.57in).
Development: Complete metamorphosis.
Food: Herbivore (adult); carnivore/parasite (larva).

This is a parasitic insect belonging to a very large family, known as the ichneumons. Their larvae live as parasites (more correctly parasitoids), either inside or on the surface of other arthropods, mostly on the larvae of other insects, killing their hosts before emerging as adult ichneumons. The sabre wasp is one of Europe's largest and it parasitizes the larvae of wood wasps. The adults are active from June to September. The female can sometimes be spotted drilling, with arched abdomen, into timber to lay its eggs on the wood wasp larvae. Sabre wasps can be found over much of Europe and also in North Africa.

Identification: Large and black with white spots and bright red legs. The antennae are long and thin, and the female's ovipositor is up to 40mm (1.57in) long.

Right: The female sabre wasp has a very long ovipositor.

Wood Ant

Formica rufa

Main habitat: Woodland.
Main region: Temperate.
Length: 4–9mm (0.16–0.35in) (workers); 9–11mm (0.35–0.43in) (queen and males).
Wingspan: 15mm (0.59in) (winged forms).
Development: Complete metamorphosis.
Food: Carnivore/omnivore.

The wood ant is one of Europe's larger ant species, and is responsible for building large mounded nests in (mainly) coniferous woods, usually in a clearing or against a tree stump. A colony may consist of as many as 100,000 individual ants which follow regular trails in search of food.

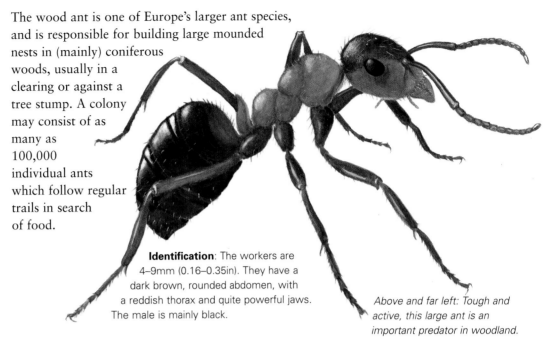

Identification: The workers are 4–9mm (0.16–0.35in). They have a dark brown, rounded abdomen, with a reddish thorax and quite powerful jaws. The male is mainly black.

Above and far left: Tough and active, this large ant is an important predator in woodland.

OTHER SPECIES OF NOTE

Yellow Ophion *Ophion luteus*
This ichneumon is quite common in Europe, and is often attracted to the lighted windows of homes on summer nights. Its larvae are internal parasites of moth larvae. The adult is yellow-brown, and has a short sting-like ovipositor which can give a slight prick.

Black Garden Ant *Lasius niger*
This is a very common ant in Europe, where it is often found in soil or in rotten tree stumps. It makes sandy domes over the entrance holes to its subterranean nest and these grow into small ant-hills. It feeds mainly on honeydew from aphids.

Ichneumon *Apanteles glomeratus*
This is a tiny ichneumon fly, only about 3mm (0.12in) long. It is black, with brown legs and long antennae and is most commonly seen flying from May through until August. It is famous as a major parasite of the larvae of the large white butterfly and helps to control the numbers of this destructive butterfly, which lays its eggs on cabbages and related plants.

Robin's Pincushion Gall Wasp *Diplolepis rosae*
This small gall wasp is best known by the galls it produces on wild (and sometimes garden) roses. The galls, formed from the plant's own tissues, look like small cushions of reddish or greenish moss, hence the common name. The gall is home to dozens of wasp larvae which spend the winter there, emerging the following spring as adult gall wasps.

Driver Ant

Dorylus nigricans

This is one of a number of ants known as driver ants, found mainly in tropical Africa, especially in the forests of west Africa and the Congo region. A single colony can contain a staggering 20 million ants, and the soldier caste have pincer-like jaws. When food is in short supply they leave the colony and form marching columns that attack and kill other insects, but also larger animals, even small mammals such as rats. There have been cases of people being attacked, although this is rare.

Below: Driver ants hunt in enormous swarms, using their numbers to immobilize prey.

Main habitat: Savanna, forest.
Main region: Tropical.
Length: 5mm (0.19in) (worker); 15mm (0.59in) (soldier); 50mm (2in) (queen); 30mm (1.18in) (male).
Wingspan: 40mm (1.57in) (male).
Development: Complete metamorphosis.
Food: Carnivore.

Identification: Soldier ants of this species are three times as large as the workers. The males, also large, are winged (below). The queen is the largest of all the driver ants.

Honey Bee

Apis mellifera

Arguably the world's most famous and important insect, because of its long association with people, and as a source of honey around the world, this fascinating species is also one of the most familiar. Honey bees live in large colonies, consisting of 2,000 males (drones), about 50,000 sterile females (workers) and just a single queen. In the wild they live in hollow trees, rock crevices and the like, but mainly they are domesticated and housed in specially designed hives. They sting only in self-defence, and the bee dies soon after stinging. They form new colonies when the queen leaves the nest in early summer, taking about half the colony with her. Honey bees feed on pollen and nectar and are therefore important for garden plants, fruit trees and certain crops.

Main habitat: Woodland, gardens, orchards.
Main region: Temperate (may have originated in South-east Asia).
Length: 12–20mm (0.47–0.79in).
Wingspan: 25mm (1in).
Development: Complete metamorphosis.
Food: Herbivore.

Identification: Rather hairy, and orange and brown. The abdomen is orange with black transverse stripes of varying width.

Left and above: The familiar honey bee is one of the most beneficial of all insects.

Buff-tailed Bumblebee

Bombus terrestris

This is one of several rather similar European bumblebees. It is quite common in gardens and the queen can be seen early in the spring, looking for a suitable site for a nest – for example in a hole in the ground, such as an abandoned mouse burrow. She will have been fertilized the previous year and hibernated through the winter. The first workers from the colony will emerge after a few weeks. Later in the year some males and new queens will be born.

Main habitat: Woodland, heathland, gardens.
Main region: Temperate.
Length: 15mm (0.59in) (worker); 20mm (0.79in) (male); 25mm (1in) (queen).
Wingspan: 28–40mm (1.1–1.57in).
Development: Complete metamorphosis.
Food: Herbivore.

Above and right: This charming bee is one of the earliest to appear in the spring.

Identification: Rather rounded, furry body with a buff-coloured tail and yellow bands on the front of the thorax and middle of the abdomen. The queen is considerably larger than the male bee and the even smaller workers. Its tongue is short so it sometimes bites through the base of flowers to get at the nectar.

OTHER SPECIES OF NOTE

Giant Carpenter Bee *Xylocopa flavorufa*
This very large bee lives in South Africa, and grows to a length of about 30mm (1.18in), with a 63mm (2.48in) wingspan. Carpenter bees nest in burrows in plant stems and feed their larvae with pollen and nectar. This species has a shiny dark body and wings, with broad yellow bands on the thorax and abdomen.

Cuckoo Bee *Psithyrus* spp.
These bees have evolved to closely resemble particular species of bumblebee, which they parasitize. They have no pollen baskets on their hind legs, and they also have darker wings. The queen enters a bumblebee nest in spring, kills the original queen and then makes the workers rear her own larvae.

Paper Wasp (Left)
Polistes gallicus
This is one of several species that make small paper nests hung from a building or branch in a sheltered site. The colonies are smaller than those of the common wasp.

Leaf-cutter Bee *Megachile centuncularis*
This dainty bee is sometimes spotted in gardens. It is a solitary species and named for its habit of cutting semicircular holes in leaves. The bee takes the pieces of leaf to line its tunnel nest.

Common Wasp

Vespula vulgaris

This active insect is well known in Europe and can sometimes be a nuisance, especially as it has a nasty sting and may be aggressive. Unlike a honey bee, a wasp usually survives after stinging. A fertile queen wasp comes out of hibernation in spring and founds a colony. This expands after the first workers hatch and is made of chewed wood – a sort of papier mâché. Wasp larvae are fed on insects while the adults feed on a range of food, such as nectar and juice from fruits.

Identification: Bright yellow and black with transparent, slightly yellow tinted wings. The head is triangular, with kidney-shaped eyes and the antennae are elbowed. The abdomen is connected to the thorax by a narrow stalk.

Main habitat: Woodland, gardens, orchards.
Main region: Temperate.
Length: 12–19mm (0.47–0.75in).
Wingspan: 30–45mm (1.18–1.77in).
Development: Complete metamorphosis.
Food: Carnivore (larva); mainly herbivore (adult).

Below: Though widely feared, wasps are useful for the garden pests they and their larvae consume.

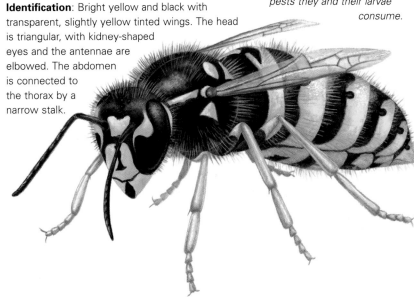

Hornet

Vespa crabro

Hornets are notorious and much feared, though they are less of a nuisance than common wasps and their sting is about the same intensity. They usually nest inside the trunks of hollow trees or rotting stumps, and the colonies contain up to about 1,000 individuals. They are much less common than wasps and are on the wing from May through August. Like those of common wasps, hornet larvae are fed the bodies of other insects, brought to them by workers. The worker hornets themselves feed mainly on nectar and sap, and also a sugary liquid which the larvae produce.

Right: Although larger than most wasps, hornets are less aggressive.

Identification: This is Europe's largest wasp. The thorax is reddish-black and rather hairy; the abdomen has dark bands near the waist, and is yellow toward the tip.

Main habitat: Woodland.
Main region: Temperate.
Length: 22–32mm (0.87–1.26in).
Wingspan: 50mm (2in).
Life cycle: Complete metamorphosis.
Food: Carnivore (larva); mainly herbivore (adult).

MILLIPEDES AND CENTIPEDES

Millipedes and centipedes are distinguishable from insects by their many legs (at least nine pairs and often many more), and by their largely uniform bodies (not divided into head, thorax and abdomen). They range in length from 0.05–30cm (0.02–12in), and they are well represented in Europe and Africa, both in gardens and also in wild habitats.

African Giant Millipede

Archispirostreptus gigas

One of the arthropod wonders of Africa, this huge millipede is also a favourite pet. It is docile, slow-moving and easy to keep. African giant millipedes grow as long as a hand and as thick as a thumb, and may live for about ten years. In the wild they live in subtropical and tropical regions of west Africa, mainly on the surface, but they also burrow to some extent. They feed on a range of foods, including fruit and leaves. If threatened, they react by curling into a spiral and they also sometimes produce a nasty chemical from pores along their body. Care should be taken when handling them as this substance can be harmful. They should be kept in a tank, with a layer of damp peaty soil, in warm conditions with a high humidity.

Identification: The body is long, black and cylindrical, with red legs and antennae. When fully grown the legs may number as many as 256.

Below: This huge millipede makes a popular and unusual pet.

Main habitat: Forest.
Main region: Tropical.
Length: 28cm (11in).
Development: Gradual.
Food: Herbivore.

Flat-backed Millipede

Polydesmus complanatus

This is a common European millipede, one of several known as flat-backed millipedes. It is often seen in leaf litter in gardens and elsewhere. Flat-backed millipedes dry out very quickly and therefore tend to stay in damp places. The female makes a domed nest in which to lay her eggs and remains with them until they hatch.

Identification: The body is distinctly flattened with about 20 segments, and is light yellow-brown. Each segment has a wing-like flange at the sides, protecting the base of each pair of legs. The antennae are angled.

Main habitat: Leaf litter.
Main region: Temperate.
Length: 20mm (0.79n).
Development: Gradual.
Food: Herbivore.

Below and above left: Flat-backed millipedes are restricted to damp sites such as underneath logs.

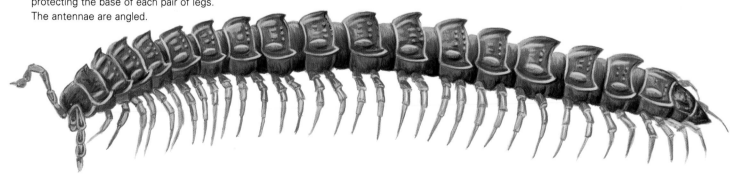

OTHER SPECIES OF NOTE

White-legged Snake Millipede *Tachypodoiulus niger*
Shiny and black, this millipede has a cylindrical body with white legs – about 50 pairs. It is common in Europe and is sometimes found indoors in damp conditions. It feeds on humus and algae, and outdoors is most common under loose tree bark, in moss or among leaf litter. It is one of the most frequently seen millipedes in Britain.

Pill Millipede (Left)
Glomeris marginata
This rather unusual millipede looks more like a woodlouse, until you count its legs – at least 17 pairs (woodlice have seven pairs). It is dark and shiny with pale edges to the plates. Pill millipedes live in the soil and leaf litter, feeding on decaying plant material. By curling up into a tight ball, pill millipedes protect their vital and vulnerable parts from attack, and reduce water loss at the same time.

Luminous Centipede *Geophilus carpophagus*
This centipede is often found among leaf litter and in damp soil, or under logs and sometimes inside sheds. Its body is very slender and pale yellow-brown, and it has more than 35 pairs of legs. It emits a weak glow and is sometimes called a 'glow-worm'. When disturbed it throws its body into tight loops. It feeds on soil invertebrates, including worms.

Yellow Centipede
Stigmatogaster subterranea
Cream-coloured and very slender, this species is common in the soil. When disturbed it twists its body into contorted shapes. It feeds on plant roots and sometimes damages crops.

Common Centipede

Lithobius forficatus

This common centipede is a swift, agile hunter. Its flexible body seems to glide along at speed. By day it hides in crevices, such as under loose bark, leaves or stones. It emerges at night to hunt for insect larvae, worms and the like, catching them using its powerful poisonous fangs. The newly hatched young have only seven pairs of legs, but soon moult and grow the full complement of 15 pairs. Although mainly a creature of the soil, this species sometimes enters houses, especially in wet weather.

Identification: Chestnut brown in colour, the jointed, bendy body has large segments, and 15 pairs of legs. The last pair of legs stick out backward and act as extra feelers; the true antennae are quite long.

Main habitat: Woodland, gardens.
Main region: Temperate.
Length: 20–30mm (0.79–1.18in).
Development: Gradual.
Food: Carnivore.

Below: Common centipedes use their long legs to run at speed and will quickly dash for cover if disturbed.

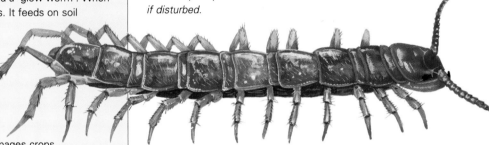

Giant African Centipede

Scolopendra spp.

There are several species of these unusually large centipedes, in Africa. This one is found in wooded savanna and open grassland, mainly in Tanzania and Kenya. It is large, fast and quite aggressive and eats crickets, cockroaches and even small mammals. It is occasionally kept as a pet, but care must be taken with this and other large centipedes as their bites can cause painful swellings.

Main habitat: Savanna.
Main region: Tropical.
Length: 23cm (9in).
Development: Gradual.
Food: Carnivore.

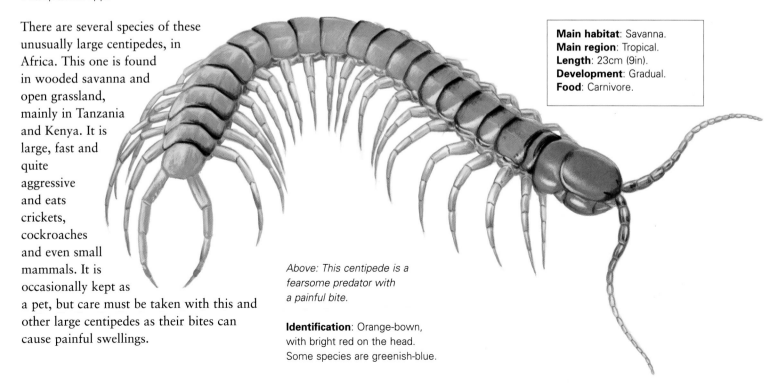

Above: This centipede is a fearsome predator with a painful bite.

Identification: Orange-bown, with bright red on the head. Some species are greenish-blue.

SPIDERS, MITES AND TICKS

Despite their unpopularity with many people, spiders have an important role to play in the balance of nature. These venomous creatures are well adapted to their predatory, carnivorous lifestyle, and they help to control many insect pests. They are common throughout Europe and Africa. Mites and ticks inhabit many locations across the region and are very numerous. Many species are parasitic.

Garden Spider

Araneus diadematus

This widespread spider is one of the most common in Europe. It is sometimes called the cross spider, because of the white cross-shaped mark on its abdomen. It belongs to a family known as orb-web spiders for their (often) orb-shaped webs. This is the spider responsible for most large, roughly circular webs found draped around garden shrubs. When an insect (for example) blunders into the web, the spider stills it with a bite and then wraps it tightly in silk threads, sometimes storing it to consume later. The spider may wait for long periods between feeds.

Identification: The male at 8mm (0.31in) is smaller than the female, 12mm (0.47in). The body is grey-brown with a line of darker patches along the back and the white cross. The legs are orange, banded brown. The webs are large, up to 40cm (16in) across.

Above and right: This spider is very common in Europe, and is often found in gardens.

Main habitat: Scrub, gardens.
Main region: Temperate.
Length: 8–12mm (0.31–0.47in).
Development: Gradual.
Food: Carnivore.

Flower Crab Spider

Misumena vatia

Crab spiders, such as this common species, have a crab-like shape and like a crab, they also tend to move sideways. The flower crab spider is a sit-and-wait, patient predator. It uses its colour as camouflage as it waits, typically on a flower, for visiting insects, which it then grabs in a quick pounce. The female usually sits on yellow or white flowers, legs spread out waiting for a suitable prey to approach.

Identification: The male is small with a dark brown prothorax and dark front two pairs of legs; the last two pairs of legs are smaller and paler. The larger female looks quite different: the whole body is either white or yellow, sometimes with a red-brown stripe on each side. However, she can change colour to some extent to match the background flower colour.

Main habitat: Flowers.
Main region: Temperate.
Length: 4mm (0.16in) (male); 10mm (0.39in) (female).
Development: Gradual.
Food: Carnivore.

Above and right: Crab spiders lurk, well camouflaged, on plants and flowers. The venom of the female (right) can kill bees and hoverflies.

Wasp Spider (Left)
Argiope bruennichi
This spider, related to the banded garden spider, is quite common in southern Europe. Its abdomen is boldly marked in black and yellow – a wasp-like warning colour. It makes large orb-webs, sometimes across tracks between bushes or small trees. It is one of Europe's largest spiders.

Baboon Spider *Mygale atra*
This large hairy spider from southern Africa belongs to the tarantula family. Its body may be up to 90mm (3.54in) long. It is brown, yellow and grey and very hairy. It lives in a silk-lined burrow and eats a range of prey, from insects to small reptiles and amphibians.

Flower Crab Spider *Thomisus onustus*
This crab spider of Europe and North Africa is coloured pink with white blotches. It often waits for its prey among the flowers of heather, where it is well camouflaged. There are also yellow forms that lurk in yellow flowers.

Blood-eating Spider *Evarcha culicivora*
This spider is found only around Lake Victoria in Kenya and Uganda. It belongs to a group known as jumping spiders. It often hunts insects on tree trunks and walls, stalking its prey. Its preferred prey is a female mosquito that has fed on mammalian blood, hence its name.

Banded Garden Spider

Argiope trifasciata

This is a common garden spider of southern Africa. The bright colours act as a warning to birds and other predators that these spiders taste nasty. It constructs a large orb-web usually among grasses, then waits, usually head downward, for insects to get entangled. Then the prey is quickly wrapped in silk and killed with a bite. The web has two zigzag bands of silk radiating outward. The function of these bands is not clear. They may help to stabilize the web. Another theory is that they may make the web more visible to birds, which avoid flying through and damaging the web.

Main habitat: Grasses.
Main region: Subtropical.
Length: 6–20mm (0.24–0.79in).
Development: Gradual.
Food: Carnivore.

Identification:
This spider is beautifully marked. The abdomen has alternating bands of yellow and black.

Right: Spiders of this genus construct very large webs.

Water Spider

Argyroneta aquatica

This amazing spider is the only spider to spend its life under the water. It does this by constructing an underwater 'diving bell' from silk – a tightly spun web that traps air. It fills the bubble by making forays to the surface and back, bringing a little more air on each trip. The spider traps air in its covering of short hairs, so looks silvery when under the water. Water spiders are found mainly in ponds, streams and shallow lakes with plentiful plant life.

Main habitat: Ponds, slow rivers.
Main region: Temperate.
Length: 8–15mm (0.31–0.59in).
Development: Gradual.
Food: Carnivore.

Identification: The water spider has a red-brown thorax and legs, and smooth grey-black abdomen. The abdomen and legs are covered in short hairs. It hunts various water creatures, dragging them back into the air-bell or to the surface to eat.

Above and right: With its ability to create an air-filled sac, under water, this spider has adapted well to life beneath the water surface.

House Spider

Tegenaria duellica (T. gigantea)

This spider is common in houses and sheds and is one of the species responsible for creating untidy cobwebs that gather dust in corners. These sheet webs may measure several feet across. This is one of Europe's larger spiders and can be alarming to come across, when for example it drops into a sink or bath. However, it is harmless. It is usually the male that wanders about. The female tends to stay with its eggs until the young spiders emerge. This long-legged spider is capable of sudden bursts of speed, which can make it all the more frightening and also quite difficult to catch. The female can live for several years and may eat the male after mating.

Identification: Large brown spider with very long, hairy legs. The male has a smaller body, but longer legs than the female. The web is a thin cobweb, with a tube-shaped base in which the spiders rest.

Main habitat: Houses, sheds, crevices in trees.
Main region: Temperate.
Length: 11–14mm (0.43–0.55in).
Development: Gradual.
Food: Carnivore.

Left and far left: Though large and fast-running, the house spider is quite harmless.

Zebra Spider

Wall Spider, *Salticus scenicus*

You may have seen this common spider sunning itself on a hot wall in the sun, or suddenly jumping away. It uses a stealthy approach and jump strategy to capture its prey, rather than a sticky web. It retreats to hide in a crevice such as between stones or under bark. It belongs to a group known as jumping spiders. The four, forward-facing eyes give it good vision to track its prey.

Main habitat: Walls, stones, bark.
Main region: Temperate.
Length: 5–6mm (0.19–0.24in).
Development: Gradual.
Food: Carnivore.

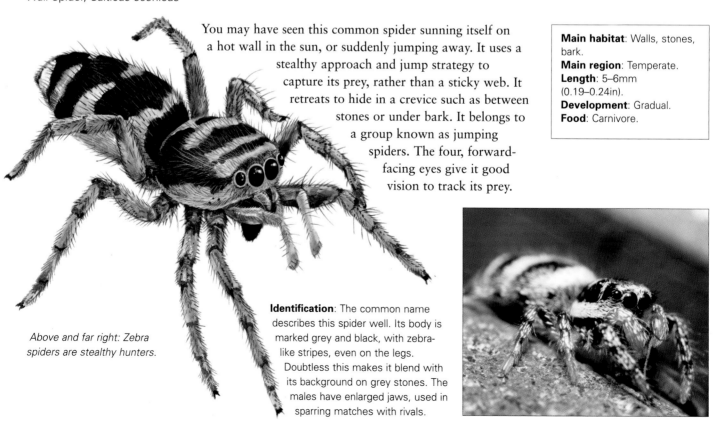

Above and far right: Zebra spiders are stealthy hunters.

Identification: The common name describes this spider well. Its body is marked grey and black, with zebra-like stripes, even on the legs. Doubtless this makes it blend with its background on grey stones. The males have enlarged jaws, used in sparring matches with rivals.

Velvet Mite

Trombidium holosericeum

Main habitat: Soil. **Main region**: Temperate. **Length**: 4mm (0.16in). **Development**: Gradual. **Food**: Parasite (larva); carnivore (adult).

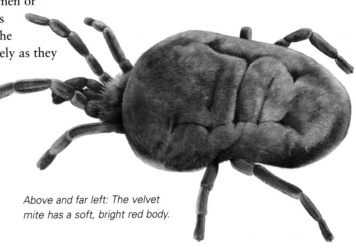

Velvet mites are relatively large and, being bright red, are quite easily spotted as they crawl over the ground. They can be found in woodland, heath and in gardens and often amble around in the open. Velvet mites lay their eggs in the soil. The larvae that hatch, also bright red in the case of this species, live as parasites on the legs and bodies of insects such as grasshoppers, or on harvestmen or spiders. After a short time as parasites, the larvae fall to the ground and begin to live freely as they mature. The adults are free-living predators, feeding mainly on smaller mites, aphids and springtails.

Identification: Large (for a mite), with soft, velvety skin and a bulbous body. The whole mite, including the legs, is a bright red colour. The legs are rather short and thin.

Above and far left: The velvet mite has a soft, bright red body.

OTHER SPECIES OF NOTE **Kite Spider** *Gasteracanthus sanguinolenta* This species is fairly common in Africa where it is usually found high in trees. The abdomen is broader than long with six pointed projections sticking out around it. The colouring is red, yellow and black, and it has short black legs. **Wolf Spider** *Pardosa lugubris* Wolf spiders are active predators that hunt down their prey by chasing it over the ground. This small European species is about 5–6mm (0.19–0.24in) long. It is dark brown with a striped thorax and lightly hairy abdomen. It can be seen in spring and summer, rushing across leaves and soil, the female sometimes with an egg sac beneath her abdomen. **Nursery Web Spider** *Pisaura mirabilis* This slender spider is about 10–15mm (0.39–0.59in) long, with long legs and a yellow central stripe on the thorax. The abdomen is grey-brown with a darker central patch, with a darker wavy border. The female spins a domed nest among grass stems into which the spiderlings hatch, hence the name. **Woodlouse-eating Spider** *Dysdera crocata* This spider is specialized to catch and eat woodlice, using its large, pincer-like jaws. These are powerful enough to give a nasty nip to any human that tries to pick it up. It lives in similar places to woodlice, in damp soil, under logs and stones, and emerges at night to hunt. Its body is about 15mm (0.59in) long, pink or reddish with a pale abdomen.

Sheep Tick

Ixodes ricinus

The sheep tick is one of the most annoying of parasites, attacking people as well as sheep. Ticks are parasitic arachnids with biting mouthparts. They sit and wait, typically atop a blade of grass and then scuttle quickly into the fur of a passing mammal, or the sock of a passing human leg. From here they make their way to the flesh, boring their mouthparts into the skin and sucking blood for several days, their abdomen expanding during this process. When full the tick drops off. Ticks can carry serious diseases, such as Lyme disease, and should be removed as soon as possible.

Main habitat: Grassland, heath. **Main region**: Temperate. **Length**: 1–3mm (0.04–0.12in). **Development**: Gradual. **Food**: Parasite.

Identification: Tiny, but expanding to the size of a pea 10mm (0.39in) when full. The body is red-brown and flat. The first-year larva has six legs, but the nymph (second year) and adult (from third year) have the full complement of eight.

Below: Sheep ticks also attack people and can be quite hard to remove.

SCORPIONS AND PSEUDOSCORPIONS

Scorpions inhabit a range of locations in tropical and warm temperate climates. They are often found in arid deserts, but are also common in humid rainforests. Africa has many species, but only a few are found in Europe (mainly in the southern regions). The smaller pseudoscorpions range from 2.5–8mm (0.09–0.31in) and are mostly found in leaf litter.

African Rock Scorpion

Hadogenes spp.

The rock scorpions of Africa can be found in rocky sites and mountains from South Africa west to Namibia and north to Tanzania. The genus includes *H. troglodytes*, the male of which is the longest scorpion in the world, at 21cm (8in). Rock scorpions have very strong clawed feet, enabling them to walk upside-down on rocks. A single scorpion may inhabit the same rocky retreat for years at a time. They are active at night, pouncing on prey that passes nearby. Although fearsome to behold, their venom is rather weak.

Identification: Very flat-bodied. Even their legs and tail are flattened, allowing them to squeeze into narrow crevices in the rocks. *H. troglodytes* ranges in colour from black to brown with yellow legs.

Right: Rock scorpions have extremely large pincers but a relatively small sting.

Main habitat: Rocks.
Main region: Tropical.
Length: 21cm (8in).
Development: Gradual.
Food: Carnivore.

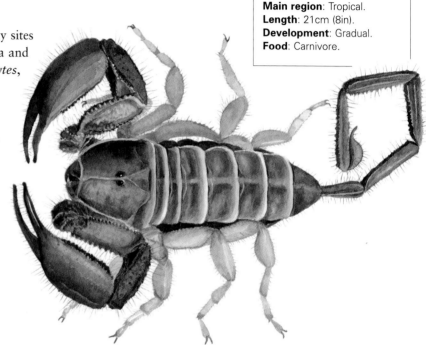

Scorpion

Parabuthus villosus

This large scorpion is one of the most dangerous in Africa. It lives in rocky and mountainous terrains and likes to hide underneath stones or timber and loose bark. It tends to be most active in the early morning or evening. Scorpions of this genus have large stings and the venom is highly toxic, making them quite dangerous. This species makes a distinctive clicking sound by scraping its sting against rough patches present at the tail base. As a general (but not entirely reliable) rule, scorpions with small claws and a large sting, like this species, are dangerously venomous while those with large claws and a small sting are less hazardous.

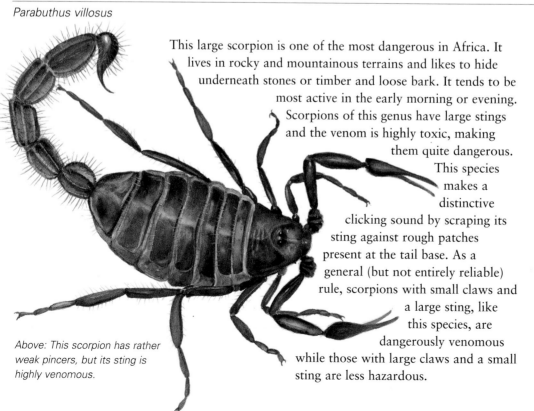

Above: This scorpion has rather weak pincers, but its sting is highly venomous.

Main habitat: Rocky sites.
Main region: Tropical.
Length: 18cm (7in).
Development: Gradual.
Food: Carnivore.

Identification: Large scorpion, growing to 18cm (7in). The body is covered in hairs and ranges from black to brown. The tail is very thick and keeled and there are rough patches on the first two tail segments. Unusually for a scorpion, this species is capable of spraying its venom more than 66cm (24in) from its sting to disable and potentially blind attackers.

Scorpion

Buthus occitanus

This is a common European species, found in southern France, Spain and Portugal; it also occurs in North Africa (Tunisia and Morocco). It lives mainly in hot, dry countryside that has little plant cover but plenty of rocks and stones. It also lives in Mediterranean forests in Spain, and occasionally wanders into houses. The sting of this species is painful but not deadly. However, closely related yellow species from Africa have a much more dangerous sting and rank among the most venomous of all of the world's scorpions.

Main habitat: Stony areas. **Main region**: Warm temperate. **Length**: 60–80mm (2.36–3.15in). **Development**: Gradual. **Food**: Carnivore.

Identification: Yellow (mainly Mediterranean) or brown (mainly African) with rather small claws and a thick tail with a large sting. The front of the carapace is granular.

Above: Common in southern Europe and North Africa, this species can deliver a painful sting.

OTHER SPECIES OF NOTE
Emperor Scorpion *Pandinus imperator* This African scorpion is found mainly in tropical west Africa, and is often kept as a pet as it is quite docile and has a fairly mild sting. It is large, black and shiny, with big pincers. The adult can reach a length of 20cm (8in). In the wild it inhabits burrows in moist forests.
Yellow-tailed Scorpion *Euscorpius flavicordis* This small scorpion is found in south-west Europe and North Africa, and has been introduced to some sites in southern England. It likes to live among stones and also in the cracks of walls. It is small and black with yellow-brown legs. Its sting is mildly venomous.
Wind Scorpion *Solpuga* spp. Classified separately from true scorpions, wind scorpions, or sun spiders, are found in African deserts and other dry tropical habitats, and in Asia and tropical America. Powerfully built, with large jaws, they prey on other arthropods and small vertebrates. They use their long front legs as feelers.
Pseudoscorpion *Chelifer cancroides* This European species is about 4mm (0.16in) long and red-brown in colour. It is very agile and moves with ease, forward, sideways and backward, in a crab-like way. It lives under bark and in birds' nests and is sometimes found in houses in dusty places. It feeds mainly on mites and book- or bark-lice. Pseudoscorpions sometimes hitch a lift on the legs of insects, thus aiding their dispersal.

Pseudoscorpion

Neobisium carcinoides

Pseudoscorpions are like miniature true scorpions, but they lack the arched and stinging tail. This common European species is tiny. It lives among leaf litter on the forest floor, or in moss, under stones or in rotten tree stumps and similar sites. Pseudoscorpions can also be found in gardens. They feed on smaller invertebrates such as mites and springtails.

Identification: It has a brown body and long front limbs with two relatively enormous pincers. Pseudoscorpions move deliberately and can walk backward.

Main habitat: Soil, leaf litter. **Main region**: Temperate. **Length**: 3mm (0.12in). **Development**: Gradual. **Food**: Carnivore.

Below and left: Pseudoscorpions are seldom spotted, as they tiny and secretive.

INSECTS OF THE AMERICAS

The Americas are home to more insects than any other landmass, largely

because of the extreme richness of the tropical regions, especially the

multi-layered habitats of the Amazonian rainforests. North America has

very varied landscapes too and many insects have also adapted well to

agricultural and other human-influenced areas. Some of the insects found

in this region are true giants – such as the beetle *Megasoma actaeon* and

the longhorn beetle *Titanus giganteus*, both from South America. At the

other extreme, the feather-winged beetle *Nanosella fungi* is one of the

world's smallest insects. The sheer numbers of insects can be impressive

too. It has been estimated that ants may account for nearly one third of

all the animal biomass in the Amazon Basin.

Above from left: A pennant dragonfly, a stag beetle and a Colorado beetle.

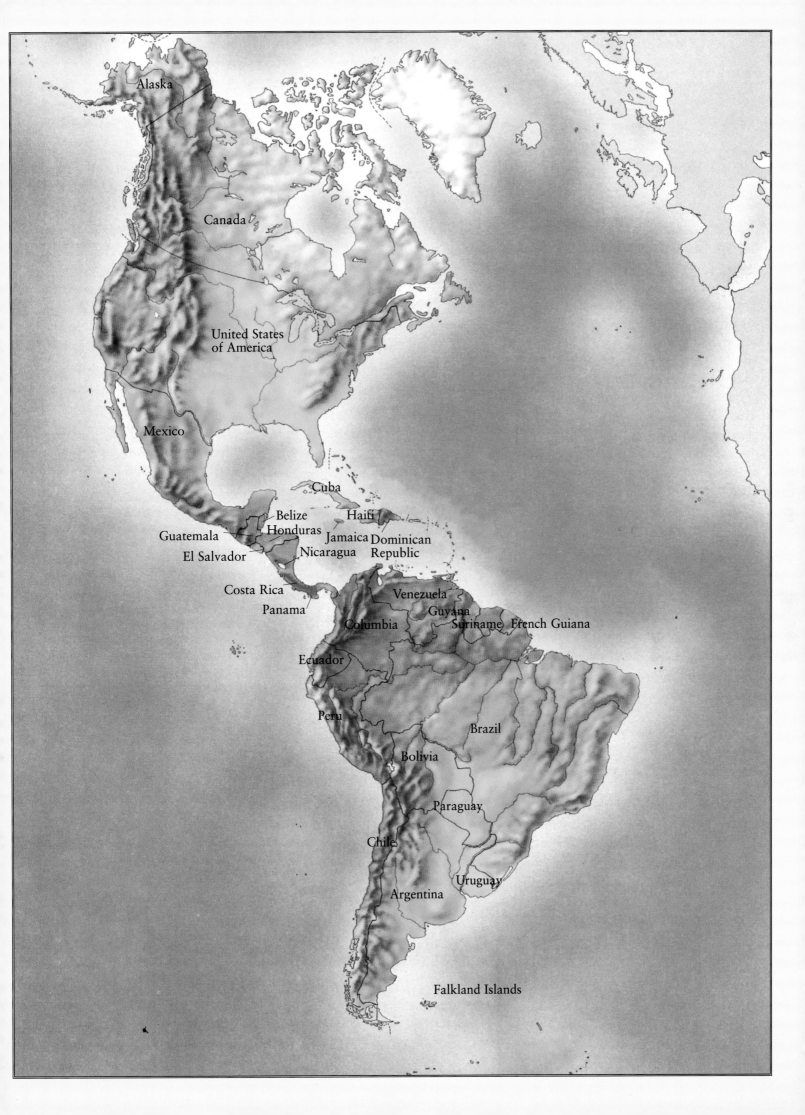

WINGLESS NEAR-INSECTS AND WINGLESS INSECTS

Bristletails and springtails are wingless true insects. Their mouthparts are very different from those of true insects and are enclosed in a pouch under the head. There are hundreds of species of proturans and diplurans, but the springtails are the largest group, with well over 2,000 species.

Seashore Springtail

Anurida maritima

Springtails are six-legged arthropods, but are not now classed as true insects. They are very abundant but because they are so small are easily overlooked, unless they occur in large numbers. This species belongs to a family called elongate-bodied springtails because their bodies are quite long. It is unusual in being aquatic or semi-aquatic, living in tidal zones, mainly on the Atlantic coast of North America.

Identification: Tiny and bluish in colour, with an obviously segmented body and stumpy legs and antennae.

Below and below right: This tiny springtail hides in holes in the rocks as the tide comes in.

Main habitat: Tidal zones of shoreline.
Main region: Temperate.
Length: 1–2mm (0.04–0.08in).
Development: Gradual.
Food: Herbivore/omnivore.

Large Slender Springtail

Pogonognathellus flavescens

Rather easier to spot because of its (relatively) huge size, this springtail belongs to a family called the slender springtails. It is one of about 36 species found in North America. It prefers to live in dark places such as beneath stones, logs and the like. The tail-like spring is normally held underneath its body. It can be released suddenly to catapult the springtail into the air.

Main habitat: Soil.
Main region: Temperate.
Length: 6mm (0.24in).
Development: Gradual.
Food: Herbivore/omnivore.

Identification: This is one of the largest of the springtails, reaching about 6mm (0.24in). Its body is shiny and brownish, and rather smooth, but with scattered clumps of hairs. The legs are fairly long, as are the antennae, the latter rather more than half the body length.

Left: Like many springtails, this species can make sudden long leaps.

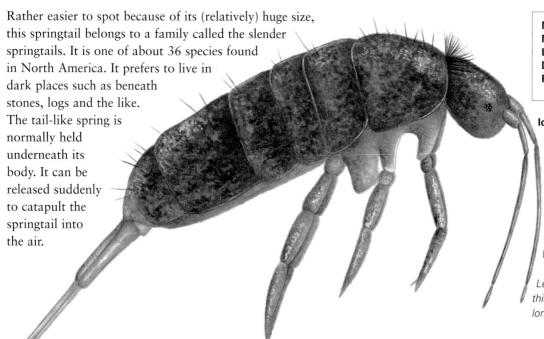

Urban Silverfish

Ctenolepisma urbana

Main habitat: Houses.
Main region: Temperate.
Length: 8–20mm
(0.31–0.79in).
Development: Gradual.
Food: Omnivore.

This is a particularly large silverfish that is not uncommon in buildings, especially in cities and other urban areas. About twice the size of the firebrat, it is a dark silvery grey with the typical silverfish three-tail projections and long narrow antennae.

Below: Silverfish seem to glide as they twist and run across the floor.

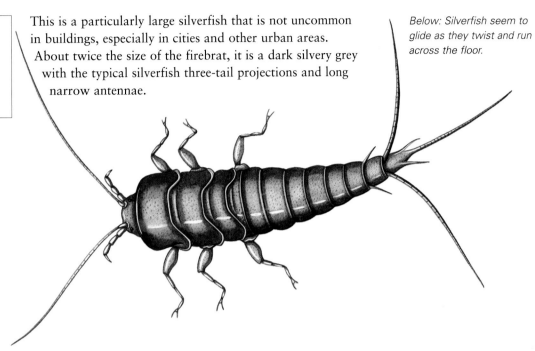

Identification: Silverfish have widely separated eyes and a three-pronged 'tail'. Brownish-grey with a bright silvery, scaly sheen, it does look rather fish-like as it scuttles rapidly across the floor, twisting its supple body as it goes. The head has compound eyes and long sensitive antennae. This silverfish grows to about 20mm (0.79in).

OTHER SPECIES OF NOTE

Smooth Springtail *Folsomia candida*
This species belongs to a family called the smooth springtails. They have smooth bodies without rough scales. This one is pale, pinkish-white and rather translucent. It is often found in soil, especially around potted plants and compost.

Smooth Springtail *Isotoma viridis*
Another smooth springtail, this species is greenish-brown in colour with a rather podgy body that has small tufts of hair. This is one of the most common American springtails, and also one of the largest, reaching a length of 6mm (0.24in).

Common Silverfish *Lepisma saccharina*
The common silverfish is usually found in cool, damp places such as food cupboards, cellars, kitchens and bathrooms, but being strictly nocturnal it is not often seen in daytime. With generally higher standards of hygiene they are not as common as they used to be in houses. It lays its eggs in cracks and feeds on food scraps such as flour, damp wallpaper and damp cloth. It can go without food for months if necessary.

Firebrat *Thermobia domestica*
This silverfish takes its name from one of its favoured habitats, the hearth of a fire or the vicinity of an oven in a bakery or kitchen. It lives on scraps of food found on the floor or in cracks. Fast-moving, firebrats disappear rapidly into the shelter of crevices when disturbed. The firebrat is less silvery than the common silverfish and it also has much longer antennae and tail filaments. It has a flattened body and runs rapidly, close to the ground. The basic body colour is yellowish with darker bands and spots.

Snow Flea

Hypogastrura nivicola

Snow fleas take their name from their habitat (being found on snow) and their habit of jumping rapidly when disturbed. Though they are a little flea-like in behaviour, they are not fleas at all, but a kind of springtail. They are tiny, bluish and clearly segmented. Their blood has a kind of natural antifreeze that stops them being damaged by the cold. The snow flea is closely related to the seashore springtail, both belonging to the largest springtail family. They are often found clustered in groups.

Below: This tiny springtail is sometimes seen in large numbers on snow patches.

Main habitat: Snow patches.
Main region: Temperate.
Length: 1–2mm
(0.04–0.08in).
Development: Gradual.
Food: Omnivore.

Identification: Very tiny springtail, belonging to a family known as elongate-bodied springtails due to their rather long abdomen. They appear like tiny specks on the snow and may be the only obvious life on this inhospitable habitat. The abdomen shows rather obvious segmentation and the legs are chunky.

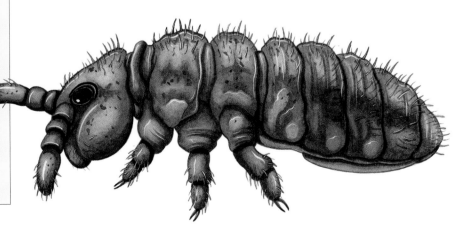

MAYFLIES AND STONEFLIES

Mayflies (Ephemeroptera) are rather delicate winged insects with short-lived adults and longer-lived aquatic larvae, known as nymphs. They have one or two pairs of thin wings and a weak, fluttery flight and are often seen around May in northern regions. Stoneflies (Plecoptera) are found in similar habitats and also have aquatic nymphs, but their adults live for two or three weeks.

Burrowing Mayfly

Hexagenia spp.

Burrowing mayflies are so-called because their larvae mostly remain hidden, burrowing into the mud at the bottom of streams and lakes. When they do emerge from the water as adult mayflies, however, they are quite easy to spot, as these are some of the largest of all mayflies. Mass hatchings of these mayflies are a feature of the Great Lakes in North America, with piles of dead and dying insects sometimes appearing in early summer.

Identification: The adult (imago) is a large mayfly with two very long tail streamers, as long as its body. The body is pale creamy-brown and the transparent wings have the typical mayfly network of small veins. The larva (nymph) is about 25mm (1in) long with obvious gills on the abdomen, three tails and curved tusk-like projections on its head.

Left: The adult burrowing mayfly is one of the region's largest mayflies.

Main habitat: Lakes.
Main region: Temperate.
Length: 30mm (1.18in) (adult).
Wingspan: 50mm (2in).
Development: Incomplete metamorphosis.
Food: Larva omnivorous; adult does not feed.

Right: A newly hatched mayfly drying on a rock.

Grey Quill

Callibaetis nigritus

Also known as the speckled spinner, this common American mayfly belongs to a family known as the small minnow mayflies (Baetidae), named after their slender, rather fish-like larvae. The adults are also small. There are more than 130 different species in this one family, which is the largest family in North America. This mayfly also emerges in very large numbers, mainly in July and August. Well known to anglers, this mayfly is the favoured food of many fish. Artificial lures mimicking this species are now also popular as a bait for fly fishing.

Right: This species of mayfly favours still and slow-moving water. Grey quills are usually found among vegetation in ponds, lakes and lazy rivers. The illustration shows a nymph.

Main habitat: Lakes.
Main region: Temperate.
Length: 15mm (0.59in).
Wingspan: 30mm (1.18in).
Development: Incomplete metamorphosis.
Food: Larva omnivorous; adult does not feed.

Identification: Nymphs are small, with three equal-length tails and heart-shaped gills. The adults have two tail filaments and the hindwings are very small. They have speckled bodies and brown patches at the front of the wings.

Golden Stonefly

Hesperoperla spp.

Main habitat: Rivers.
Main region: Temperate.
Length: 15mm (0.59in).
Wingspan: 40mm (1.57in).
Development: Incomplete metamorphosis.
Food: Larva carnivorous; adult does not feed.

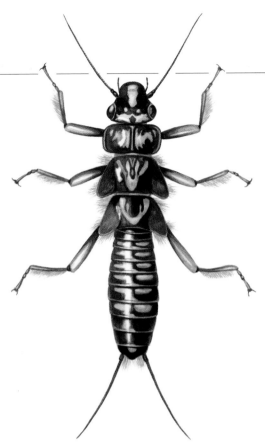

This genus of stonefly is quite common, especially in the western USA. The larvae live in flowing water where they feed mainly on the larvae of other insects such as midges. This and related species are known by anglers as 'golden stones' as some species have a distinctly golden sheen. They emerge from the water in the spring and waste no time in trying to find a mate. The eggs are laid shortly afterward, then the adults die.

Identification: The larva (nymph) is about 25mm (1in) long with rather a large head and generally flat body, and two long cerci. The adult is darker, with a glossy sheen and long, rather broad wings.

Left and far left: The larva of this species has a flattened body, which helps it to move beneath the gravel and stones on the bottom of the streams and rivers where it lives.

OTHER SPECIES OF NOTE

Flat-headed Mayfly e.g. *Cinygma* spp.
The flat-headed mayflies (family Heptageniidae) are medium-sized mayflies that have rather thin-bodied larvae with large, flattened heads. This is an adaptation to living under stones in fast-flowing rivers and streams.

Minnow Mayfly
Baetis spp. (Left)
Like the grey quill, these minnow mayflies have rather active, fish-like larvae. The adult insects are quite small. *Baetis* minnow mayflies are widespread across North America. Several species have more than one breeding event every year, so, despite their short adult lifespan they can often be seen over rivers.

Rolled-wing Stonefly e.g. *Megaleuctra* spp.
These stoneflies (family Leuctridae) take their name from the wings of the adults, which are rolled around the sides of their bodies. This genus is found around wet patches in upland areas. There are around 45 species of rolled-wing stonefly in North America.

Winter Stonefly e.g. *Capnia* spp.
Winter stoneflies (family Capniidae) often emerge in late winter or early spring, and may even be active while there is still snow on the ground. The nymphs live under rocks and gravel in streams feeding on plant material; the adults browse mainly on algae.

Giant Stonefly

Pteronarcys dorsata

As its common name suggests, this is a large stonefly. It is fairly common in North America, especially in the north and east, where the adults tend to emerge in spring and early summer (mainly from March through June). The larvae crawl about at the bottom of streams where they feed mainly on decaying leaves.

Identification: The adults are dark, blackish-brown but are often a pink colour when they first emerge, giving them the local name of 'salmon flies'. They have long antennae and two tail filaments (cerci). Like all stoneflies, the wings are held flat over the body. Most of their egg-laying activity occurs after dark. Hatching may occur at dawn or dusk.

Right: The giant stonefly is one of the largest species; its larvae (right) inhabit running water.

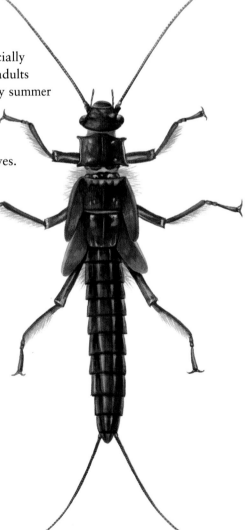

Main habitat: Rivers.
Main region: Temperate.
Length: 30mm (1.18in).
Wingspan: 75mm (2.95in).
Development: Incomplete metamorphosis.
Food: Larva omnivorous; adults do not feed.

DRAGONFLIES

Dragonflies are large, active insects with large compound eyes and a flexible neck. The order contains about 6,000 species. They have two pairs of large, rather stiff, usually transparent wings. The wings have a network of fine veins. The thorax is short and broad, with three pairs of long, hairy, clawed legs, and the long segmented abdomen is usually rather slender.

Flame Skimmer

Libellula saturata

This stunningly beautiful dragonfly is quite common in the south-west USA and Mexico. It is sometimes known as the firecracker skimmer and can look rather like a streak of red flame as it shoots past. The larvae (nymphs) live in thick mud at the bottom of ponds and streams.

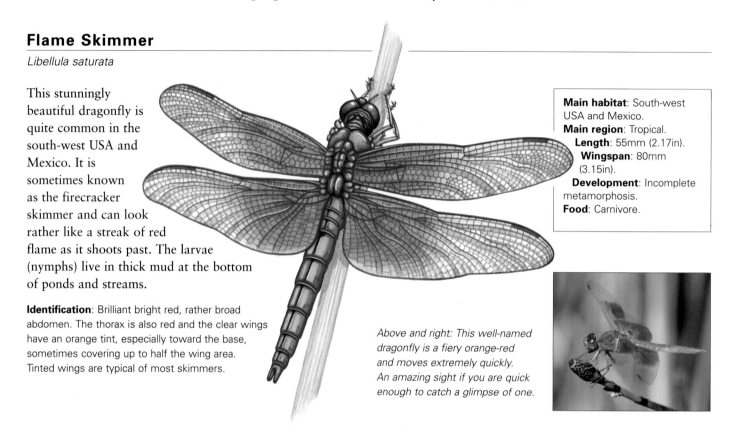

Identification: Brilliant bright red, rather broad abdomen. The thorax is also red and the clear wings have an orange tint, especially toward the base, sometimes covering up to half the wing area. Tinted wings are typical of most skimmers.

Main habitat: South-west USA and Mexico.
Main region: Tropical.
Length: 55mm (2.17in).
Wingspan: 80mm (3.15in).
Development: Incomplete metamorphosis.
Food: Carnivore.

Above and right: This well-named dragonfly is a fiery orange-red and moves extremely quickly. An amazing sight if you are quick enough to catch a glimpse of one.

Eastern Amberwing

Perithemis tenera

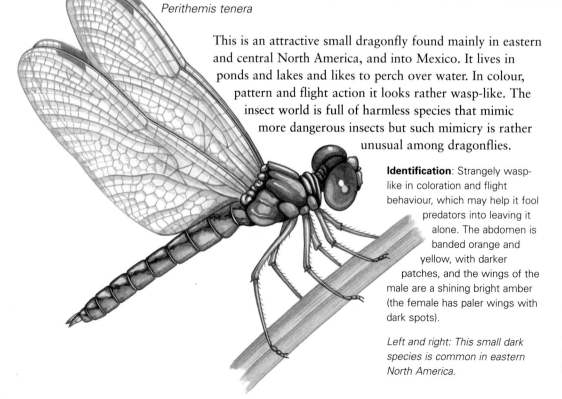

This is an attractive small dragonfly found mainly in eastern and central North America, and into Mexico. It lives in ponds and lakes and likes to perch over water. In colour, pattern and flight action it looks rather wasp-like. The insect world is full of harmless species that mimic more dangerous insects but such mimicry is rather unusual among dragonflies.

Main habitat: Ponds, lakes and slow streams.
Main region: Temperate.
Length: 22mm (0.87in).
Wingspan: 30mm (1.18in).
Development: Incomplete metamorphosis.
Food: Carnivore.

Identification: Strangely wasp-like in coloration and flight behaviour, which may help it fool predators into leaving it alone. The abdomen is banded orange and yellow, with darker patches, and the wings of the male are a shining bright amber (the female has paler wings with dark spots).

Left and right: This small dark species is common in eastern North America.

Green Darner

Anax junius

Main habitat: Small ponds. **Main region**: Temperate, subtropical. **Length**: 76mm (3in). **Wingspan**: 10cm (4in). **Development**: Incomplete metamorphosis. **Food**: Carnivore.

This is one of the most common of North American dragonflies, and also one of the largest and fastest. It is found throughout most of the USA, north to southern Canada and also south to Costa Rica. It tends to migrate, often in large groups.

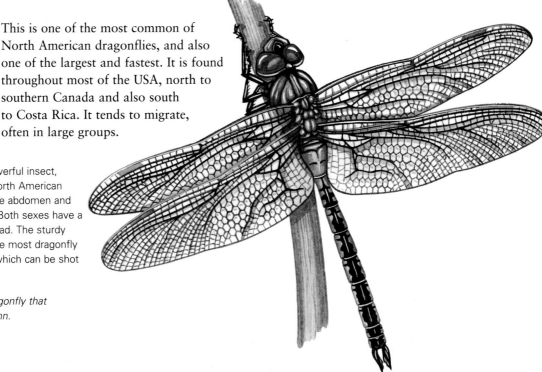

Identification: This is a large and powerful insect, indeed it is one of the largest of all North American dragonflies. The male has a bright blue abdomen and green thorax; the female is browner. Both sexes have a black spot in the middle of the forehead. The sturdy larva is also bright green in colour. Like most dragonfly larvae it has large, underslung jaws, which can be shot out telescopically to capture prey.

Right: The green darner is a large dragonfly that migrates in flocks in spring and autumn.

OTHER SPECIES OF NOTE
Regal Darner *Coryphaeschna ingens* This is one of the largest of North America's dragonflies. It is found mainly in and around swamps in the south-east, and west as far as eastern Texas. Its abdomen is banded brown and green and the thorax has bold diagonally alternating stripes of the same colours. The male has bright blue-green eyes.
Common Sand Dragon *Progomphus obscurus* This dragonfly belongs to a family known as the club-tailed dragonflies (Gomphidae). The abdomen is club-shaped at the tip and in this species is yellow and brown. It is common in the eastern and central states of the USA, near rivers and streams, often at sandy creeks. Most other sand dragons are tropical species.
Prince Basket-tail *Epitheca princeps* This is a large dragonfly found in south-east Canada and the eastern USA, around slow streams and ponds. It is a very active species, hunting on the wing for long periods. It is mainly brownish, with large green eyes. Its main distinctive feature are the wings with their large brown blotches, three on each wing.
Black Saddlebags *Tramea lacerata* This is a very distinctive dragonfly. A large skimmer, with broad hindwings, it has a dark body and very dark, blackish saddlebag-shaped patches at the base of each hindwing. These make it very easy to identify even in flight. This common species sometimes gathers in flocks – unusual behaviour for a dragonfly.

Dragonhunter

Hagenius brevistylus

This large dragonfly belongs to a family known as clubtails, from the enlarged, rather bulging abdomen tip. This species is the largest of North American clubtails. Well-named, it is an aggressive and fierce hunter, catching other insects, even other dragonflies in flight. Its favoured habitats are forest rivers, mainly in the east. When perched, this impressive dragonfly tends to hold its body horizontally.

Main habitat: Forested rivers. **Main region**: Temperate **Length**: 80mm (3.15in). **Wingspan**: 95mm (3.74in). **Development**: Incomplete metamorphosis. **Food**: Carnivore.

Identification: This insect is easily identified by its large size, combined with bright yellow-and-black coloration. Note the broad yellow diagonal stripes on the sides of the thorax and the large, widely separated eyes.

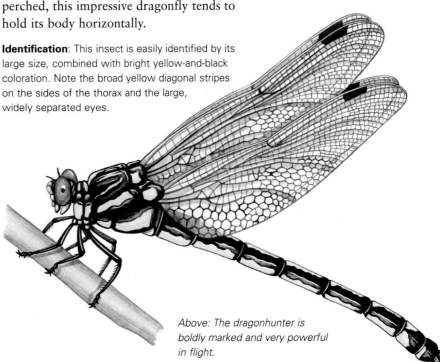

Above: The dragonhunter is boldly marked and very powerful in flight.

Halloween Pennant

Celithemis eponina

This is a small dragonfly found commonly near ponds and marshes, mainly in eastern North America as far north as south-east Canada. It belongs to the skimmer family (Libellulidae), many of which have coloured patches on their wings. The halloween pennant's wings are strikingly patterned – making it most distinctive in flight. The name 'pennant' comes from the habit of these dragonflies to sit waving, flag-like, atop a tall stem. The 'halloween' part of the name is derived from its yellow-orange colouration. Its breeding ground is over ponds and other still water, but it can be found foraging over fields. Halloween pennants are in season all year around. Mating usually takes place in the morning, with males waiting for females to be attracted to them.

Identification: A fairly small dragonfly, with a yellow-brown body. The wings are the most distinctive feature, being brownish in colour, with darker diagonal stripes.

Left and below right: The unusual striped wings are characteristic of this species.

Main habitat: Ponds and streams.
Main region: Temperate.
Length: 35mm (1.38in).
Wingspan: 60mm (2.36in).
Development: Incomplete metamorphosis.
Food: Carnivore.

Wandering Glider

Pantala flavescens

Identification: The abdomen is conical and yellow in colour, with a dark central stripe. The broad wings are transparent and finely veined. The hindwings are broader than the forewings. The large eyes are reddish.

The wandering glider is another skimmer. It has very broad wings and is extremely adept in flight, spending long periods on the wing, hunting other flying insects and also prospecting for a suitable breeding site, such as a small pond. This dragonfly often covers long distances, flying powerfully and gliding periodically, making full use of the prevailing wind. It lays its eggs in ponds that dry out and the larvae can grow and hatch into adults in a short period – often as short as five weeks. They can also survive a dry period in the mud, then appear again with fresh rains.

Left and right: Broad, powerful wings help this dragonfly cover a wide area when in flight.

Main habitat: Temporary rainwater pools.
Main region: Warm temperate and tropical.
Length: 45mm (1.77in).
Wingspan: 80mm (3.15n).
Development: Incomplete metamorphosis.
Food: Carnivore.

DAMSELFLIES

Damselflies are rather daintier than dragonflies; they tend to have a more fluttering, less direct and slower flight pattern. They are generally smaller than dragonflies and most hold their wings upright over their backs when at rest. Their larvae (nymphs), like those of dragonflies, are aquatic. There are about 130 species in North America.

Great Spreadwing

Archilestes grandis

This is one of America's largest and most unusual damselflies. Unlike most damselflies, spreadwings hold their wings out sideways when at rest, but rather more angled than dragonflies. Seen from the side, it has rather a hunchbacked appearance. The species often hangs downward from a twig or stem. Its range extends from Colombia and Venezuela through Central America and into the USA as far as the Great Lakes. The nymph of the great spreadwing is also rather distinctive. It too is long-bodied, and it has three feathery 'tails'.

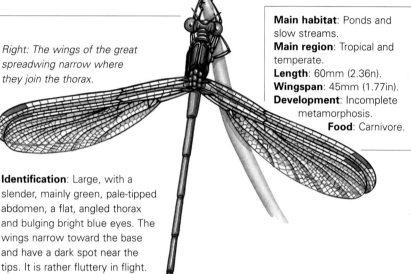

Right: The wings of the great spreadwing narrow where they join the thorax.

Identification: Large, with a slender, mainly green, pale-tipped abdomen, a flat, angled thorax and bulging bright blue eyes. The wings narrow toward the base and have a dark spot near the tips. It is rather fluttery in flight.

Main habitat: Ponds and slow streams.
Main region: Tropical and temperate.
Length: 60mm (2.36n).
Wingspan: 45mm (1.77in).
Development: Incomplete metamorphosis.
Food: Carnivore.

OTHER SPECIES OF NOTE

Familiar Bluet *Enallagma civile*
This is a common damselfly from southern Canada southward. It breeds in ponds, rivers, swamps and saltmarshes. It is small and dainty with an abdomen (male) striped pale blue and black and transparent wings. The female is less colourful.

Desert Firetail *Telebasis salva*
A splendid tiny damselfly that is fairly common in the south-west of the USA and ranges south to northern South America. Only about 25mm (1in) long, it is a bright fiery red in colour (male), all over its abdomen, thorax and eyes. The female is rather paler.

American Rubyspot *Hetaerina americana*
This damselfly is found in southern Canada and most of the USA. The abdomen is brown with yellow longitudinal stripes. The most striking features are the shiny red upper thorax and the deep red wing bases (male). The female is a shiny bronze-green. They are on the wing from June to September and prefer sunny riverbanks.

Common Spreadwing *Lestes disjuncta*
This common large damselfly is found from Alaska and Canada south to Florida and California. It breeds in weedy ponds and slow streams and flies from July to October. The thorax is browny-black with a central yellow stripe and the wings are relatively short, narrowing at the base.

Ebony Jewelwing

Calopteryx maculata

This is one the finest of the damselflies, and well-named, for its body and wings have a dark, metallic sheen, making it instantly recognizable as it flutters around streamside vegetation in mid-summer, mainly in eastern North America. When at rest the wings are folded together above the body. The abdomen and the large eyes are a shiny blue-green.

Identification: The wings of this damselfly are broad and rounded and, in the male, are deep ebony black in colour. The female also has dark wings, but usually rather smoky and with a small white spot. The narrow abdomen is shiny blue-green in the male and dark brown in the female.

Right: The dark, almost black, broad wings are characteristic of the male. Note the rather hunched thorax of this species.

Main habitat: Streams and rivers, often near trees.
Main region: Temperate.
Length: 45mm (1.77in).
Wingspan: 60mm (2.36in)
Development: Incomplete metamorphosis.
Food: Carnivore.

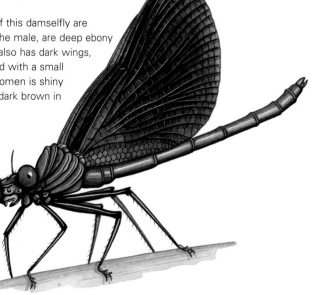

COCKROACHES AND TERMITES

Cockroaches (Blattodea), often called roaches for short, are rather low-slung insects with flattened, oval bodies and rather a leathery texture, some being quite shiny. They live in crevices and mainly emerge to feed at night on organic scraps. Most are very swift runners. Termites (Isoptera) are social insects, living in colonies. Some species damage timber and are considered pests.

American Cockroach

Periplaneta americana

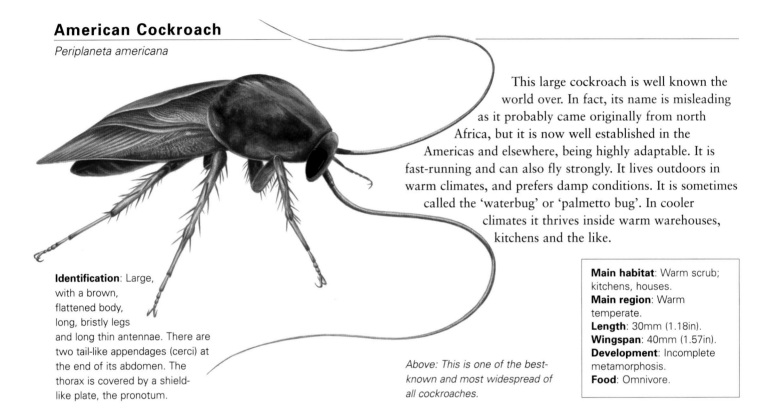

This large cockroach is well known the world over. In fact, its name is misleading as it probably came originally from north Africa, but it is now well established in the Americas and elsewhere, being highly adaptable. It is fast-running and can also fly strongly. It lives outdoors in warm climates, and prefers damp conditions. It is sometimes called the 'waterbug' or 'palmetto bug'. In cooler climates it thrives inside warm warehouses, kitchens and the like.

Identification: Large, with a brown, flattened body, long, bristly legs and long thin antennae. There are two tail-like appendages (cerci) at the end of its abdomen. The thorax is covered by a shield-like plate, the pronotum.

Above: This is one of the best-known and most widespread of all cockroaches.

Main habitat: Warm scrub; kitchens, houses.
Main region: Warm temperate.
Length: 30mm (1.18in).
Wingspan: 40mm (1.57in).
Development: Incomplete metamorphosis.
Food: Omnivore.

Death's Head Cockroach

Blaberus craniifer

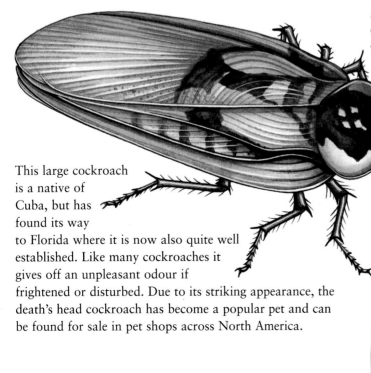

Left: The death's head cockroach is large, with a rather shiny body and wings.

Main habitat: Scrub; houses.
Main region: Tropical.
Length: 40–58mm (1.57–2.28in).
Wingspan: 60mm (2.36in).
Development: Incomplete metamorphosis.
Food: Omnivore.

This large cockroach is a native of Cuba, but has found its way to Florida where it is now also quite well established. Like many cockroaches it gives off an unpleasant odour if frightened or disturbed. Due to its striking appearance, the death's head cockroach has become a popular pet and can be found for sale in pet shops across North America.

Identification: It has a brown-and-cream sturdy body, with long wispy antennae. However, its most distinctive feature is the pattern on the pronotum (the head-shield part of the thorax) that looks a little like a skull or mask, giving this cockroach its common name.

Drywood Termite

Incisitermes spp.

Main habitat: Dry scrub.
Main region: Subtropical, tropical.
Length: 4mm (0.16in).
Wingspan: 10mm (0.39in) (winged adult).
Development: Incomplete metamorphosis.
Food: Herbivore.

This kind of termite (family Kalotermitidae) makes its colonies inside wood, and an infestation may turn the timber gradually into a dust-like powder, with damaging results to wooden buildings. There are many species in the southern USA and in Central and South America and about 350 species worldwide. In parts of South America they also damage the woody parts of certain crops such as coffee. Winged adults are produced in the autumn and these swarm on hot days and fly off to infect fresh timber.

Identification: These termites are mostly rather small. The workers are creamy white, with a grub-like appearance. The soldier caste is larger, with a yellow body and an orange-brown head armed with powerful jaws.

Right: The illustration shows a winged form of the adult termite.

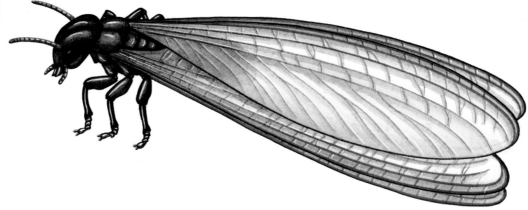

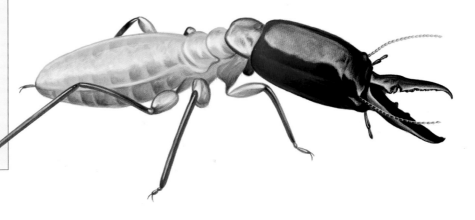

OTHER SPECIES OF NOTE

Colossus Cockroach *Megaloblatta longipennis*
This magnificent cockroach is one of the giants of the insect world, reaching a length of about 90mm (3.54in) and with a wingspan an amazing 18cm (7in). The body is orange-brown and the wings are long and rather broad. It is found mainly in tropical Central America and northern South America, notably in Colombia.

Green Banana Roach *Panchlora nivea*
This is one of the prettiest of all cockroaches. The nymphs are brown like most cockroaches, but the adults turn a bright, vivid green. They are active, flying readily and climbing well. This cockroach is found in Cuba, Mexico, Texas and Florida.

Brown-hooded Cockroach
Cryptocercus punctulatus
This small cockroach is found in highland regions of North America, including the Appalachian mounain range. It is rather unusual in living, termite-like, in social family groups in rotting wood. It has a rather rounded, shiny brown body.

Formosan Subterranean Termite
Captotermes formosanus
This is a species of termite that has become established in the southern USA, notably in southern California and also around New Orleans, where it attacks wooden buildings. It is an introduced species, which originated from the island of Taiwan (formerly known as Formosa, hence its common name).

Dampwood Termite

Zootermopsis spp.

These are regarded as rather primitive termites as their colonies consist of only two castes: soldiers and reproductive adults. There is no worker caste, unlike in many other termites. Their colonies can be very large, with up to 3,500 individuals and they prefer moist wood, such as logs or rotting stumps. They sometimes attack wooden fence posts and the like and can cause considerable damage to timbers in humid or damp locations. They are found mainly in western North America, although a few species occur farther east.

Main habitat: Damp wood.
Main region: Temperate.
Length: 15mm (0.59in).
Wingspan: 55mm (2.17in) (winged adult).
Development: Incomplete metamorphosis.
Food: Herbivore.

Identification: The young stages (nymphs) are pale and grub-like, but instead of remaining in this state these grow into either soldiers or reproductives.

Below and right: The soldier caste has large pincer-like jaws.

MANTIDS AND EARWIGS

Mantids (Mantodea) are mainly tropical. The group contains mostly quite large carnivorous species with claw-like grasping front legs used for catching their prey. There are about 2,000 species found throughout the warmer regions, of which about 20 are found in North America, with many more in South America, especially in the tropics. Earwigs (Dermaptera) number some 1,900 species.

Dead-leaf Mantis

Acanthops falcata

Many mantids are camouflaged, either by colour or by shape, or both. This remarkable mantis is so cryptically coloured and shaped that it easily escapes detection among the leaves of the tropical forests of South America. The male is fully winged and can fly, whereas the female's wings are unusable. Like most mantids it sits still and waits for prey to come within striking range.

Identification: The young have very long, rather curvy abdomens. They moult every couple of weeks and reach maturity after about six moults. The adults are amazing, with brown shiny wings that are flattened and crumpled, with many veins, making them look very like a dead leaf.

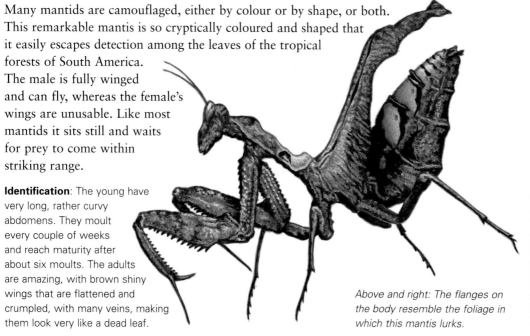

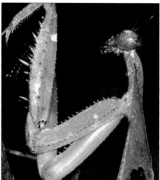

Above and right: The flanges on the body resemble the foliage in which this mantis lurks.

Main habitat: Forest.
Main region: Tropical.
Length: 50mm (2in).
Wingspan: 60mm (2.36in) (male).
Development: Incomplete metamorphosis.
Food: Carnivore.

Carolina Mantis

Stagmomantis carolina

This rather dainty mantis is quite common in eastern North America, west to Utah. The male flies readily and is sometimes attracted to lights at night. The female, however, is flightless and tends to spend much of her time sitting on twigs or flowers. This species consumes many smaller insects including garden pests and is therefore sometimes used as a biological control. It is often kept as a pet and is easy to feed and breed.

Identification: The male is a bright apple-green colour, the female rather speckled and grey-green or grey. Females can grow to about 70mm (2.76in), males usually only to about 50mm (2in). The wings of the male are functional, but those of the flightless female are small and bud-like.

Main habitat: Woodland.
Main region: Temperate.
Length: 70mm (2.76in).
Wingspan: 60mm (2.36in) (male).
Development: Incomplete metamorphosis.
Food: Carnivore.

Below: Often seen on flowers, the male is bright green.

Ring-legged Earwig

Euboriella annulipes

Main habitat: Leaf litter, roots, etc.
Main region: Worldwide.
Length: 15mm (0.59in).
Development: Incomplete metamorphosis.
Food: Omnivore.

Identification: The nymphs look very like the adults, with a dark brown head and abdomen, and yellow-grey pronotum. The whitish legs have a dark ring around the femur – hence the common name. They may have one or two bands. Adults are dark brown and wingless and the females are a little larger than the males.

This earwig is widespread in the Americas, Hawaii and elsewhere, although it may have originated in Europe. For example, it is the most common earwig pest in Florida. However, it is also useful in the garden as it consumes other insect pests. In warm regions it lives outdoors but can also survive happily inside greenhouses in colder climates. Under warm conditions it may have two or more generations each year.

Above right: Although generally useful, this species sometimes damages plant roots.

OTHER SPECIES OF NOTE

Grizzled Mantis *Gonatista grisea*
This American mantis is rather small and easily overlooked, especially as its camouflage is rather effective. The wings, body and legs are patterned in a mottled creamy-brown, which blends well with leaves and bark. Only the male is winged. It is found from Cuba and Florida north to South Carolina.

Minor Ground Mantis *Litaneutria minor*
This mantis is even smaller, at only about 25mm (1in). An active species, it clambers about in grasses and other low vegetation and is found from Florida west to the Great Plains.

Linear Earwig *Doru lineare* (Left)
This earwig is common in North America. The abdomen is shiny and dark brown and the head bright chestnut. The hindwings are folded and protected by the creamy-yellow elytra (hardened forewings). The legs are also yellowish in colour.

Riparian (Tawny) Earwig
Labidura riparia
This is another species that lives in coastal habitats, and is also found on riverbanks, mainly in the southern and eastern USA. It can protect itself by exuding a nasty-smelling chemical. It has a yellow-brown body and long cerci.

Maritime Earwig

Anisolabis maritima

This earwig is rather unusual in that it lives close to the sea, its usual habitat being in the tidal zone of the shore. It has a very wide distribution in many parts of the world and may, in fact, be originally native to Europe and Asia, and was probably spread by ships. However, it is now common on both coasts of North America. Maritime earwigs hide under driftwood and other debris, and emerge at night to hunt for other small insects and crustaceans. The females look after their eggs and young in an underground nest.

Identification: A large wingless earwig, the males are usually larger than the females. Males have thick rather curved cerci; those of the female are straighter. The body is shiny and dark brown, broadening toward the rear. The rather sturdy legs are pale red.

Main habitat: Seashore.
Main region: Temperate.
Length: 30mm (1.18in).
Development: Incomplete metamorphosis.
Food: Carnivore.

Above: This coastal earwig forages on the shore at night, feeding mainly on smaller invertebrates.

GRASSHOPPERS

Grasshoppers are mainly very active insects with powerful hind legs for jumping and are known for their often loud 'songs' produced by rubbing movements either of their legs or wings. There are about 22,000 species, showing many variations on the basic body theme. Most grasshoppers belong to the large family Acrididae with about 7,000 species (630 in North America).

Horse Lubber

Teniopoda eques

The lubber grasshoppers are some of the largest of the North American grasshoppers. They have plump, bulky bodies and a tendency to exude a nasty-smelling fluid as a defence when threatened. Many, including this species, raise their wings as a display, revealing bright colours on the hindwings. Horse lubbers are found mainly in the south-west of the USA and may march about in large numbers.

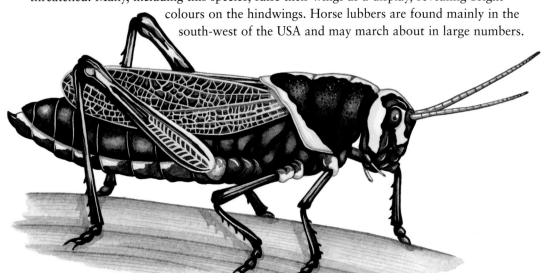

Main habitat: Grassland.
Main region: Subtropical.
Length: 60mm (2.36in).
Wingspan: 30mm (1.18in).
Development: Incomplete metamorphosis.
Food: Herbivore.

Identification: Large, with rather short antennae and long, powerful hind legs. The body is dark blackish or greenish-brown and there are bright yellow markings on the head and sides. The head is rounded in outline with orange-and-black striped antennae. The forewings have a greenish net-like pattern and the delicate hindwings are pink, and visible in flight.

Left: In the nymph stage, horse lubbers often move around in groups.

Green Fool Grasshopper

Acrolophitus hirtipes

This common species has a wide range in North and Central America, from Canada south to Mexico. It has rather a weak, fluttery flight, with its long legs dangling, behaviour that perhaps gave rise to its common name.

Identification: Medium-sized grasshopper with a pale green background colour, overlain with paler stripes. The head is sharply angled slanting downward and it has a characteristic hump on its back, below the neck. In the south of its range it may have leopard-like spots. The hindwings are yellow with a broad black band.

Main habitat: Grassland.
Main region: Temperate.
Length: 40mm (1.57in).
Wingspan: 40mm (1.57in).
Development: Incomplete metamorphosis.
Food: Herbivore.

Left: This species has very long legs, which it trails during flight.

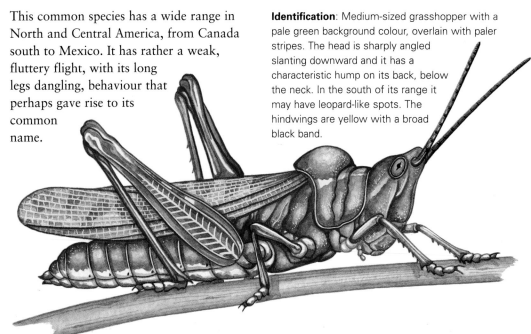

Carolina Grasshopper

Dissosteira carolina

Main habitat: Grassland and wasteland.
Main region: Temperate.
Length: 35mm (1.38in).
Wingspan: 60mm (2.36in).
Development: Incomplete metamorphosis.
Food: Herbivore.

Sometimes called the Carolina locust, this is a large species, common in much of Canada and the USA. It is not very fussy as to habitat, being found in grassland, along roads and tracks and also in quarries. It is extremely well camouflaged when resting, but very easily spotted in flight when the large, colourful hindwings are prominent. The males indulge in hovering display flights. It is sometimes a pest of crops such as sweetcorn, cotton and potatoes.

Identification: Brown and locust-like, with short antennae and mottled grey-brown forewings and body. The hindwings are broad and brightly patterned – mainly black, with pale yellow edging.

Right: The dark, pale-edged hindwings are obvious in flight.

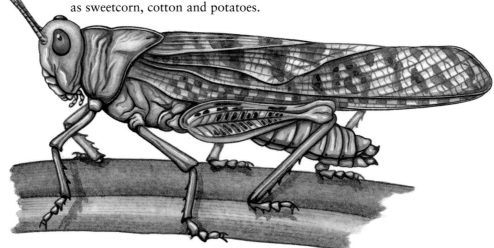

OTHER SPECIES OF NOTE

South-eastern Lubber *Romalea guttatus*
This is a large grasshopper found mainly in the south and east of the USA, south to Florida. It is about 60mm (2.36in) long and bright orange in colour. Its hindwings are bright red and are used to flash as a warning. It can also produce an unpleasant liquid from its thorax.

Toothpick Grasshopper *Achurum* spp.
This genus is unusual in that the body is pencil-thin and both head and wingtips are extended and narrowly pointed, hence the common name. The hindwings are very small. Toothpick grasshoppers are found from the southern USA south into Mexico and Central America.

Blue-winged Grasshopper *Leprus intermedius*
This is another cryptically coloured and patterned grasshopper, until it flies, when its colourful hindwings are revealed. These are mainly bright blue, with a broad black margin. This species inhabits mainly rocky and grassy sites in the south-west USA.

Pallid-winged Grasshopper
Trimerotropis pallidipennis
This is a destructive species found from south-west Canada, south to Argentina, in grassland, deserts and on wasteland. Periodically its numbers explode and the young nymphs especially can do serious damage to crops. A strong flier, it sometimes appears in large numbers around street lights. It can be very common, especially in desert habitats.

Rainbow Grasshopper

Dactylotum bicolor

This is one of the prettiest of all grasshoppers and also quite widespread, occurring across much of the western USA and north Mexico. The adults are most active from June through October and feed on a range of plants, though they are not a pest of crops. They can be very abundant. This species is also known as the pictured grasshopper or barber-pole grasshopper.

Main habitat: Poor grassland.
Main region: Temperate, subtropical.
Length: 35mm (1.38in).
Wingspan: 10mm (0.39in).
Development: Incomplete metamorphosis.
Food: Herbivore.

Identification: Unmistakable short-winged grasshopper. Small, with short antennae and long, powerful hind legs. Its most striking feature is its bright coloration – the whole body is patterned with alternating bands of red, blue, white and yellow, hence the common name.

Below: The bright warning colours of this grasshopper make it look freshly painted.

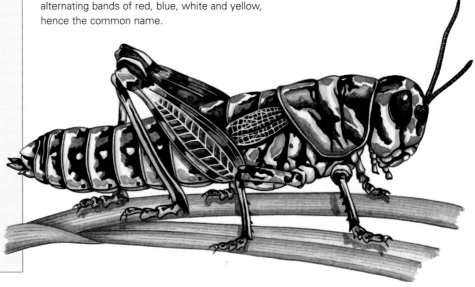

CRICKETS

The largest cricket families are the bush crickets (Tettigoniidae) with some 6,000 species, and the true crickets (Gryllidae) with about 4,000 species. In America the members of the Tettigoniidae are usually called katydids; this name comes from the call of one species, the true katydid. There are many species from both families throughout the Americas.

True Katydid

Pterophylla camellifolia

This familiar large bush cricket takes its common name from the repeated calls which sound a little like 'katy did, katy didn't'. Both sexes call to each other, mainly on hot summer nights, often from high in the branches of trees, so they are not always easy to spot. Both sexes can fly, but rather weakly, and they prefer to clamber. Common in North America, mainly south of the Great Lakes.

Identification: Like most bush crickets, it is leafy green in colour, and well camouflaged among foliage. The forewings are broad and leaf-like, adding to the cryptic effect, and the antennae are long and narrow. The female has a large, sickle-shaped ovipositor.

Main habitat: Trees and bushes.
Main region: Temperate.
Length: 45mm (1.77in).
Wingspan: 40mm (1.57in).
Development: Incomplete metamorphosis.
Food: Herbivore.

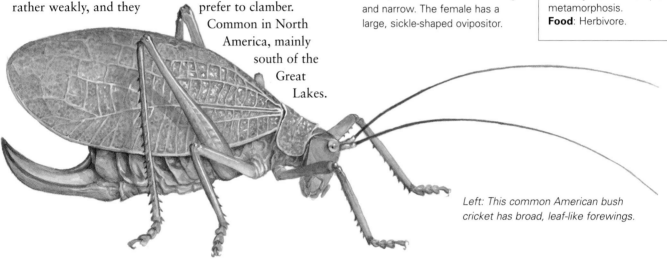

Left: This common American bush cricket has broad, leaf-like forewings.

Camel Cricket

Tropidischia xanthostoma

This giant among crickets inhabits dark places, such as wells, under bridges, caves and in shady damp forests in the Pacific coastal region from California north to British Columbia. The common name comes from its hump-backed appearance. This is a noctural insect that scavengers for its food.

Main habitat: Dark, damp sites.
Main region: Temperate.
Length: 30mm (1.18in) (body only).
Development: Incomplete metamorphosis.
Food: Herbivore/omnivore.

Left: The spider-like camel cricket lives in damp, dark habitats.

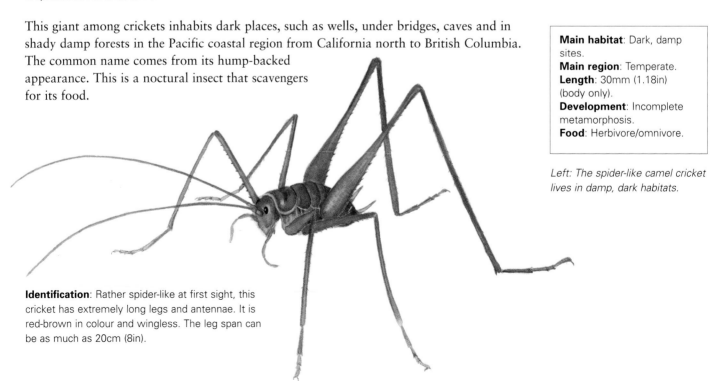

Identification: Rather spider-like at first sight, this cricket has extremely long legs and antennae. It is red-brown in colour and wingless. The leg span can be as much as 20cm (8in).

Grig

Cyphoderris spp.

Grigs, or hump-winged grigs, are sturdy crickets with very plump bodies and reduced wings. Like many crickets they have a characteristic call, in this case usually a high-pitched trill. They are most active at night and tend to spend the day sheltering under stones. They are found mainly in the Pacific north-west, in evergreen forests.

Below: Grigs are more often heard than seen and emerge mainly at night.

Identification: Very fat and dumpy, with tiny or reduced wings, medium-long antennae and sturdy legs. The body is rather shiny and grey-brown and speckled.

Main habitat: Trees.
Main region: Temperate.
Length: 30mm (1.18in).
Development: Incomplete metamorphosis.
Food: Herbivore.

OTHER SPECIES OF NOTE

Ommatopia pictifolia
This predominantly rainforest-dwelling insect is a remarkable leaf-mimic. A dingy-brown bush cricket, it is found in tropical South America, notably in Brazil. Its entire body resembles a dead leaf, especially its forewings, which even have leaf-like veins and discoloured patches. When it flies it displays coloured eye-spots on the hindwings.

Spiny Katydid *Panacanthus cuspidatus*
This remarkable katydid lives in the rainforests of Central and South America. Its mainly green body is covered with sharp spines which form a very effective protective barrier against the predators which might otherwise eat it, such as monkeys and bats. The male has a high-pitched song to attract females.

Snowy Tree Cricket *Oecanthus fultoni*
This small cricket is rather slim and bright green and looks like a green lacewing. It is found in most of North America and is sometimes called the 'thermometer cricket', as its rate of chirping speeds up with increasing temperature.

Camel Cricket *Ceuthophilus* spp.
Camel crickets are large, hump-backed insects with very long, narrow antennae and long, spiny, powerful hind legs. They are usually found in damp habitats such as caves and cellars, and emerge at night to hunt for their prey. This is a very large genus with dozens of different species.

Jerusalem Cricket

Stenopelmatus spp.

This group of wingless crickets contains at least 40 species in North and Central America. They are some of the strangest of all crickets, with their rather bloated, dumpy shape and nocturnal burrowing lifestyle. In North America they are found mainly in the west, especially in California and Mexico, where they often live in sand-dunes. They are related to the wetas of Australia and New Zealand.

Right: Despite their nickname of 'potato bug', these insects feed on dead organic matter and other insects.

Main habitat: Dry sites.
Main region: Subtropical.
Length: 45mm (1.77in).
Development: Incomplete metamorphosis.
Food: Carnivore.

Identification: Large and bulky, with a plump body and legs, a large head and long, narrow antennae. The head has powerful jaws which can inflict a nasty bite. The head, legs and thorax are orange-brown and the abdomen is darker.

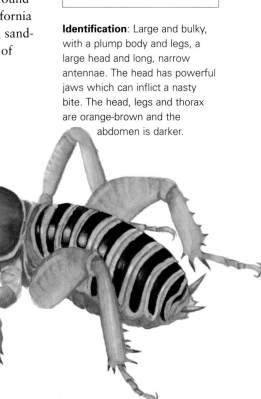

STICK AND LEAF INSECTS

Stick and leaf insects (Phasmida) are among the world's most remarkable invertebrates, with amazing camouflage. Most of the 2,500 or so known species live in warm or tropical regions. Stick insects have long, thin bodies and resemble twigs. North America has about 35 species, most of which are wingless and hence rely entirely on their camouflage to escape predators.

Peruvian Fern Stick Insect

Oreophoetes peruana

This pretty wingless stick insect from South America is unusual in a number of ways. It is brightly coloured, as a warning of its unpalatable taste; it eats ferns; and it can discharge a foul-smelling liquid from its thorax if disturbed. The latter has been shown to deter spiders, cockroaches, ants and frogs. Despite this habit they make good pets and are easy to feed.

Identification: The young are yellow with black stripes, the adult females are orange-yellow and the adult males are an amazing bright red. The antennae are unusually long and the legs thin and delicate.

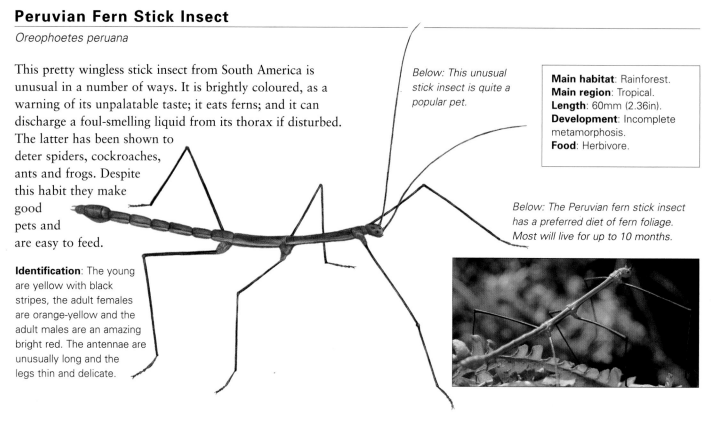

Below: This unusual stick insect is quite a popular pet.

> **Main habitat**: Rainforest.
> **Main region**: Tropical.
> **Length**: 60mm (2.36in).
> **Development**: Incomplete metamorphosis.
> **Food**: Herbivore.

Below: The Peruvian fern stick insect has a preferred diet of fern foliage. Most will live for up to 10 months.

Giant Walkingstick

Megaphasma dentricus

This large wingless stick insect is a familiar common species in North America, especially in the south-east and mid-west. In fact it is North America's longest insect. Found on a wide range of plants in woods and grassland, it is also rather fond of grape vines. Males are rare and the females produce more than 100 eggs, dropped on to woodland floor.

Identification: The body is long and slender, the male being smaller than the female; size range is about 6.5–10.8cm (2.5–4in). The antennae are long and the adults are green or reddish-brown, with yellowish legs, and sometimes have a red stripe along the back.

> **Main habitat**: Trees and shrubs.
> **Main region**: Warm temperate.
> **Length**: 10cm (4in).
> **Development**: Incomplete metamorphosis.
> **Food**: Herbivore.

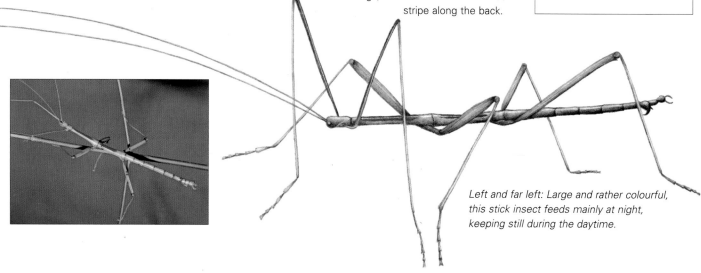

Left and far left: Large and rather colourful, this stick insect feeds mainly at night, keeping still during the daytime.

Northern Walkingstick

Diapheromera femorata

Main habitat: Forest.
Main region: Temperate.
Length: 90mm (3.54in).
Development: Incomplete metamorphosis.
Food: Herbivore.

This is the most common stick insect in North America, mainly found in the east and south, as far north as south-east Canada. Although not usually very easy to see, sometimes its populations explode and then it is not hard to find. The favoured habitat is broad-leaved or mixed forest where it feeds on many species including oak and black cherry.

Identification: The body is very thin and twig-like, and light brown in colour (rather greenish in the female). The males are smaller than the females. The legs are also very thin, especially the outer sections. The antennae are very long.

Right and far right: This is the stick insect most likely to be spotted in eastern North America.

OTHER SPECIES OF NOTE

Western Short-horned Walkingstick
Parabacillus hesperus
As its name implies, this stick insect has short antennae (horns), but its legs are long and spindly like those of most other stick insects. The western short-horned walkingstick favours dry habitats and is mainly found in the south and west of the USA, including Arizona.

Colorado Short-horned Walkingstick
Parabacillus coloradus
This stick insect is closely related to the above species and grows to about 70mm (2.76in) long. It has a brown body with pale stripes on its head. As well as Colorado it is found in a range of southern states including Texas and New Mexico, mainly in upland grassy sites.

Timema spp.
These are small stick insects that look at first sight more like caterpillars or earwigs than true stick insects. They are usually green with stout bodies, rather short, stumpy legs and short antennae. They are found mainly in the south where they feed on the leaves of shrubs such as California lilac (*Ceanothus*).

Trinidad Twig *Ocnophiloidea regularis*
This small stick insect comes from the Caribbean island of Trinidad. It is cream-coloured with black stripes along its sides. The Trinidad twig is quite often kept as a pet. It can be fed on the leaves of a wide variety of plants including bramble, rose and hawthorn.

Two-striped Walkingstick

Anisomorpha buprestoides

This stick insect is most common in the south-east, especially in Florida, west to Texas. It feeds on the leaves of trees and bushes (including roses) and is most common in the autumn, which is when the females deposit their eggs in the soil. It has a number of alternative common names including 'devil-rider' and 'musk-mare'. The first refers to the fact that the male often rides on the back of the female when mating, and the second to the fact that they can secrete a musky-smelling acidic spray if disturbed. This defensive compound is squirted from glands in the thorax region of the back and is painfully irritating to the eyes and the mucous membranes.

Main habitat: Trees and bushes.
Main region: Warm temperate.
Length: 40mm (1.57in) (male), 70mm (2.76in) (female).
Development: Incomplete metamorphosis.
Food: Herbivore.

Identification: The body is broad (for a stick insect) and brown with long black stripes. The antennae are quite long. It may be found in aggregations with several pairs or individuals in darkened sights such as under bark. It is a noctural insect.

Left: Females of the species lay up to 10 eggs in holes in the ground.

BUGS

This group are the true bugs (Hemiptera). True bugs are a large group, with more than 82,000 species found worldwide. They are the most varied order of insects with incomplete metamorphosis. Both nymphs and adult bugs have mouthparts that are modified for piercing and sucking, whether this be for feeding on sap from plants or fluid from animals.

Periodical Cicada

Magicicada spp.

These bugs are among the most remarkable of all insects. As well as being large and noisy, like most cicadas, they appear in numbers on regular cycles of either 13 or 17 years. No-one knows why they evolved this bizarre strategy – possibly as some kind of predator or disease avoidance. The northern and eastern species such as *M. septendecim* are on a 17-year cycle, while the southern and central species such as *M. tredecim* are on the 13-year cycle. The emergences are local and so do not occur simultaneously throughout their range. The females lay their eggs in the twigs of trees, and when their numbers are large this can cause much damage. The nymphs live underground before emerging years later. The adults hatch in summer and may be a nuisance with their loud songs and also when they die in huge numbers.

Right: The large cicadas in this genus undergo periodical mass hatchings.

Identification: Periodical cicadas are large bugs with transparent wings, dark bodies and bright red eyes. The wings are shiny and have net-like red veins. Like most cicadas, the front legs are more powerful than the others and rather claw-like, helping them to cling firmly to the bark of trees.

Main habitat: Woodland.
Main region: Temperate.
Length: 45mm (1.77in).
Wingspan: 70mm (2.76in).
Development: Incomplete metamorphosis.
Food: Herbivore.

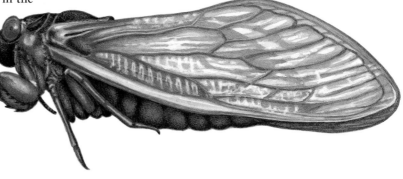

Hieroglyphic Cicada

Neocicada hieroglyphica

This small cicada is common in the south-east USA where it is associated with trees, mainly oaks. The song of the male is heard from early spring and is often described as a sequence of high-pitched whining calls. The males call in small groups, from morning until evening. As with all cicadas, the amazingly penetrating 'songs' are produced from vibrations of drum-like organs on the sides of the thorax.

Identification: A pretty species, with a greenish-yellow body, patterned in script-like scribbles (hence the name). Some individuals are creamy-yellow in background colour.

Main habitat: Trees.
Main region: Warm temperate.
Length: 30mm (1.18in).
Wingspan: 50mm (2in).
Development: Incomplete metamorphosis.
Food: Herbivore.

Left: This is a medium-sized cicada and one of the earliest in the year to start calling.

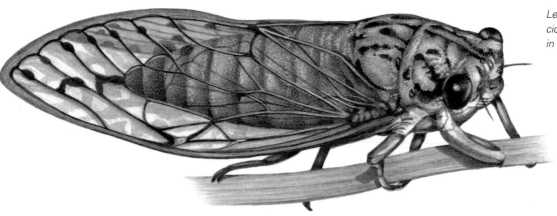

Peanut-head Bug

Fulgoria laternaria

Main habitat: Rainforest.
Main region: Tropical.
Length: 8cm (3in).
Wingspan: 12cm (4.5in).
Development: Incomplete metamorphosis.
Food: Herbivore.

Identification: The extended head is prominent and shaped like the shell of a peanut. The huge nose is hollow. At rest this bug is cryptically coloured with bark-like patterns, but the hindwings have bright shiny eye-spots to confuse any predator. Peanut-head bugs are able to produce noise by tapping their hollow nose on tree bark, for example.

This tropical bug, from the Amazon rainforests, is weird indeed. It is related to plant-hoppers but looks rather different, with its bizarre head extension. Members of this group are also known as 'lantern bugs', as their heads were once wrongly reported to glow in the dark. Another name for this species is 'alligator bug' as the head looks a little like that of a crocodilian. It feeds by sucking the sap from trees. One theory for its appearance is that the shape mimics that of a lizard. When attacked it can release a nasty chemical.

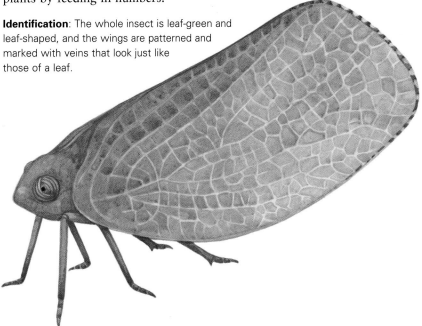

Below: This tropical insect is surely one of the strangest of all bugs.

OTHER SPECIES OF NOTE
Cactus Dodger *Cacama valvata*
This cicada lives in the south-west of the USA, in dry sites, often on cactus plants. Its common name comes from its habit of flitting from one plant to another as it sings. Most other cicadas buzz for hours on end from the same spot.

Dog Day Cicada *Tibicen* spp.
The many cicadas in this genus are commonly heard in the 'dog days' of high summer and are widespread in North America. Unlike the periodical cicadas, they emerge regularly each year. However, as with most cicadas, the nymphs live underground for more than a year.

Two-lined Spittlebug *Prosapia bicincta*
This is a common spittlebug, with a mainly eastern distribution in North America. Spittlebug nymphs protect themselves by exuding a spit-like foam of waste and mucus. They leap with great agility and are also called frog-hoppers.

Thorn Bug *Umbonia crassicornis* (Below)
These bugs belong to the group known as tree hoppers. Many of them are excellent mimics of thorns, and this species is one of the best. It has a tall, green pointed extension on its thorax and it sits tight against a twig, looking amazingly

like one of the plant's own thorns. The effect is heightened when a group of bugs sit close to each other.

Plant-hopper

Acanalonia spp.

These bright green plant-hoppers are often extremely well camouflaged, and they can be quite hard to spot as they rest on plant twigs, looking rather like buds or small leaves. The wings are held angled like a roof over the body. Their resemblance to leaves has presumably evolved under pressure of predation by insect-eating animals such as birds. There are many rather similar species found on a range of host plants. The adults appear in summer and sometimes damage plants by feeding in numbers.

Identification: The whole insect is leaf-green and leaf-shaped, and the wings are patterned and marked with veins that look just like those of a leaf.

Main habitat: Trees and shrubs.
Main region: Temperate.
Length: 10mm (0.39in).
Wingspan: 15mm (0.59in).
Development: Incomplete metamorphosis.
Food: Herbivore.

Below: This bug is an excellent leaf-mimic as it sits at the side of a green shoot.

Pea Aphid

Acyrthosiphon pisum

This very small bug is one of a large family of plant-sucking insects that are serious pests of crops, in this case mainly of members of the pea family, including alfalfa. Like many aphids, it can reproduce at amazing speed and each female can produce up to eight nymphs each day, without the need of males, and up to 15 generations each year. Eventually a winged generation is produced and then the bugs can fly off to infect another field of crops. Males appear in the autumn. After mating, the females lay eggs that overwinter. The pea aphid causes large agricultural losses and is of major economic significance.

Main habitat: Fields, crops.
Main region: Temperate.
Length: 3mm (0.12in).
Wingspan: 5mm (0.19in).
Development: Incomplete metamorphosis.
Food: Herbivore.

Identification: Tiny and bright green, with (depending on the generation) transparent wings. The legs and antennae are thin and long.

Left: Like most aphids, the pea aphid can breed rapidly and damage plants.

Cochineal Scale

Dactylopius coccus.

Main habitat: Cactus leaves.
Main region: Tropical.
Length: 2–4mm (0.08–0.16in).
Wingspan: 3mm (0.12in) (male only).
Development: Incomplete metamorphosis.
Food: Herbivore.

Identification: The tiny, aphid-like adults are rarely seen. The scales are small, fluffy patches on the leaves of the cactus, reddening with the dye where they have been disturbed.

These tiny bugs are among the smallest found in the order Hemiptera. They are almost microscopic and the adults are rarely seen. The visible part is usually the scale itself, created by the insect as protection. In this case the scale is found on prickly pear cacti, notably in Mexico. This insect is the source of cochineal dye, a bright red chemical used to colour a range of products from drinks and sweets to cosmetics. The Aztecs used this as a dye and for painting. The dye itself is a chemical produced by the insects as a defence.

Right: The adult male has long tail filaments, but the adults are rarely seen.

Left: Scale damage can be extensive.

Giant Mesquite Bug

Thasus spp.

The bugs of this genus are large and quite colourful, especially the nymphs. They are found on the leaves and branches of mesquite trees (*Prosopis*), on which they feed, often with many bugs clustered close together. However, they do not seem to harm the trees to any great extent. They suck sap from the shoots and also attack leaves and seed pods. They range north from Mexico to Arizona and New Mexico.

Identification: Mainly red, white and grey. The adult has a long body and its chunky legs are patterned grey and red. The hind legs of the male have enlarged femurs. Their wings are black with yellow veins. The nymphs are very bright, with a fiery red body and white spots.

Main habitat: Mesquite trees.
Main region: Warm-temperate, tropical.
Length: 35–40mm/ (1.38in–1.57in).
Wingspan: 80mm (3.15in).
Development: Incomplete metamorphosis.
Food: Herbivore.

Left: Giant mesquite bugs sometimes come to lighted windows at night.

Harlequin Bug

Murgantia histrionica

Main habitat: Crops. **Main region**: Temperate, tropical. **Length**: 10mm (0.39in). **Wingspan**: 15mm (0.59in). **Development**: Incomplete metamorphosis. **Food**: Herbivore.

This bug has a wide range from Mexico north to New England. One of a group known as shield bugs or stink bugs, it is well named as its wings and thorax carry a bright pattern. It is a major pest of crops such as cabbages, sprouts, radishes and mustard, and also attacks tomatoes, beans and asparagus, among other plants. It sucks the sap from the plants so that they dry out, wilt and die.

Above: This bug does considerable damage to crops, notably cabbages.

Identification: Easily identified by the yellow-and-greenish-black or reddish-and-black wavy pattern on the thorax and wings. The nymph is similarly patterned, though wingless. Like other shield bugs, the body looks flat and shield-like in shape and the antennae are quite long. The timespan from egg to adult is approximately 50 to 80 days, with several nymph stages between, with each instar becoming more boldly coloured.

OTHER SPECIES OF NOTE

Giant Bark Aphid *Longistigma caryae*
This is an unusually large aphid, reaching more than 6mm (0.24in) in length. Its body has black spots and the wings and legs are also black. Its favoured habitat is the bark of a range of tree species. It occurs mainly in the southern and eastern USA.

Grape Phylloxera *Phylloxera vitifoliae*
This tiny, gall-forming bug is native to North America. It is notorious for having almost destroyed European vines, in particular in France. Vines in America had largely evolved resistance to it, but it was introduced to Europe by mistake. It produces blister-like galls on vine leaves.

Thread-legged Bug *Emesaya* spp.
These bugs bear a striking resemblance to stick insects but are nevertheless true bugs, related to assassin bugs. Like assassin bugs, they use their needle-like mouthparts to pierce their prey and suck out the body fluids. Their front legs have grasping claws with which they grab smaller insects.

Cotton Stainer *Dysdercus suturellus*
This is a colourful bug, sometimes called the 'red bug', due to its bright red body. Its other common name comes from the fact that it often leaves a stain on cotton, on which it feeds. Cotton stainers also feed on other crops including citrus. Tropical insects, they are also found in the southern states of the USA, especially southern Florida.

Florida Stink Bug

Euthyrhynchus floridanus

This is one of the larger stink bugs of the region, found mainly in the south-east USA, notably Florida, and in Central America. This species is a predator of other insects, using its piercing mouthparts to suck the juices from its prey. Unlike many of its pest relatives, this species is beneficial, feeding on many species that damage plants.

Main habitat: Crops, gardens. **Main region**: Tropical, temperate. **Length**: 12–17mm (0.47–0.67in). **Wingspan**: 18mm (0.71in). **Development**: Incomplete metamorphosis. **Food**: Carnivore.

Below: Large and colourful, this bug feeds mainly on other insects.

Identification: Males are about 12mm (0.47in) long, females up to 17mm (0.67in). The pattern above is distinctive. The head, thorax and elytra are dark, and these contrast with bright yellow-orange patches on the back and around the margin.

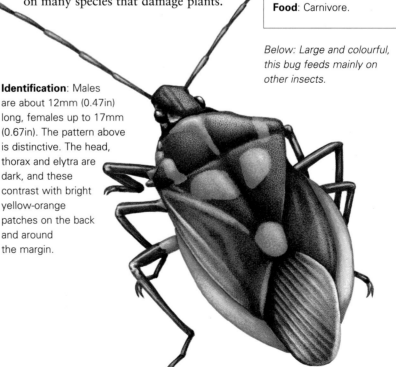

Large Milkweed Bug

Oncopeltus fasciatus

This large bug is widespread and common in North America. It is associated with its favoured plant, milkweed. It is sometimes seen in large numbers, with many individuals clustered close together, often on buildings. Like many brightly coloured insects, the colours warn predators that these bugs are not worth eating. They taste bad because their bodies store the toxins from the milkweed plants on which they feed, notably on the seed pods. The females lay up to 2,000 eggs on the milkweed plant, and the nymphs hatch after about a month. They are found from Brazil to Texas and Florida and as far north as Massachusetts.

Main habitat: Fields.
Main region: Temperate-tropical.
Length: 12mm (0.47in).
Wingspan: 30mm (1.18in).
Development: Incomplete metamorphosis.
Food: Herbivore.

Identification: Boldy patterned in orange-red and black, with long, dark legs and medium length antennae. Even the eggs are red, and the nymphs are a bright orange-red, but lack the dark wings of the adults.

Left and right: This bug gains protection from the poisons it stores from its food plant.

Chinch Bug

Blissus leucopterus

The chinch bug is a very small bug, well known mainly because of the damage it causes to crops such as cereals and sorghum, especially in the American Midwest. It also feeds on lawn grasses. The adults overwinter and become active in spring when they attack plant tissues, often at the base of the stem. There can be three generations each year and numbers can build up rapidly. In some areas a subspecies of this bug, the hairy chinch bug (*B. l. hirtus*), is quite common. Rather than damaging commercial crops, this form of the chinch bug attacks lawn grasses, leaving them very patchy and dry. The false chinch bug (*Nysius raphanus*), found mainly in western North America, is sometimes found inside houses, especially in hot weather. It can also cause damage to plants if present in large numbers.

Main habitat: Crops, grasses.
Main region: Temperate.
Length: 4mm (0.16in).
Wingspan: 6mm (0.24in).
Development: Incomplete metamorphosis.
Food: Herbivore.

Identification: The adult is small and mainly black, with silvery wings. The nymphs are orange at first, turning darker, and have a broad white stripe.

Left: The chinch bug is very small, but can cause severe damage to crops.

Giant Water Bug

Lethocerus americanus

Main habitat: Warm
temperate, tropical.
Main region: Central and
southern North America.
Length: 45–65mm
(1.77–2.56in).
Wingspan: 60mm (2.36in).
Development: Incomplete
metamorphosis.
Food: Carnivore.

Identification: Long, flat brown
body, the second and third pairs
of legs have a fringe of hairs and
are used for swimming. The
front pair are modified as pincers.
The wings are well developed.
These bugs often fly towards
light at night.

This fierce aquatic bug is one of the largest members of the
order. It spends its whole life in the water where it feeds on
a range of prey from other invertebrates to small fish and
frogs. The strong front legs are hooked at the tip and work
like pincers to grasp their prey, after which
the needle-like
mouthparts inject
enzymes that
dissolve the
internal
tissues. The
bug then
sucks out the
juices from
the victim. Giant
water bugs breathe
at the water's surface
through tubes at the tip of
their abdomen.

*Below: Giant water bugs
are some of the largest of all
bugs and can be rather scary
to encounter.*

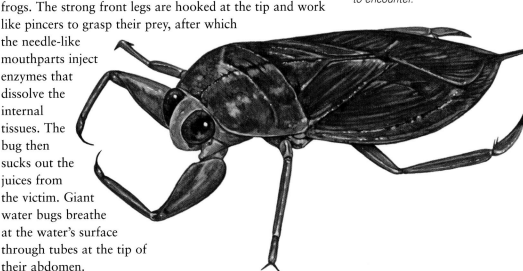

OTHER SPECIES OF NOTE

Damsel Bug *Nabicula subcoleoptrata*
This strange bug has short wings and its body
is shaped rather like an ant, which it mimics.
It is found in grassland and is an active predator
of aphids, caterpillars and other small
arthropods. Damsel bugs are shiny and black
with yellow legs.

Wheel Bug *Arilus cristatus*
This interesting large bug belongs to the family
known as assassin bugs, named for their habit
of attacking and killing other insects, even those
larger than themselves. The bite is more painful
than a bee or wasp sting. The adult wheel bug
has a serrated wheel-like crest on the thorax and
grows to about 30mm (1.18in). The nymph has a
bulging red abdomen.

Toad Bug *Gelastocoris* spp.
This small bug is very well named. It is squat,
brown, lives in the muddy shores of ponds and
streams and its skin is bumpy. Its head is also
rather toad-like, with large, bulging eyes. Toad
bugs creep about at the edge of the water
waiting to catch and feed on smaller insects.

Water Scorpion *Ranatra fusca*
Also known as a water stick insect, this aquatic
bug has a thin, streamlined body and long,
spindly legs. It grasps its prey using its claw-like
front legs and breathes at the surface through
long tubes at the tip of its body. Water scorpions
are quite common in slow-flowing rivers
but are hard to spot as they move slowly and
hide in the vegetation.

Common Water Strider

Gerris remigis

This is one of several bugs that have
perfected the art of living on the surface of
the water. Their thin legs have complex
rows of microscopic hairs that allow them
to rest on the surface of the water without
breaking the surface tension. They can
therefore skate along, and can even jump
and land again without
getting wet. They use their
front legs
to catch
prey by
suddenly
darting
forward
in a
surprise
attack.

Main habitat: Ponds and
streams.
Main region: Temperate.
Length: 12mm (0.47in).
Wingspan: None.
Development: Incomplete
metamorphosis.
Food: Carnivore.

*Above: Water striders can
glide with ease across the
water surface.*

Identification: The body is dark brown
and narrow, and the middle and back
pairs of legs are long and stilt-like; the
wings are absent. Their eyes are large
and bulging, giving them keen vision.
The shorter front legs are claw-like and
used to catch prey.

CADDIS FLIES, LACEWINGS, ANTLIONS AND SCORPION FLIES

The aquatic larvae of caddis flies are as well known as the adults. North America has about 1,260 species, mainly in mountainous regions. Lacewings and antlions have net-veined wings. Scorpion flies are named for the scorpion-like tail of the male of some species.

Large Caddis Fly

Hydatophylax argus

This rather big caddis fly is considered by many to be one of America's most beautiful species. It belongs to a large family of caddis flies that are most frequent in northern latitudes where the larvae build cases and live in the mud of rivers, streams and ponds, feeding on plant material.

Below: This unusually large caddis fly breeds in cool rivers.

Identification: The body and upper legs of the adult are bright yellow and the long antennae are dark, as are the prominent eyes. The wings are its most attractive feature – large and patterned in grey and silver in a lace-like network of veins with a group of pearly spots in the middle.

Main habitat: Rivers.
Main region: Temperate.
Length: 25mm (1in).
Wingspan: 50mm (2in).
Development: Complete metamorphosis.
Food: Larva herbivorous; adult does not feed.

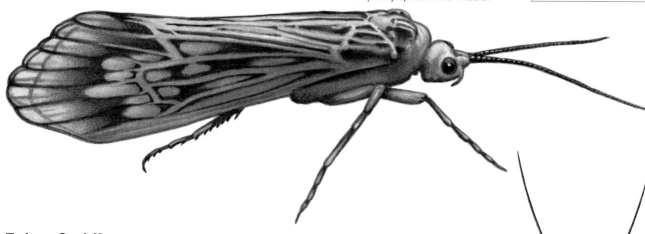

Zebra Caddis

Macrostemum zebratum

Main habitat: Flowing fresh water.
Main region: Temperate.
Length: 17mm (0.67in).
Wingspan: 30mm (1.18in).
Development: Incomplete metamorphosis.
Food: Larva omnivorous; adult does not feed.

This is another caddis fly that is common in the north-east and upper Midwest. It is on the wing in late June and early July. Adults rarely eat. The zebra caddis belongs to a family known as the net-spinning caddises, from their habit of weaving net-like larval webs. The larva spins a web and hides there, often using the web as a net to catch debris as it flows past.

Identification: The adult zebra caddis has a pretty pattern of dark brown and beige stripes on its wings, giving it a zebra-like look. The long antennae are as long as the body and the wings extend beyond the tip of the body. Like most caddis flies, the adult resembles a small moth, but the antennae are longer and narrow and the wings are covered in soft hairs. Like a moth, this caddis fly holds its wings roof-like over its body when at rest. It is a poor flier. It is also attracted to lights at night. As well as using their silk nets for snaring edible particles, the larvae also use them to help anchor themselves to the riverbed so they are not swept downstream. Females of the species attach strings of eggs to vegetation over water or directly over water so that the larvae, when they hatch, drop straight into the water. The larvae pupates under the water surface, but it takes approximately a year for the egg to develop through its various stages into the adult. The species goes through complete metamorphosis. Once an adult the expected life span is a month.

Right: This species is more prettily marked than many caddis flies.

Brown Lacewing

Hemerobius spp.

Main habitat: Woodland and gardens.
Main region: Temperate.
Length: 5–8mm (0.19–0.31in).
Wingspan: 8–12mm (0.31–0.47in).
Development: Complete metamorphosis.
Food: Carnivore (prefers soft-bodied insects including aphids and mealy bugs).

These are small- to medium-sized brown lacewings. About 60 North American species are known from northern Florida to the Caribbean. Some species are used as biological controls for pests of economically important crops due to their rapid reproduction rates, the healthy appetites of both adults and larvae, and the relatively long adult lifespans. The larvae are sometimes called aphid wolves, as they are so good at hunting aphids.

Identification: Wing venation is important in identification of individual species. The wings are rather broad and beige or brown in colour. The larvae are slender with a small head. The pupa rests in a silk cocoon for about two weeks before emerging as an adult.

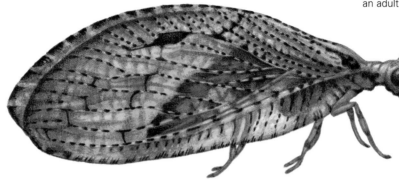

Left: Despite their name, colour is not a reliable characteristic for identification. Their bodies and their wings are very hairy.

OTHER SPECIES OF NOTE

Beaded Lacewing *Lomamyia* spp.
These lacewings are grey-brown, and look rather like brown lacewings, but their wings have upturned hooks at the tips. The wings of the females have specialized scales that exude a liquid. Their larvae feed on drywood termites.

Tree Antlion *Glenurus gratus*
This antlion is quite large, with a wingspan of about 10cm (4in). It looks a bit like a dragonfly in flight, but its flight is weaker and floppier. The carnivorous larvae live in burrows or in tree holes, hence the common name. The tips of its four wings have dark markings.

Wasp Mantisfly *Climaciella brunnea*
This mantisfly is found over most of North America. It is unusual in that it mimics the paper wasp (*Polistes*). Its body is orange-brown with bright yellow rings. It looks very like a wasp until you notice the mantis-like front limbs.

Scorpion Fly *Panorpa nuptialis*
Scorpion flies are very odd insects, with a long, downcurving face and the male with a sting-like enlarged end to the abdomen. They are harmless. This is one of the largest species, with an orange body and wings striped with broad dark bands. It can be seen in open woodland.

Dustywing family *Coniopterygidae*
These very small neuropterans are also known as waxflies due to the pale waxy covering on their bodies. Only about 3mm (0.12in) in length they look very like whiteflies until inspected closely. Their larvae are predators of aphids and mites.

Mantisfly

Mantispa sayi

Mantisflies (family Mantispidae) are curious relatives of lacewings. Very unusual in appearance, they look like a cross between a lacewing and a praying mantis. These predatory insects derive their common name from their mantid-like front legs, which they use to catch their prey. This species is one of the most common mantisflies in eastern North America. Mantisflies are often attracted to lights at night, where they hunt for other insects similarly drawn to the light. The larvae feed on the egg sacs of spiders and may hitch a ride on a spider to find their food.

Main habitat: Woods, scrub.
Main region: Temperate.
Length: 15mm (0.59in).
Wingspan: 40mm (1.57in).
Development: Complete metamorphosis.
Food: Larva parasitic; adult carnivorous.

Identification: Rather grotesque, with the body and wings similar to a lacewing, but with a very long neck and front legs with sharp, serrated, grasping claws.

Below: The front legs of a mantisfly have large pincer-like claws.

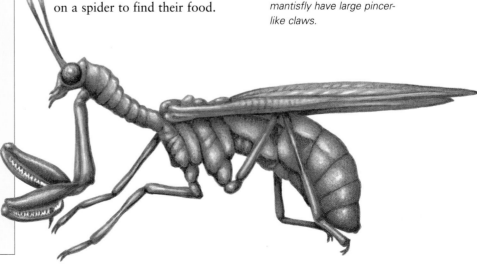

BEETLES

With over 300,000 known species, beetles (Coleoptera) are by far the largest of the insect orders. In North America alone there are an estimated 24,000 species, and in South America there are countless more, especially in the tropical forests. They are found in all sorts of habitats and occur in a wide variety of shapes and sizes.

Beautiful Tiger Beetle

Cicindela formosa

This is a large, active, predatory beetle. This species is one of several in the genus. They like to emerge in warm sunny places and their long legs give them great speed as they hunt for their prey. They have large eyes and sharp jaws. The adults are most active in late May and June through to August and September. They like open sites with sparse vegetation and other areas of dry sandy soil. They inhabit central and north-east North America.

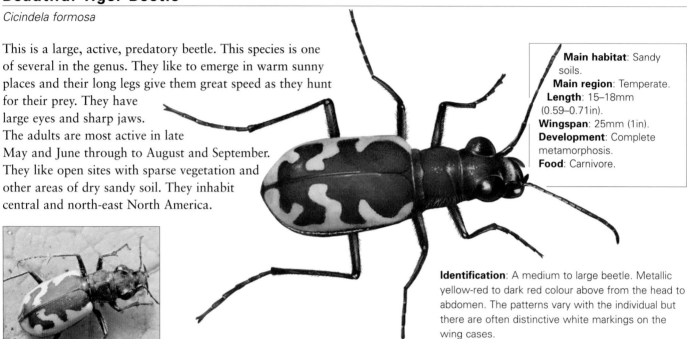

Main habitat: Sandy soils.	
Main region: Temperate.	
Length: 15–18mm (0.59–0.71in).	
Wingspan: 25mm (1in).	
Development: Complete metamorphosis.	
Food: Carnivore.	

Left and above: The bulging eyes of this species are typical of tiger beetles.

Identification: A medium to large beetle. Metallic yellow-red to dark red colour above from the head to abdomen. The patterns vary with the individual but there are often distinctive white markings on the wing cases.

Caterpillar Hunter

Calosoma scrutator

Another fast-running predator, this species is generally a nocturnal feeder, preying on soft-bodied larvae including caterpillars. It is also known as the 'fiery searcher'. As a ground beetle, the adults and larvae normally feed at ground level but unusually, both also sometimes climb trees to find prey. The adults are relatively large and capable of emitting noxious odours in self-defence. The larvae pupate in the soil before maturing as adults, which can overwinter and live for up to three years. They rarely fly.

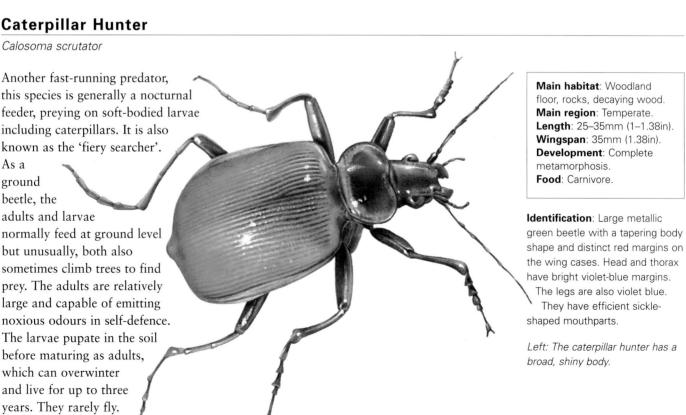

Main habitat: Woodland floor, rocks, decaying wood.	
Main region: Temperate.	
Length: 25–35mm (1–1.38in).	
Wingspan: 35mm (1.38in).	
Development: Complete metamorphosis.	
Food: Carnivore.	

Identification: Large metallic green beetle with a tapering body shape and distinct red margins on the wing cases. Head and thorax have bright violet-blue margins. The legs are also violet blue. They have efficient sickle-shaped mouthparts.

Left: The caterpillar hunter has a broad, shiny body.

Colorado Potato Beetle

Leptinotarsa decemlineata

Main habitat: Weeds, crops.
Main region: Warm
temperate, tropical.
Length: 10mm (0.39in).
Wingspan: 20mm (0.79in).
Development: Complete
metamorphosis.
Food: Herbivore.

Identification: Round-bodied
with bright yellowish-orange wing
cases, each with five bold brown
stripes along their length.

Also known as the ten-striped spearman, this
sturdy, dumpy beetle is a notorious pest of
potato crops. It also feeds on plants
related to potatoes including
tomatoes, aubergines and the native
'buffalo bur'. This is a serious pest
with up to three generations
possible per crop-growing season
and a developed
resistance to
pesticides. It is found especially
in Mexico (where it is native)
to south-western parts of the
USA, including
Colorado. The final
stage larvae pupate in the soil.

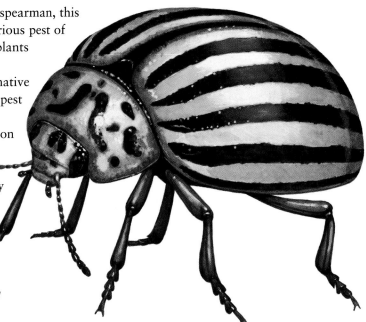

*Left and right: This brightly coloured
chunky beetle is a serious pest of
potato crops.*

OTHER SPECIES OF NOTE

Ground Beetle *Pterostichus* spp.
This genus of ground beetle contains several
other common American species. They are
common in forests and some can also be seen
in gardens where they like to lurk underneath
stones. They are very shiny and black with
narrow grooves along the elytra.

False Bombardier *Galerita* spp.
This ground beetle genus is widespread in
Central and North America. Large and blue-black,
its members have a brown pronotum and legs
and a black, rather narrow head. The favoured
habitat is under stones and leaves, and they are
sometimes found around houses.

Golden Tortoise Beetle
Charidotella sexpunctata
This leaf beetle is brightly coloured like many
other members of its large family. Its body is
rounded and shiny and remarkably can change
colour. It does this by changing the fluid
content of tiny cavities in the cuticle, becoming
more or less iridescent. The usual colour is a
bright, shiny greenish-gold. It is commonly found
on the morning glory plant.

Three-lined Potato Beetle *Lema trivittata*
This pretty beetle is another serious pest of
potatoes. The genus is mainly tropical or
subtropical. This species is small, only
about 6mm (0.24in) long, with a rectangular
body, bright yellow with three bold black stripes.
Both the larvae and the adult beetles feed on
plant foliage.

Cottonwood Leaf Beetle

Chrysomela scripta

Another boldly marked leaf beetle, this
species is a pest of willows, poplars, aspen
and alder plantations, especially on younger
trees. Breeding occurs from spring onward
when the insects take flight from their
ground-based wintering quarters to lay eggs
on fresh leaf growth. Once hatched the
young collect in groups and feed together
causing a lot of damage in concentrated
areas. Both young and adults feed on
vegetation. Lady beetles are a natural
predator of this pest. With six to eight
generations possible on plants in the
southern USA, the large population numbers
are also tackled with pesticides.

*Below: This leaf beetle is a common
American species.*

Main habitat: Bark, forest
floors, leaf litter.
Main region: Temperate.
Length: 6mm (0.24in).
Wingspan: 12mm (0.47in).
Development: Complete
metamorphosis.
Food: Herbivore.

Identification: A medium-sized
beetle with a black head and
thorax and yellow wing cases
marked with broken black stripes
and blotches. The margins of the
thorax can be yellow or red. The
larvae are up to 12mm (0.47in)
long, dark to light brown with
obvious white marked glands
on their sides, which emit a
defensive scent if the larva
is disturbed.

Ironclad Beetle

Zopherus nodulosus

This is one of several species from this flightless genus, found mainly in the south-west USA, notably in western Texas, and also in Mexico and south as far as Venezuela. The common name comes from the fact that these beetles have a very tough exoskeleton. The larvae feed on rotting wood, though the adults tend to feed on lichens growing on oak trees, where their colouring gives them camouflage. In Mexico some species of this slow-moving genus are caught and used as jewellery.

Main habitat: Trees, wooden houses.
Main region: Warm temperate.
Length: 12–30mm (0.47–1.18in).
Wingspan: None.
Development: Complete metamorphosis.
Food: Lichenivore.

Identification: Extremely hard exoskeleton, patterned with irregular black and white patches. The mouthparts are adapted for chewing and the legs are sturdy.

Far left: The ironclad beetle gains protection from its tough, patterned body.

Cottonwood Borer

Plectrodera scalator

The adults of this splendid longhorn beetle are medium to large and coloured cream and white. This species is a pest of cottonwood, poplar and willow trees. The larvae are difficult to see as they spend much of their time feeding within the roots of the plant. This affects the strength of the tree, often causing it to topple over. Adults overwinter and then emerge to breed during late spring and summer. The adults feed on the tree's bark, leaves and shoots. Found in the USA, mainly east of the Rockies.

Main habitat: Trees.
Main region: Temperate.
Length: 25–38mm (1–1.49in).
Wingspan: 45mm (1.77in).
Development: Complete metamorphosis.
Food: Herbivore.

Identification: A large long-horned beetle with long, segmented antennae, which are almost the same length as the body. The body is black and narrow, tapering toward the rear and is patterned with creamy white areas.

Right: This active beetle is easy to spot, being large and boldly patterned.

Long-jawed Longhorn Beetle

Trachyderes mandibularis

Main habitat: Trees and shrubs.
Main region: Warm temperate, tropical.
Length: 25mm (1in).
Wingspan: 40mm (1.57in).
Development: Complete metamorphosis.
Food: Herbivore.

This large longhorn beetle is also known as the 'horse-bean longhorn'. It is found from Florida to California and south to Honduras. It is most commonly spotted between July and September and the larvae bore into hackberry, figs and tamarisk, as well as citrus fruits and horse-beans. Adults are most active during the day and may be found near wounded trees oozing sap.

Identification: Rather an impressive longhorn beetle, with striped antennae longer than its body and which are curled outward toward their tips. The head and thorax are black, and the elytra yellow or orange and black.

Left: The curved antennae of this longhorn beetle are longer than its body.

American Carrion Beetle *Necrophila americana*
The American carrion beetle is one member of a family of beetles (Silphidae) that are nearly always found around decaying corpses. They do a great job in helping to recycle the remains of dead animals. Their larvae feed on the corpses while the adults catch maggots. The adult beetle is flat-bodied with dark elytra and a yellow pronotum with a central dark patch.

Harlequin Beetle *Acrocinus longimanus*
This decorative longhorn beetle lives in tropical America. Its common name reflects the delicately marked body, which is a medley of intricate black, white and orange shapes. Both sexes have extremely long antennae, and the male also has very long front legs with which he guards the female.

Milkweed Longhorn Beetle *Tetraopes* spp.
This genus of longhorn beetles feeds on milkweed plants (*Asclepias*) from which they store poison in their bodies. Their bright red colours act as a warning of their unpleasant taste and deter attack. Various species are found in Central America, Mexico and the USA, north to Canada.

Red-lined Carrion Beetle *Necrodes surinamensis*
Often seen at lighted windows at night, this widespread beetle is attracted to carrion, where it feeds on maggots. Both the larvae and adults of this species hunt the same prey. The adults have ridged wing cases with a line of red dots toward the rear.

Titanic Longhorn Beetle

Titanus giganteus

In flight this huge longhorn beetle is the size of a bird. They have powerful jaws and can deliver a painful and dangerous bite, and they also have sharp claws. They have even been reported to be capable of biting through a pencil, or a finger! This huge beetle lives in the rainforests of northern South America – notably in Brazil, Colombia, Ecuador and Peru. They seem not to feed as adults and though heavy they can fly, often being attracted to lights at night. The larvae feed on decaying timber.

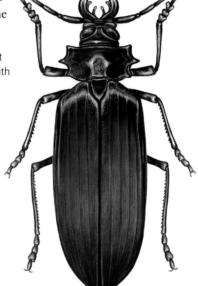

Below: This huge insect is one of the largest known.

Identification: Arguably the world's largest beetle, measuring up to 17cm (7in), and with a wingspan of 30cm (12in). The body is elongated oval and has a reddish-brown sheen. The wings are a pale orange colour. The long antennae curve backward.

Main habitat: Trees and shrubs.
Main region: Tropical.
Length: 17cm (7in).
Wingspan: 30cm (12in).
Development: Complete metamorphosis.
Food: Herbivore (larva).

American Burying Beetle

Nicrophorus americanus

This is one of about 30 species of the family Silphidae in North America. Burying beetles take their name from their habit of storing scavenged food underground. Once much more common, the American burying beetle was listed as an endangered species in 1989, the decrease in numbers being possibly due to the use of the agricultural chemical DDT affecting the availability of food, combined with loss of habitat. It now occurs over only 10 per cent of its original range. It has been recorded recently from Nebraska, Rhode Island, Oklahoma, Arkansas and South Dakota (previously recorded in 35 states), with reintroductions attempted in Ohio and Massachusetts. Unusually for beetles, both sexes jointly raise their offspring by staying with them and providing them with a food store, usually the carcass of a small animal or bird, buried alongside the eggs. The American burying beetle is the largest scavenging beetle in North America. A strong flier, this nocturnal beetle can travel up to 1km (0.6 mile) when foraging at night. Its preferred habitat is linked to the availability of food.

Main habitat: Soil.
Main region: Temperate.
Length: 20–35mm (0.79–1.38in).
Wingspan: 60mm (2.36in).
Development: Complete metamorphosis.
Food: Carnivore/omnivore.

Identification: A strikingly marked large beetle – a shiny black body marked with irregular areas of bright orange on the head shield and face, orange tips on the large antennae and four scalloped orange-red markings on the wing cases. The orange colour distinguishes it from other carrion beetles.

Above left: Burying beetles perform a useful task by recycling waste.

Giant Stag Beetle

Lucanus elaphus

This majestic beetle is found mainly in the eastern states of North America, in woods and forests where it lays its eggs in rotting wood such as logs or stumps. The larvae feed on the decaying wood and emerge as adults in the summer. The adult feeds mainly on plant sap. Male stag beetles use their antler-like mandibles in displays and fights over female mating rights. The males lock their mandibles together and joust and jostle with the aim of turning the opponent on to his back. Stag beetles fly well and are sometimes attracted to lights at night.

Identification: The adult male has huge mandibles that look like the antlers of a male deer, hence the common name. The mandibles can be as long as the body of the beetle itself. Those of the female are shorter. The antennae are elbowed and the body is a glossy black.

Above right and right: The massive antler-like jaws of the male give this splendid beetle its name.

Main habitat: Woodland.
Main region: Temperate.
Length: 40mm (1.57in) (excluding 'antlers').
Wingspan: 60mm (2.36in).
Development: Complete metamorphosis.
Food: Herbivore.

Rain Beetle

Pleocoma spp.

Rain beetles are unusual in a number of respects. Squat and dumpy, they spend a lot of time in burrows in the ground and do not feed as adults, having no digestive tract and feeble mouthparts. The larvae live in the soil feeding on plant roots and may take as long as 12 years to reach maturity. The name comes from the tendency of male rain beetles to fly when it is damp or wet, often at dawn or dusk in autumn and winter. They seek out the females in their burrows and mate, dying soon afterward. They are found mainly in western North America.

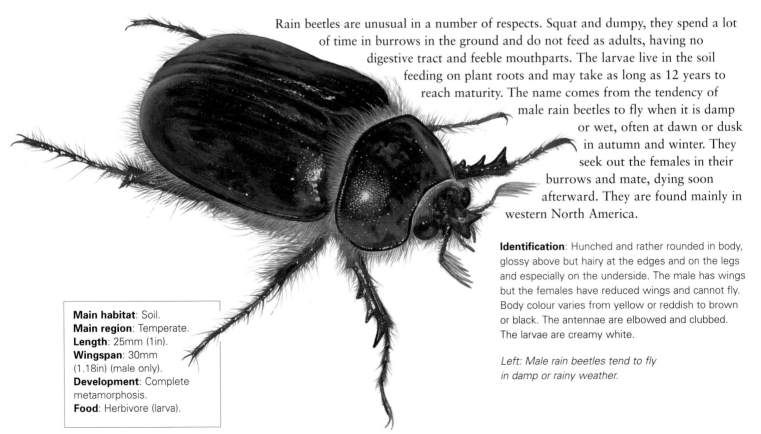

Identification: Hunched and rather rounded in body, glossy above but hairy at the edges and on the legs and especially on the underside. The male has wings but the females have reduced wings and cannot fly. Body colour varies from yellow or reddish to brown or black. The antennae are elbowed and clubbed. The larvae are creamy white.

Left: Male rain beetles tend to fly in damp or rainy weather.

Main habitat: Soil.
Main region: Temperate.
Length: 25mm (1in).
Wingspan: 30mm (1.18in) (male only).
Development: Complete metamorphosis.
Food: Herbivore (larva).

Eastern Hercules Beetle

Dynastes tityus

Main habitat: Woodland.
Main region: Temperate.
Length: 60mm (2.36in).
Wingspan: 90mm (3.54in).
Development: Complete
metamorphosis.
Food: Herbivore.

*Right: Although not as large as
some of the tropical species,
this Hercules beetle is an
impressive insect.*

This splendid large beetle is related to a number of mainly tropical species, but this one ranges north to New Jersey and Indiana and west to Texas. Like other members of the family, its larvae feed on rotting wood. The adults look scary but they are harmless vegetarians, feeding mainly on fruit and sap. They are not uncommon in mixed woodland and are often attracted to lights at night. The larvae develop among tree leaf debris.

Identification: Large and glossy, with long legs. The most remarkable features are the projections on the head and thorax – long spine-like extensions. The female lacks these projections. The thorax and abdomen are yellowish or olive green with brown spots, and the legs are darker. The larva is a large whitish grub.

OTHER SPECIES OF NOTE

Rhinoceros Beetle *Megasoma actaeon*
This large rhinoceros beetle is common in the rainforests of South and Central America, often seen clambering up a tree trunk using its sharp curved claws. Although not the longest beetle, this mouse-sized insect is one of the most bulky and may measure 70mm (2.76in) across and 12cm (4.5in) long. As well as being raised in vivaria, this species is often kept as a novelty pet by local children. It is a gentle insect and harmless to people, and its armour gives it good protection from predators such as birds. It likes to eat fruit, such as bananas.

Elephant Beetle *Megasoma elephas*
This is another rhinoceros beetle, similar to *M. actaeon*. It differs from that species in having a covering of fine yellowish hairs all over its body. The elephant beetle is found in Central America and also in southern Mexico.

Horned Passalus *Odontotaenius disjunctus*
This common beetle is often found in woods in the eastern USA. Like many large woodland beetles it breeds in rotting timber. This species is colonial and the various members of the colony communicate with calls made by rubbing parts of their bodies together. The larva is a rather shapeless white grub.

Green June Beetle *Cotinis nitida*
This pretty beetle belongs to the scarab or dung-beetle family and is quite often seen, as it flies by day. Its body is a bright metallic green. The larvae feed on the roots of grasses.

Hercules Beetle

Dynastes hercules

This giant tropical beetle (also known as the 'rhinoceros beetle') has enormous horns, which are used by rival males in combat. It is one of the largest of all beetles. Hercules beetles are found in the rainforests of Central and South America, where they trundle about in search of fruit. The largest males may reach a length of 17cm (7in). The larvae live in rotting wood for up to two years and can reach 11cm (4.5in) and weigh 120g (4.25oz).

Main habitat: Rainforest.
Main region: Tropical.
Length: 16cm (6in).
Wingspan: 10cm (4in).
Development: Complete
metamorphosis.
Food: Herbivore.

Identification: The projections on the head of the adult male are amazing – sometimes longer than the rest of the body. The females may have a longer body but they lack the horns.

Below and below left: The huge, horn-like projection on the male's head is used in jousting contests with rivals.

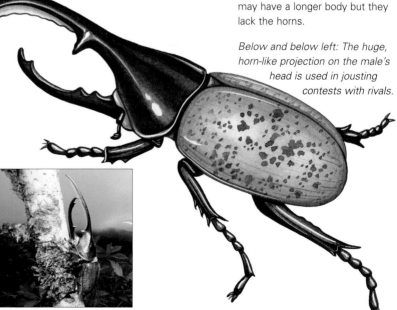

Convergent Lady Beetle

Hippodamia convergens

This beetle belongs to the large family of ladybirds (Coccinellidae), also known as ladybugs and lady beetles. The members of this family are mostly colourful with rounded, spotted carapaces. This is one of the most common species, found across much of North, Central and South America. Like other ladybirds it is a voracious predator and consumes large numbers of garden pests such as aphids. For this reason it is often sold as a biological control agent.

Identification: Head and thorax are black with white markings. The domed elytra are orange with several black spots, though newly emerged beetles lack the black spots at first. The larvae are dark and grow to about 7mm (0.28in).

Main habitat: Fields, gardens, crops.
Main region: Temperate, tropical.
Length: 4–7mm (0.16–0.28in).
Wingspan: 12mm (0.47in).
Development: Complete metamorphosis.
Food: Carnivore.

Above: Like many ladybirds, this beetle is a helpful destroyer of garden pests.

Right: The prominent black and white head is common to all lady birds of this species, but the number of spots is variable up to a maximum of 13 spots.

Firefly

Photinus pyralis

Fireflies, also known as lightning bugs, are able to produce their own light by a chemical process called bioluminescence. Some species in the family (Lampyridae) only glow as larvae or as wingless forms, while others flash as they fly. In all cases the flashes serve as signals to attract a mate. This species is a common firefly of North America, mainly in the east. The males twinkle as they fly in the evenings, signalling with a yellow light to females that lurk in the vegetation below by flashing about every five seconds. When a female responds with a flash he may descend to find her. Adults and larvae are active carnivores, feeding on other invertebrates.

Below: Fireflies use chemically produced light to flash signals in the dark. They can be seen in the early evening flying low over grass emitting flashes. The flash that they emit has a significant yellow colour. Male and females emit different signals and males will only respond to flashes from females.

Main habitat: Fields, hedges.
Main region: Temperate.
Length: 15mm (0.59in).
Wingspan: 25mm (1in) (male only).
Development: Complete metamorphosis.
Food: Carnivore.

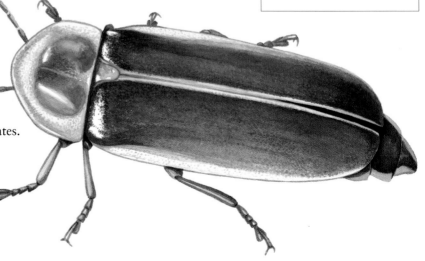

Identification: The elytra are rather long and narrow, and brown with yellow markings. The thorax is paler with an orange centre. The male produces light from his lower abdomen. Inside there is a layer of tissue that acts rather like a mirror to reflect the light outward through the transparent layers below.

Palmetto Weevil

Rhynchophorus cruentatus

Main habitat: Palms.
Main region: Sub-tropical.
Length: 20–30mm
(0.79–1.18in).
Wingspan: 60mm (2.36in).
Development: Complete
metamorphosis.
Food: Herbivore.

Identification: The adult is brightly marked with varying amounts of black and red. The rostrum is rather short and down-curved and the elytra have obvious longitudinal ridges. Males and females can be distinguished from each other by the texture of the rostrum (snout). On males this is bumpy and on females it is smooth.

The weevil family (Curculionidae) is huge, with more than 50,000 species. They typically have their mandibles at the tip of a long 'nose', the rostrum. The palmetto weevil is the largest weevil in North America, where it is found from Florida to South Carolina and west to Texas. It feeds mainly on various species of palm, including saw palmetto and cabbage palm, but will also attack dead and dying trees and other stressed palms, where it will cause considerable damage.

Left: Large by weevil standards, this species feeds mainly on small palms.

OTHER SPECIES OF NOTE

Firefly *Photuris* spp.
These fireflies are common, mainly in wet fields. They produce a green or greenish-yellow light. This genus of firefly has a cunning way of finding food. The females sometimes trick the males of *Photinus* fireflies by mimicking the flashes of females of that genus. Then when the male of the 'wrong' species flies down to find a mate, he becomes a tasty meal instead.

Red Flat Bark-beetle *Cucujus clavipes*
This beetle is a striking bright red, especially on the elytra and inside legs. Its whole body is very flattened which aids it as it squeezes and burrows beneath the bark of trees. The larva feeds under the bark of trees such as maple, ash and poplar. It is found mainly in the west of North America, north to Alaska. It has a kind of natural antifreeze and can survive the winter by hibernating at very low temperatures.

Cotton Boll Weevil *Anthonomus grandis*
This is a notorious pest of cotton, responsible for heavy cotton yield losses every year. A native of Mexico where it feeds on wild cotton plants, it has spread to attack cotton fields in the southern USA and also in South America. It is small, about 6mm (0.24in), and grey, with quite a long snout. It is controlled by chemical sprays and by biological agents such as fire ants and parasitic wasps.

New York Weevil *Ithycerus noveboracensis*
This handsome large weevil is all grey with black and white markings along its ridged abdomen. It grows to a length of 18mm (0.71in) and its snout is rather short and broad.

Spruce Beetle

Dendroctonus rufipennis

This modest-looking little beetle is extremely destructive and causes huge amounts of damage each year to commercial forests of spruce. It belongs to the bark-beetle family (Scolytidae), which includes the beetle responsible for spreading Dutch elm disease. The spruce beetle has regular population explosions and then kills large numbers of spruce trees, especially in the north-west of North America. An outbreak typically lasts for between two and five years. The adults are on the wing from May through July. The damage is caused by the beetles burrowing into and under the bark, weakening the trees and allowing in infections.

Main habitat: Spruce forest.
Main region: Temperate.
Length: 4mm (0.16in).
Wingspan: 6mm (0.24in).
Development: Complete
metamorphosis.
Food: Bark.

Identification: A small beetle with a black head and thorax and brown, rough elytra. The legs are rather broad toward the base and the domed head carries a pair of short, clubbed antennae. It can be identified by the patterns left in the bark of affected trees.

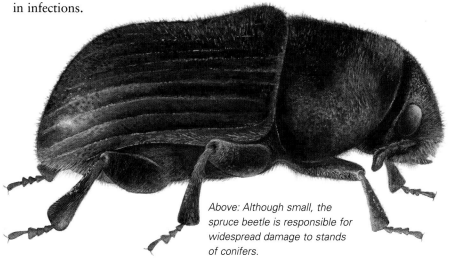

Above: Although small, the spruce beetle is responsible for widespread damage to stands of conifers.

Cigarette Beetle

Lasioderma serricorne

This species is commonly known as the cigarette beetle as the adults feed on stored cigarettes, cigars and dried tobacco leaves, among other food. The beetles may also consume books. Its larvae feed on a wider range of foodstuffs including pepper, grain, raisins, spices (even chillies) and dried fish. A pest in many households, it is hard to control due to its ability to survive on such a broad range of foods. It has even been known to eat pyrethrum powder with no apparent ill effects. It also has a fast reproductive cycle, producing from three to six broods in a season.

Identification: A small oval beetle covered in silky yellow-brown hairs. More common in the winter and autumn but persists all year round. The adults are strong fliers. The long thin larvae are a little shorter than the adults, with a white body, a yellow head and brown mouthparts.

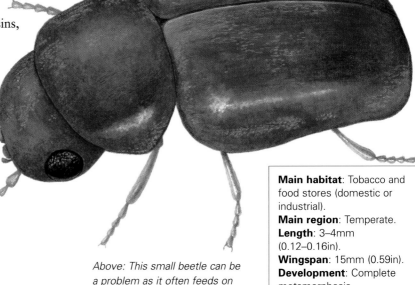

Above: This small beetle can be a problem as it often feeds on stored food.

Main habitat: Tobacco and food stores (domestic or industrial).
Main region: Temperate.
Length: 3–4mm (0.12–0.16in).
Wingspan: 15mm (0.59in).
Development: Complete metamorphosis.
Food: Omnivore.

Giant Palm Borer

Dinapate wrightii

The giant palm borer gets its name from the plant it lives on, its large size and its ability to tunnel into the palm trunk. The mainly nocturnal adults chew holes to live in and as a safe place to rear young. This impressive beetle lives mainly on the trunks of the California fan palm (*Washingtonia filifera*) in the south-west USA.

Identification: A very large beetle, which grows from 30–50mm (1.18–2in) long. The body is shaped like a cylinder, dark brown above and hairy underneath. The hairs are golden in colour. The wing cases are textured with slight ridges and form points toward the rear of the body. The beetle's head faces downward from beneath a shield-like domed carapace.

Main habitat: Palm trees.
Main region: Tropical.
Length: 30–50mm (1.18–2in).
Wingspan: 60mm (2.36in).
Development: Complete metamorphosis.
Food: Herbivore.

Below left: This large beetle has a distinctively domed thorax.

Giant Green Water Beetle

Dytiscus marginicollis

Main habitat: Semi-permanent salt- and freshwater ponds.
Main region: Temperate
Length: 26–33mm (1–1.3in) long.
Wingspan: 50mm (2in).
Development: Complete metamorphosis.
Food: Carnivore.

This large water beetle swims strongly using the paired action of its hind legs. It is an active predator, catching underwater prey such as fish and invertebrate larvae. It can adapt to live in either salt- or freshwater ponds. The adult beetles are attracted to bright lights at night. They mate in late autumn or early spring and are most active from May to August. Its main range is from British Columbia to Manitoba.

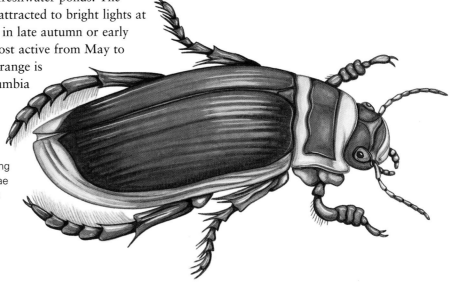

Identification: A medium-sized reddish-brown or metallic green beetle with an elongated body. The adults have a chevron pattern on the head and yellow margins to the wing cases and the shield covering the thorax area. The antennae have a yellow base and red tips. The front and middle legs are either yellow or pale red.

Right: Though aquatic, this large water beetle sometimes flies to seek new ponds.

OTHER SPECIES OF NOTE

Large Chestnut Weevil *Curculio caryatrypes*
This rather pretty weevil has beautiful mottled and striped body markings of black and fawn and a rounded body shape. Its rostrum is very long, sometimes longer than the rest of the body. Its mouthparts are situated at the end of its rostrum. In some areas it is a troublesome pest of chestnut trees.

Rufous Soldier Beetle *Cantharis rufa*
This soldier beetle is common in all of North America as well as in Europe and Northern Asia. It is uniformly red-brown in colour with small black eyes. It is a member of the Cantharidae family which are also known as 'leather wings'. Long bodied, and with strong flight, the adults are often found on flowers and they are important pollinators.

Violet Rove Beetle *Platydracus violaceus*
This beetle, as its name implies, has a rather beautiful violet coloration, with a metallic sheen. It has a long body and fast scuttling behaviour, and can be mistaken for an earwig. However, rove beetles are distinguished from earwigs by their short wing cases.

Fire-coloured Beetle *Dendroides cyanipennis*
This is a stunning soldier beetle mimic with a black body, head, antennae and startling orange red legs and thorax. It, like others in its genus, has beautiful large, feathery antennae, and very large eyes. It is one of the Pyrochroidae family, which curiously can contain an irritant chemical (cantharidin) and may cause skin blistering.

Giant Water Scavenger Beetle

Hydrophilus triangularis

This is a widespread species found from Ontario to western Florida, California and Texas. A silken egg case is laid and attached to floating vegetation or debris. The larvae develop underwater and take a wide range of prey including snails, fish and insect larvae. Despite their name they are predatory insects. Once fully developed, the larva leaves the water to pupate in nearby soil, emerging as an adult after 17–21 days. Adult beetles may overwinter and emerge in spring to repeat the breeding cycle the following summer. These beetles are a favourite food of bullfrogs.

Main habitat: Deep water with vegetation.
Main region: Temperate-subtropical.
Length: 27–40mm (1.06–1.57in).
Wingspan: 80mm (3.15in).
Development: Complete metamorphosis.
Food: Carnivore.

Below: This splendid water beetle is one of the region's largest.

Identification: This large oval-shaped diving beetle has a violet to green metallic shiny black colouring above, and is slightly browner beneath. Antennae, feeding palps and 'feet' (ends of legs) also tend toward brown. Incomplete ridges and dotty textured markings occur along the length of the wing cases. The hind legs are built for swimming, with long spurs midway down the length. Larvae are longer than the adults, 40–60mm (1.57–2.36in), and yellow-brown, with a flattened oval head shape.

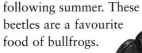

FLIES

Although not the best loved of insects, flies (Diptera) are nonetheless one of the most fascinating of the insect orders. Flies are very diverse, with about 120,000 known species, and they exist in virtually all habitats. In many temperate countries, flies make up about a quarter of all insects. The fly fauna of North America is roughly 17,000 species.

Cluster Fly

Pollenia rudis

This common fly is widespread in North America, although it may originally have been introduced from Europe. It frequently hibernates inside houses, sometimes clustered in groups and it may emerge at almost any time of the year, mistaking centrally heated rooms for spring weather. It lays its eggs on the surface of soil and the newly hatched maggots seek out earthworms on which they feed. Although this species can be irritating, it is not known to spread diseases.

Identification: Similar to, but a little larger than, the common housefly. The body is yellow-grey with black markings and the thorax has a dense covering of yellow hairs.

Right and far right: Cluster flies often gather inside houses in the autumn and winter.

> **Main habitat**: Fields, gardens, houses.
> **Main region**: Temperate.
> **Length**: 8mm (0.31in).
> **Wingspan**: 15mm (0.59in).
> **Development**: Complete metamorphosis.
> **Food**: Carnivore (larva); omnivore (adult).

Shiny Bluebottle-fly

Cynomya cadaverina

The common name bluebottle-fly probably comes from the 16th-century word 'bot' meaning 'little maggot'. These and related flies are thought to be responsible for the transmission of numerous diseases and pathogens, so although small they have a wide impact on other animals, including humans. This fly is widespread from Alaska to Georgia. The larva feeds on carrion and other decaying matter. This and related flies often arrive first at a corpse to lay their eggs.

> **Main habitat**: Fields.
> **Main region**: Temperate.
> **Length**: 8mm (0.31in).
> **Wingspan**: 15mm (0.59in).
> **Development**: Complete metamorphosis.
> **Food**: Carnivore (larva); omnivore (adult).

Identification: The abdomen is metallic blue-green with black and grey markings and about 6mm (0.24in) long. The head has a covering of grey or yellow-brown short hairs. It has large red-brown compound eyes.

Left: This species is at its most prolific in spring when the weather is cool.

Virginia Flower-fly

Milesia virginiensis

Main habitat: Flowery sites.
Main region: Temperate.
Length: 20mm (0.79in).
Wingspan: 40mm (1.57in).
Development: Complete metamorphosis.
Food: Herbivore (adult).

Identification: Large black compound eyes and clear bluish-grey wings with black venation. Short yellow antennae projecting between the eyes. The predominantly yellow, narrow and tapering body is marked with stripes and patterns in black with some rusty patches on the abdomen and 'thighs'.

A brightly coloured yellow-and-black fly which closely resembles a wasp (yellow jacket). The Virgina flower-fly is a member of the syrphinae subfamily. This cleverly disguised mimic is referred to as the 'news bee' due to its habit of hovering in front of people as if to impart 'news'. It is also called the yellowjacket hoverfly. These attractive hoverflies prefer sunny areas, such as open woodland or forest edges, and also frequent gardens and parks. The adult feeds on pollen. In the USA it is found from the Rockies east to Florida. It is on the wing from May to September.

Left: This attractive flower-fly is commonly seen hovering where there are flowers nearby.

OTHER SPECIES OF NOTE

Horsefly *Goniops chrysocoma*
This horsefly has a rounded yellow-and-black stripey body, furry fawn thorax and small yellow head with large eyes. The wings are mostly dark grey. It is common in eastern wet wooded areas. The female circles its prey before biting it to feed on its blood. The favoured prey is usually deer, but humans also get bitten. The male feeds on nectar.

Tachinid Fly *Gymnosoma fuliginosa*
This is one of seven North American fly species of the family Tachinidae that parasitize stink and shield bugs to raise their young. This fly has a rounded red-brown body with a number of black oval markings above. The wings are clear grey, with orange patches near the body. Its large compound eyes are red-brown. Adult flies are often found on the flowers of wild carrot. Unusually, eggs are laid on the outside of the host and the larva enters the bug through gaps in its body segments. While inside, it feeds on body fat and may finally emerge without actually killing the host. However, it usually leaves the bug weakened or sterile.

Bee Fly *Anthrax georgicus*
This insect has a dark furry body and mottled grey wings. A member of the family Bombyliidae, it is a bee mimic that lays its eggs in nests of solitary bees. Its larvae eat the food stores and grubs of their host. The adults hover in flight and possess a long proboscis that is used to sip nectar. This species occurs mainly in the south-west of North America.

Mydas Fly

Mydas clavatus

These large black insects belong to the Mydidae family, which include some of the largest flies known. There are about 50 species in North America, mainly in the west. The adult has a relatively short lifespan so is rarely seen. It tends to prefer open country and often selects hot sandy places. This species mimics a wasp and might at first sight be mistaken for a hymenopteran, or possibly a robber fly. The larvae feed on beetle grubs in decaying wood, while the adults feed on pollen.

Below: The mydas fly has orange markings on the abdomen and orange feet.

Main habitat: Open country.
Main region: Temperate-subtropical.
Length: 10–60mm (0.39–2.36in).
Wingspan: 50mm (2in).
Development: Complete metamorphosis.
Food: Carnivore (larva); herbivore (adult).

Identification: Shiny black body with short hairs and black wings. Small black head with distinctive clubbed antennae. The legs are fairly long and the rear pair are rather powerful. There is a distinctive orange base to the abdomen.

Inland Floodwater Mosquito

Aedes vexans

This mosquito is widespread from southern Canada south across the USA and in many other American countries. It is cosmopolitan around the globe and considered a nuisance since it helps to spread disease. It may be found around virtually any area of still water, including small temporary pools and flooded areas, and in fact has been known to travel up to a 10 mile radius from its source of water. The eggs overwinter in water or on damp ground that is prone to flooding. The larvae have a horizontal posture in the water. The adults are active during the night, and feed mainly at dusk or by day in the shade.

Above: This species is one of the most troublesome and widespread of all mosquitoes.

Main habitat: Still water.
Main region: Temperate and tropical.
Length: 8mm (0.31in).
Wingspan: 10mm (0.39in).
Development: Complete metamorphosis.
Food: Carnivore (female); herbivore (male).

Identification: Attractive when magnified, males having feathery antennae and with black-and-white body patterns. They show an arched, springy body posture with long hind legs that seem to push the body up in the middle. The thorax is rather a golden brown with grey scales. The legs have pale 'knees'. The relatively small head faces down. Adult females feed on humans and animals, but the males feed on nectar.

Non-biting Midge

Orthocladius spp.

This genus contains about 30 species in North America. It is a poorly understood group and work is still being carried out on the identification of many species. This genus belongs to the family of non-biting midges (Chironomidae) which are harmless to people. They are an important source of food for birds. Mainly associated with flowing waters and some still waters including lakes and ponds, they are also found over moist soil. These midges frequently gather in large swarms, typically in the evenings and sometimes over water. Such swarms are usually of male insects which dance up and down to attract the females. The males use their feathery antennae to help them locate the females.

Left: These midges are often abundant, but are quite harmless.

Main habitat: Wetlands.
Main region: Temperate.
Length: 10mm (0.39in).
Wingspan: 10mm (0.39in).
Development: Complete metamorphosis.
Food: Omnivore (larva).

Identification: Midges have a more flattened appearance than mosquitoes, having a low-slung resting body posture with forelegs facing forward. Their wings are shorter than their body. They swarm in suitable conditions (varies between species, affected by light, humidity and temperature) in order to mate and complete their short adult life. They can also have distinct feathery antennae.

Giant Cranefly

Holorusia rubiginosa

Main habitat: Moist areas including damp woodland and stream edges.
Main region: Temperate.
Length: 38mm (1.49in).
Wingspan: 70mm (2.76in).
Development: Complete metamorphosis.
Food: Herbivore (larva).

Identification: Large red-brown or olive-green cranefly with olive wings spanning up to 70mm (2.76in) and distinctive white striped markings on the brownish thorax. The red abdomen is pointed in adult females and bulging in adult males, both having a thin 'waist'.

This cranefly is one of the largest flies in existence. Although large it is delicate in appearance, with long thin legs. The larvae live and grow at the edges of streams, in water, moist soils or leaf mould. Adults are attracted by lights so are often found inside houses. They are harmless to people as they possess no biting mouthparts because they do not need to feed in their brief adult life. This species is found mainly in western North America, from California to British Columbia.

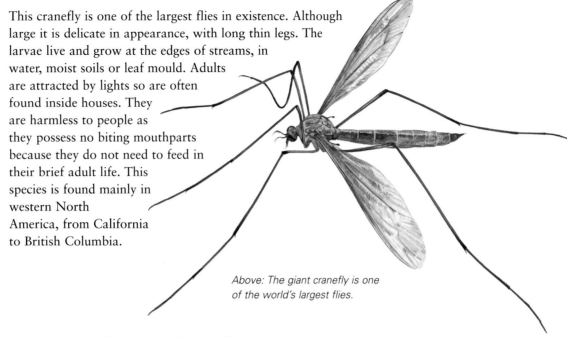

Above: The giant cranefly is one of the world's largest flies.

OTHER SPECIES OF NOTE

Northern House Mosquito *Culex pipiens*
A common mosquito that lays its eggs in rather beautiful geometrical floating rafts on the water surface. The resulting aquatic larva can survive in very shallow water depths so it is hard to eradicate potential habitats for it. It is a serious vector of disease in the north-west, where birds and horses are the main hosts. Only adult females bite to feed on blood. Females also sometimes hibernate inside houses.

Cranefly *Hexatoma albitarsis*
This cranefly has long legs and a delicate appearance. The overall colouring is dark. The larva of this species lives underwater in streams with sand or gravel bottoms, though some related species can tolerate still water. The larvae feed on small worms and insect larva in the same habitat. *Hexatoma* is a large genus with 35 species in North America.

Winter Cranefly, *Trichocera* spp.
These flies are common, especially in eastern North America. Most species are about half the size of a typical cranefly. Winter craneflies fly on sunny days in late autumn or early spring, although as their name indicates they can also be seen swarming in winter in some regions. The larvae feed by scavenging on decaying plants.

Wood-boring Cranefly *Ctenophora* spp.
These craneflies are rather wasp-like, with a shiny black, yellow or red body. The females have a long pointed ovipositor, which they use to deposit eggs under dead wood. Most species are shorter-legged and broader-bodied than typical craneflies.

Phantom Cranefly

Bittacomorpha clavipes

This rather unusual fly looks a little like a normal cranefly, but has a number of distinctive features. It lives in damp sites near ponds and marshes and is a slow flier. The larva lives in the water, feeding on swamp debris, and it breathes through a thin extensible tube at the tip of the abdomen. The adults mate on the wing, often with the male trailing behind in the air during copulation.

Right: This cranefly earns its name through its tendency to float in the air rather than fly.

Main habitat: Wetlands.
Main region: Temperate.
Length: 12–16mm (0.47–0.63in).
Wingspan: 20mm (0.79in).
Development: Complete metamorphosis.
Food: Herbivore (larva).

Identification: This cranefly has distinctive black-and-white banded legs, broader toward the tips as if it is wearing socks. In fact these are filled with air and increase the insect's buoyancy, perhaps helping it to drift as it wafts along in the breeze.

American Horsefly

Greenhead, *Tabanus americanus*

Horseflies are renowned for their painful bites and tenacious pursuit of prey. They follow large dark objects, which can include vehicles. Horseflies have unusual compound eyes which are either brightly patterned or a solid colour such as green, giving this species the alternative name of 'greenhead'. Only the females bite, feeding on the blood of large mammals (including humans); the males feed on nectar. The aquatic larvae eat mainly other insects. Males can be distinguished by the fact that their eyes join in the middle, whereas female eyes have a gap between them. Often found close to ponds, swamps and marshes. Widespread in North America, from Newfoundland to Florida, west and south to Texas and northern Mexico, and north to the Canadian Northwest Territories.

Main habitat: Damp woodland and forest.
Main region: Temperate.
Length: 20–28mm (0.79–1.1in).
Wingspan: 40mm (1.57in).
Development: Complete metamorphosis.
Food: Carnivore (larva and adult female).

Identification: The most obvious feature of the American horsefly is the large eyes which are a solid bright green, set in a grey to tan head with reddish-brown antennae. It has a long, tapering blackish-brown body, with short, grey-white hairs at the margins, with transparent grey wings that extend past the tail at rest. The body is also hairy in places.

Left: Female horseflies inflict painful bites using their razor-sharp mouthparts.

Giant Robber Fly

Promachus hinei

Robber flies are cunning and speedy hunters, preying on other insects. They tend to lie in wait and spring out to catch their prey as it flies by, relying on the element of surprise, speed and accuracy. *Promachus* is Latin for 'fighter in the front ranks'. This species is also known as 'bee killer' due to its favoured meal. The giant robberfly grows to an impressive 50mm (2in) long. It is found mainly in the central states of the USA. It often catches other insects during active flight, even those larger than itself.
It makes a loud buzzing noise in flight.

Main habitat: Meadows, fields.
Main region: Temperate.
Length: 50mm (2in).
Wingspan: 65mm (2.56in).
Development: Complete metamorphosis.
Food: Omnivore (larva); carnivore (adult).

Identification: This species has a pale tapering body with dark incomplete stripes and a hairy grey-brown domed thorax. The head is rather flattened and faces forward, with black compound eyes. The male has long dark claspers on the rear end, extending beyond the wings at rest.

Left: Robber flies are powerful predators of other insects.

Lovebug

March Fly *Plecia nearctica*

Main habitat: Grassland.
Main region: Temperate.
Length: 6mm (0.24in) (male);
8mm (0.31in) (female).
Wingspan: 12–16mm
(0.47–0.63in).
Development: Complete
metamorphosis.
Food: Herbivore (larva); adult
does not feed.

Identification: Small black flies
with opaque black wings, black
legs and a bright red thorax. The
male is smaller than the female in
overall size but has much bigger
bulging compound eyes, which
are also black in both sexes.

This small fly
swarms in
huge numbers
to breed in
late April and
May and again in
August and September.
It is mainly found in southern
North America. It is sometimes seen swarming
in March, hence its other common name.
Its body contains an acid, which
makes it unpalatable to predators. It
also creates problems for the human population as the acidity
can penetrate car paint and even chrome. This insect gets its
nickname of 'lovebug' from the fact that the adults do not
physically separate once mating has started, flying paired until
the female is ready to lay her eggs. The larva sometimes causes
damage to turf and may also be spread via transport of
horticultural products.

*Above: Lovebugs can be a
nuisance when they emerge
in large numbers.*

OTHER SPECIES OF NOTE

Deer Fly *Chrysops* spp.
There are around 100 species in this genus that
occur in North America. Generally smaller than
the horseflies, they often have spotted eyes – a
useful tip when trying to identify them. Many
have longitudinal stripes in golden yellow and
black on a medium-sized hairy body. The wings
are a mixture of bold clear and black markings.
Deer flies live mainly in wet woodland.

Snipe Fly *Chrysopilus ornatus*
Snipe-flies belong to the family Rhagionidae
with over 30 species occurring in the woodlands
of North America. The larvae are found mainly
in decaying wood, whereas the long-legged
adult snipe flies hunt among foliage for other
insects. *C. ornatus* is a small to medium fly of
delicate appearance. It has a tapering black-and-
yellow striped body, large brown compound
eyes and clear grey wings marked with some
dark patches.

Windfall Maple Fly *Macroceromys americana*
This is a fly which mimics hymenopteran
sawflies and ichneumon wasps. It is from the
family Xylomyidae, which has ten other species
in North America. The larva resides on logs
under bark, predating other insects or eating
decomposing plant material. The adult likes to
live on forest edges. The appearance of this
species is wasp-like, with a bright yellow-and-
black body and legs along with a bright red
abdomen. The wings are dark and the antennae
have the appearance of piercing mouthparts,
projecting forward between the eyes.

Picture-winged fly

Idana marginata

This is a pretty insect, with attractively
marked wings, hence its common name.
A close relative of the fruit flies, it
looks a little similar in overall
appearance, being small, stout
and rounded. This species
sometimes feeds at sap flows in
north-east America and south-
east Canada.

Main habitat: Woodland.
Main region: Temperate.
Length: 5mm (0.19in).
Wingspan: 15mm (0.59in).
Development: Complete
metamorphosis.
Food: Herbivore.

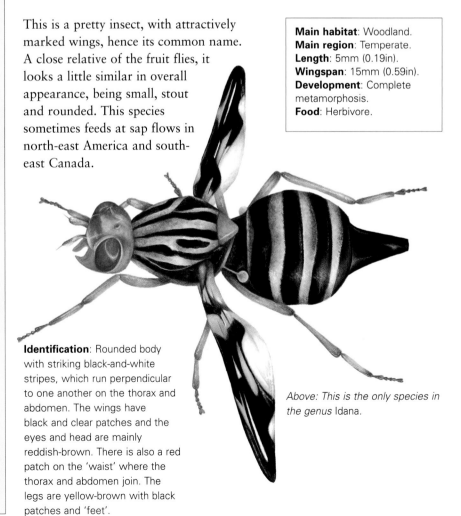

Identification: Rounded body
with striking black-and-white
stripes, which run perpendicular
to one another on the thorax and
abdomen. The wings have
black and clear patches and the
eyes and head are mainly
reddish-brown. There is also a red
patch on the 'waist' where the
thorax and abdomen join. The
legs are yellow-brown with black
patches and 'feet'.

*Above: This is the only species in
the genus* Idana.

BUTTERFLIES AND MOTHS

A truly cosmopolitan group, butterflies and moths (Lepidoptera) are found in every habitat where there is vegetation. Butterflies are some of the most colourful of all insects. Of the 180,000–200,000 species, wingspans range hugely, from 30–320mm (1.18–13in). There are about 10,000 moth species in North America alone, and some 700 butterflies, while the figures for South America are even bigger.

Monarch

Milkweed Butterfly, *Danaus plexippus*

This is one of the most familiar butterflies of North, Central and South America. The caterpillars metabolize toxic chemicals from their food plant milkweed (*Asclepias*), making both them and the adults poisonous to predators. Both advertise this fact with bright warning colours. The adults are unusual in migrating south for the winter. The two main groups in North America either fly from the east to certain Mexican pine forests or from the west to Californian eucalyptus forests. The butterflies stay in communal frost-free roosts before returning north the following spring to lay their eggs.

Identification: Both sexes are bright orange-red with black wing borders and wing veins. The wingtips are patterned with paler red 'windows' and black and white spots thought to mimic false 'eyes' to scare predators. Females have thicker black wing veins and males possess a swollen pouch on their hindwings. The body is black. The butterfly has modified forelegs, giving it the appearance of having only four legs instead of the usual six.

Main habitat: Open country.
Main region: Temperate and tropical.
Length: 40mm (1.57in).
Wingspan: 86–124mm (3.39–5in).
Development: Complete metamorphosis.
Food: Herbivore.

Above: The Monarch is one of the region's best-known butterflies.

Zebra Swallowtail

Eurytides protesilaus

This beautiful butterfly belongs to the swallowtail family (Papilionidae), which contains many splendid species, most of which are very boldly patterned and have long tail-like extensions to their hindwings. It is found over much of South and Central America, south to Argentina and north to Mexico.

Main habitat: Forest.
Main region: Tropical.
Length: 40mm (1.57in).
Wingspan: 80–100mm (3.15–4in).
Development: Complete metamorphosis.
Food: Herbivore.

Identification: The body is black above and white below. The wings are large and deep; mainly pearly white, with dark stripes. The forewings each have a black margin, one long black stripe and five shorter black stripes. The tails on the hindwings are rather long and streamer-like. The hindwings have a scalloped rear margin with blue and red spots. Including the tails, the wings are deeper than they are wide.

Left: Like many of the swallowtails, this is a striking species, with tail streamers.

White Witch

Ghost Moth, Great Owlet Moth, *Thysania agrippina*

This splendid moth has the largest wingspan of any moth worldwide, reaching up to 30cm (12in) across. It occurs in southern American states such as Texas and farther south in the tropical forests of Costa Rica and Brazil. Its wings are camouflaged to make it seem invisible to predators while resting on tree trunks. It is relatively slow in flight.

Identification: Large, rather narrow-winged moth with mottled brown patterning, which creates excellent camouflage on tree bark. Its wings have a wonderful silky appearance and may be either silvery or have a golden sheen.

Right: The white witch is a magnificent insect with a huge wingspan.

Main habitat: Tropical and subtropical forest.
Main region: Temperate-tropical.
Length: 65mm (2.56in).
Wingspan: 30cm (12in).
Development: Complete metamorphosis.
Food: Herbivore.

OTHER SPECIES OF NOTE

Black Witch *Ascalapha odorata*
Another huge moth, with a wingspan exceeding 20cm (8in). It occurs mainly in the tropics, notably Mexico and the Caribbean, but sometimes moves north, even (rarely) as far as Canada. It is beautifully marked with mottled, camouflaged patterns of grey and pale browns. Records exist of it arriving in large numbers as 'fallout' from hurricanes on the shores of Mexico. With its large dark brown wings it is rather bat-like. There are characteristic eye-spots on the forewings, each shaped like a number nine.

Faithful Beauty *Composia fidelissima*
This is a small- to medium-sized day-flying moth with a wingspan of around 45mm (1.77in). It is stunningly beautiful, being predominantly black with areas of metallic blue on the hindwings. All wing borders are spotted with white ovals and the forewings also have a red marking close to the head. The body colouring echoes that of the wings, being black and metallic blue, with small white spots. Adults like hardwood habitats and the young pink caterpillars feed on spurge plants. It is found in the Caribbean and north to southern Florida.

Painted Lichen Moth *Hypoprepia fucosa*
A widespread small, brightly coloured moth found in Canada and the USA. Its wings are streaked blue-grey with orange-yellow and pink markings and a narrow black wing border. The thorax and abdomen are grey. It has two small red 'shoulder patches' on the forewings, a yellow head, and black eyes and antennae.

Luna Moth

Moon Moth, *Actias luna*

Main habitat: Woodland.
Main region: Temperate, tropical.
Length: 30mm (1.18in).
Wingspan: 8–11.5cm (3–4.5in).
Development: Complete metamorphosis.
Food: Herbivore (larva).

This superb moth occurs from Canada south to Mexico. In the north of the range, for example in Canada, it produces only a single generation each year, whereas farther south there may be up to three generations. Although sometimes known as a silkmoth due to its large, silky woven cocoon, it is unrelated to the commercial silkmoths of Asia (genus *Bombyx*). The larval food plants include the leaves of trees such as hickory, walnut and paper birch. The larvae take as little as two weeks to pupate before they emerge as beautiful large moths. The adult's only purpose is to mate – it does not eat and only lives for about a week.

Identification: The caterpillar is plump and lime-green with a pale line along each side of the body. It grows up to 90mm (3.54in) long. The adult moth is pale lime-green. All the wings have pale coloured borders.

Left: The luna moth has eye-spots in the centre of each wing.

WASPS, BEES, ANTS AND SAWFLIES

Wasps, bees, ants and sawflies (order Hymenoptera) are highly specialized insects, with chewing mouthparts and four membranous wings. In North America there are an estimated 17,000 species and many times more than this in South America, especially in the tropical regions. The total number of hymenopteran species in the world has been estimated at 280,000.

Pigeon Tremex

Tremex columba

This is a horntail wasp of forested areas, found mainly east of the Rockies and also south-west to California. Females lay their eggs in hardwood using their pointed drilling organ (ovipositor) projecting from the rear of the abdomen. Many people mistake these for stings and fear these insects, which are actually completely harmless. The larvae feed on decaying wood for up to two years. The female aids this process by adding a little wood-rotting fungus to each egg tunnel from a specialized pouch she possesses to store fungal spores. This accelerates the natural rotting process of the wood and benefits the developing brood.

Identification: Yellow-and-black striped cylindrical body with red-brown and black thorax, head and eyes. The wings are dark, the forewings longer than the hind pair. The legs are yellow. The female possesses an obvious pointed ovipositor on her abdomen for laying eggs in wood. The male also has a horny projection on his rear, though smaller than that of the female.

Main habitat: Woodland.
Main region: Temperate, subtropical.
Length: 25–38mm (1–1.49in).
Wingspan: 40–45mm (1.57–1.77in).
Development: Complete metamorphosis.
Food: Herbivore.

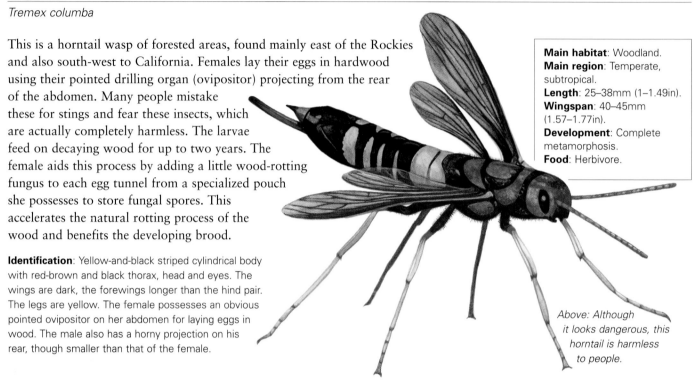

Above: Although it looks dangerous, this horntail is harmless to people.

Giant Ichneumon

Megarhyssa atrata

This is a large ichneumon wasp which raises its young on the larvae of wood-boring horntail wasps. The females have a very long ovipositor, which they use to penetrate deeply into the wood where the wood-boring larvae of horntails are growing. She locates the larvae by tapping the wood with her antennae. She then lays an egg on or near the larva and her young feed on the horntail wasp larvae until they emerge as adult ichneumons to repeat the cycle. The adult wasps do not feed. This species is widespread in North America, especially in the east, from Canada south to Florida.

Identification: A small black wasp of slender build with a tiny waist and an extraordinary long, narrow ovipositor (female) up to four times as long as the body. The head is yellow with a black mask-like band around black eyes. It has slender pale legs and antennae.

Main habitat: Forest.
Main region: Temperate.
Length: 35mm (1.38in) (male); 38mm (1.49in) (female), 13cm (5in) (ovipositor).
Wingspan: 60mm (2.36in).
Development: Complete metamorphosis.
Food: Carnivore (larva).

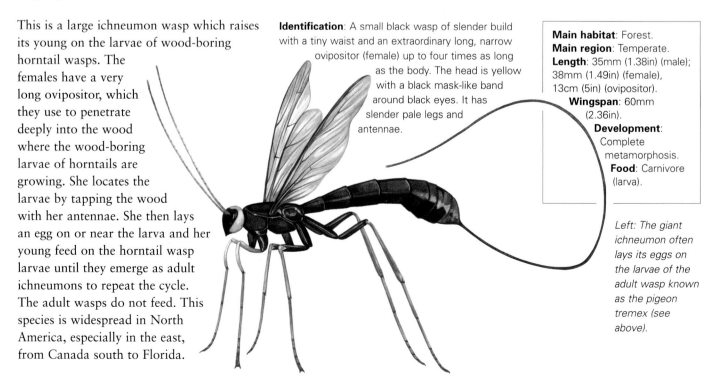

Left: The giant ichneumon often lays its eggs on the larvae of the adult wasp known as the pigeon tremex (see above).

Leaf-cutter Ant

Atta spp.

This fascinating ant lives in large colonies within tropical forests, notably in the Amazon region. Leaf-cutter ants construct huge nests, only a fraction of which are visible above ground as mounds of earth. Colonies may consist of up to three million individuals. The ants co-operate with one another to collect and transport huge amounts of leaf material into the nest. With the collected leaves they create 'fungus gardens' and in due course feed on the resulting fruiting fungus. Ants of this kind are increasingly being recognized as ecologically important in forests due to their leaf recycling activities. These insects are also known as 'parasol ants' as they hold the leaf portions high above their body on their journey to the nest.

Below: These ants harvest leaves in tropical rainforests.

Main habitat: Forest.
Main region: Tropical.
Length: 7–8mm (0.28–0.31in).
Wingspan: None.
Development: Complete metamorphosis.
Food: Herbivore.

Identification: Typical ant shape, with a slim waist, elongated body and long thin legs capable of running very fast. The head is large and reddish in colour, with well-adapted tough mouthparts for cutting and holding the large quantities of plant material that they transport to their nests. The antennae are long and elbowed.

OTHER SPECIES OF NOTE

Wood Wasp *Xiphydria* spp.
This genus includes six species of North American wood wasps. These cylindrical wasps are similar in appearance to horntails but smaller and generally slighter in build. Their reproduction is also similar, with females boring holes in decaying wood in which they place some wood-rotting fungus before laying an egg. Maple wood is chosen in preference to others. *Xiphydria maculata* is a common species found in the north-east USA, extending to southern Canada.

Elm Sawfly *Cimbex americana*
One of four species in this genus occurring in North America. As well as elm it will live and feed on maple, alder, birch, poplar, basswood and willow trees. The adults are predominantly black with yellow, clubbed antennae, yellow feet and transparent brown wings. The white larvae closely resemble caterpillars and have black spots along each side of the body. They feed on the leaves of the trees in which they live and pupate in fibrous spun cocoons.

Ichneumon Wasp *Acrotaphus wiltii*
This ichneumon is a parasitoid of orb spiders. The female stings the spider, causing it to be paralysed for a short time, during which she takes the opportunity to lay an egg on its body. Her egg then hatches and the wasp larva develops on the living spider. The adult wasp is dazzling orange-yellow and black in colour. Its body is characteristically slim and delicate with long legs and antennae.

Giant Hunting Ant

Pachycondyla villosa

This large ant occurs in tropical South and Central America, north to southern Texas. It is rather common in Costa Rica and Brazil and also occurs south as far as northern Argentina. It is also known as the 'hairy panther ant'. It likes to nest in a wide range of habitats including wet and dry native forests and also plantations where it nests in soil, tree stumps and logs, as well as in bromeliad plants. It is active during the day, hunting other insects for food. If a nest is disturbed, the worker ants become aggressive and defend it fiercely.

Below: Ants nest in cavities in dead wood, or knots of dead branches of live trees.

Main habitat: Forest.
Main region: Tropical.
Length: 15–17mm (0.59–0.67in).
Wingspan: None.
Development: Complete metamorphosis.
Food: Carnivore.

Identification: A large blackish ant with golden hairs and reddish coloration on the legs. It has a large head relative to the rest of body, with long mobile antennae. Large dark eyes and huge serrated mouthparts for catching prey. Females are larger than the males, being 17mm (0.67in) in length.

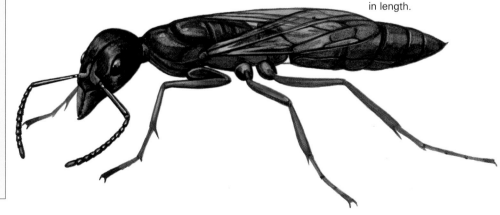

Valley Carpenter Bee

Xylocopa varipunctata

Carpenter bees include some of the largest bees. This medium-sized species is unusual in having different coloured males and females. These solitary bees nest in old and decaying wood including unpainted man-made structures. Their name comes from their habit of creating tunnels in wood to construct egg chambers, often leaving tell-tale piles of sawdust below the entrance hole. A ball of nectar and pollen is placed inside each chamber and the female lays an egg on each ball. Male bees can be very aggressive and territorial but are harmless despite their size and are less often seen, being rather short-lived. They are important pollinating insects in Arizona and California.

Main habitat: Valleys and foothills.
Main region: Temperate.
Length: 18–20mm (0.71–0.79in).
Wingspan: 45mm (1.77in).
Development: Complete metamorphosis.
Food: Herbivore.

Identification: Typical medium-sized bee with obvious body hairs used for collecting pollen, though not as hairy as bumblebees. Females are black with a violet or bronze metallic sheen but the males look quite different, being buff or golden in colour – rather like a flying teddy bear!

Right: Carpenter bees have sturdy, hairy bodies.

Miner Bee

Anthophora abrupta

This North American bee is also sometimes known as the 'moustached mud bee'. This refers to the moustache-like area on the face of the male, used to help spread pheromones to attract the female bee. The female builds a clay tunnel up to 75mm (2.95in) over her chosen nest entrance. Sometimes the tunnel and nest are recycled from previous years. These bees are important pollinators of the wide range of plants on which they feed. The miner bee could be mistaken for a bumblebee, but it is paler yellow and black.

Main habitat: Clay banks and walls.
Main region: Temperate.
Length: 12–17mm (0.47–0.67in).
Development: Complete metamorphosis.
Food: Herbivore.

Identification: Medium-sized bee, fairly round in shape and superficially similar to a bumblebee. Hairy body, with pale thorax, black head, abdomen and legs. Clear wings with dark veins and large black compound eyes.

Above: Miner bees are smaller and faster in flight than bumblebees.

Cicada Killer

Sphecius spp.

Main habitat: Well-drained soil.
Main region: Temperate, tropical.
Length: 40mm (1.57in).
Wingspan: 75mm (2.95in).
Development: Complete metamorphosis.
Food: Carnivore (larva); herbivore (adult).

This large wasp has a life cycle that runs parallel to the cicada species it predates. Females dig burrows nearly 1m (3ft) long into the ground and site a nest chamber at the end. Cicadas are caught and paralysed by the wasp and brought back to the chamber where the female wasp lays an egg on it. The female is busy during her short adult life of about a month, digging nest tunnels in dry grassy sites. She digs not just one but many tunnels and nest chambers, into each of which she lays an egg. The larvae feed on the collected insects, then grow and overwinter safely underground until ready to emerge the following July to breed. A number of species are found in North, Central and South America. The adults, although fierce-looking, are fairly harmless to humans, and feed on pollen and nectar.

Identification: Large black-and-yellow wasp covered in small dark hairs. Clear brown wings and large brown compound eyes. Black antennae with many segments. The brown legs have spurs to aid digging.

Far left and above: This wasp specializes in catching cicadas on which to raise its larvae.

Bald-faced Hornet

White-faced Hornet, *Dolichovespula maculata*

Main habitat: Trees and bushes.
Main region: Temperate.
Length: 12–14mm (0.47–0.55in); 20mm (0.79in) (queen).
Wingspan: 24mm (0.94in); 33mm (1.3in) (queen).
Development: Complete metamorphosis.
Food: Carnivore (larva); omnivore (adult).

Identification: A black-and-white hornet with greyish wings. The 'white' or 'bald-faced' description comes from a pale patch on the head between the eyes and mouthparts. Males and females are similar to look at but have subtle differences. Males have slightly longer antennae (an extra segment) and also an additional segment to their abdomen, marked with white, giving them a few more pale patches than the females.

This is a large social wasp found living in colonies. This species has a wide distribution across North America but is absent from the driest regions of the Mid west and south-west. Bold-faced hornets build beautiful rounded paper nests suspended from trees or bushes, which inside contain hexagonal cells for raising their young. The larvae are fed on pre-chewed insects while the adults consume nectar and fruit as well as other insects. Fertilized queens usually overwinter, and then emerge to found a new colony the following spring.

Right: This large hornet is common in North America.

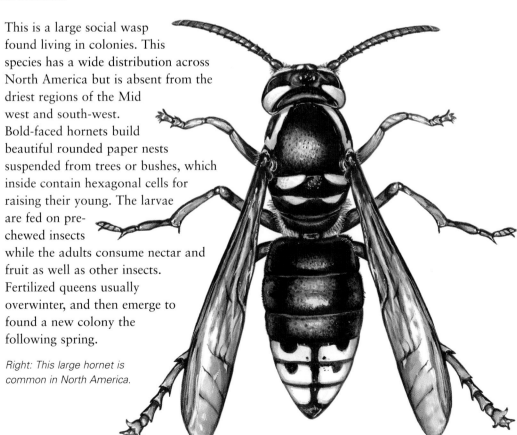

MILLIPEDES AND CENTIPEDES

Millipedes and centipedes are distinguishable from insects by their many legs (at least nine pairs and usually many more), and by their uniform bodies (not divided into head, thorax and abdomen). There are approximately 15,000 species, ranging in length from 0.05–30cm (0.02–12in). The Americas boast some of the largest members of both groups.

House Centipede

Scutigera coleoptrata

Like other members of this centipede order (Scutigeromorpha) this species is fast-running and very agile, using its speed to hunt insects such as cockroaches, moths, flies and crickets, and also spiders. House centipede jaws contain an immobilizing venom, and they can inflict a painful bite if handled, but luckily this species is not particularly aggressive. As the common name suggests, it is found in houses, where it prefers damp sites such as pantries or cellars. Now found in America and elsewhere, this centipede is probably native to southern Europe.

Main habitat: Houses, gardens.
Main region: Temperate, subtropical.
Length: 40mm (1.57in).
Development: Gradual.
Food: Carnivore.

Identification: Mainly yellow-brown, this centipede has three dark stripes along its back. The 15 pairs of legs increase in length toward the rear, and are distinctly banded.

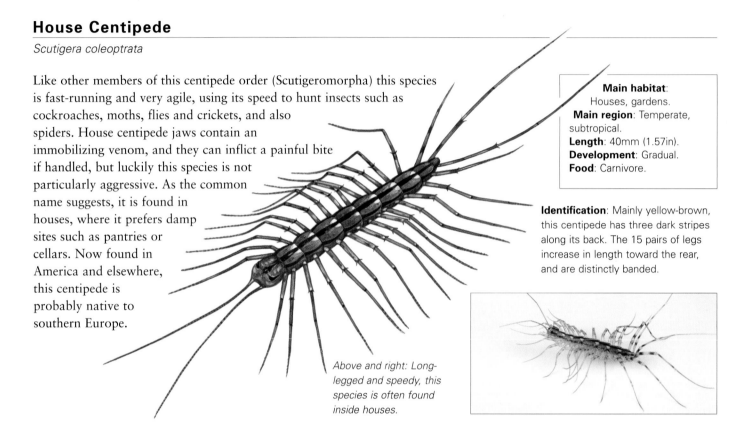

Above and right: Long-legged and speedy, this species is often found inside houses.

Giant Desert Centipede

Scolopendra heros

One of the world's largest centipedes, this species lives mainly in dry scrubland and deserts in Mexico and the south-west USA. Very active and quite aggressive, it hunts for insects and even vertebrates such as small rodents and lizards, and can capture creatures larger than itself. The last pair of legs are modified for grasping, and also stick upward, mimicking antennae, which may confuse a predator into attacking the wrong end. This species lives mainly on the surface but occasionally retreats into burrows. Its venomous bite is very painful and, although not deadly to humans, can cause swelling and pain for up to two days. Poison is held in special glands at the base of the fang-like front claws and injected into wounds when the centipede bites. There are a number of subspecies with differing colour patterns, the red-headed form being the most colourful. The giant desert centipede is sometimes kept as an unusual pet. However, this is a dangerous species and should not really be handled.

Right: A large centipede with a very painful poisonous bite.

Main habitat: Desert.
Main region: Tropical–subtropical.
Length: 16–20cm (6–8in).
Development: Gradual.
Food: Carnivore.

Identification: Very large, with 20 pairs of evenly sized legs and long, slender antennae. The red-headed subspecies has a black shiny body, yellow legs and a bright red head; the blue-tailed form is yellowish, and bluish toward the rear; the black-headed form is red or orange with a black head.

OTHER SPECIES OF NOTE

Amazonian Giant Centipede

Scolopendra gigantea

From tropical South America comes this largest of all centipedes, at more than 30cm (12in) in length. An active carnivore it can catch prey as large as birds, lizards, bats and rodents. It has one very unusual hunting method – sometimes hanging from the walls of caves, then striking at and killing passing bats. Also known as the 'Peruvian giant yellowleg centipede', it lives in north and west South America and also on Trinidad and Jamaica. Its body is reddish or maroon and the legs are yellow. The venom of this species is quite powerful and can cause swelling and fever in humans unfortunate enough to suffer its bite.

Eastern Red Centipede

Scolopocryptops sexspinosus

This medium-sized centipede is the most common species in eastern North America. It grows to a length of about 75mm (2.95in) and is quite easy to spot as its body, legs and antennae are bright orange-red. As with other members of the order Scolopendromorpha, it is an agile hunter with narrow antennae held out in a V shape and antennae-like hind legs, adapted for grasping. Its favoured habitats are underneath rocks, logs and bark. Like many centipedes, it delivers a painful bite.

Leggy Millipede *Illacme plenipes*

Famous for having the most legs of any animal – as many as 750 – this rare species is known only from a small area in central California. It is cream-coloured, with a long, thin, flexible body and grows to about 15mm (0.59in) (male) and 33mm (1.3in) (female). The females have more legs than the males.

Yellow-spotted Millipede

Harpaphe haydeniana

This millipede belongs to the order of flat-backed millipedes (Polydesmida), and has a flat rather than domed back. It is found in the temperate rainforests of North America's Pacific north-west, ranging from California north to Alaska. Its body is black with bright yellow spots along each side and it grows to about 50mm (2in) long. It feeds mainly on humus on the forest floor. It is sometimes called the almond-scented millipede because it exudes a scent of bitter almonds (hydrogen cyanide) when disturbed. This, combined with its warning colours, gives it some protection from predators such as shrews or beetles, which might otherwise eat it.

American Giant Millipede

Narceus americanus

Identification: A long, tubular, shiny, purplish-brown body with 182 slender, evenly sized red legs. The dorsal segments are edged in red. Two large compound eyes. When disturbed it may curl into a tight spiral and exude a nasty-smelling chemical – often enough to deter attack.

Main habitat: Woodland, gardens.
Main region: Temperate.
Length: 10cm (4in).
Development: Gradual.
Food: Herbivore.

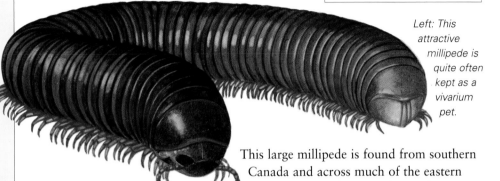

Left: This attractive millipede is quite often kept as a vivarium pet.

This large millipede is found from southern Canada and across much of the eastern USA. It patrols the ground, mainly at night, and forages among the leaf litter for decaying vegetable matter and fungi. American giant millipedes are quite slow-moving and so frequently fall prey to rodents, amphibians and lizards, although they can defend themselves by producing a noxious fluid. Mated females lay up to 300 white eggs, placed in a shallow hollow. These hatch after a few weeks as tiny young, with only three pairs of legs at first. The number of legs increases steadily as they moult and grow.

Texas Striped Millipede

Orthoporus ornatus

This is one of a number of species known as desert millipedes, being found mainly in dry and desert habitats in North and South America. The Texas striped millipede occurs mainly in Texas, Arizona and New Mexico. Although it lives in arid regions, it cannot withstand extreme heat and seeks out humid places, emerging from burrows at night, especially after rain. It feeds mainly on plant material but sometimes scavenges dead animals. It can live for about ten years. The young are born in spring. This is another popular pet species and is fairly easy to keep, using peat-based soil that retains humidity well. It can be fed on fruit and vegetables and leafy greens.

Below: This is a mainly nocturnal species, often found in deserts.

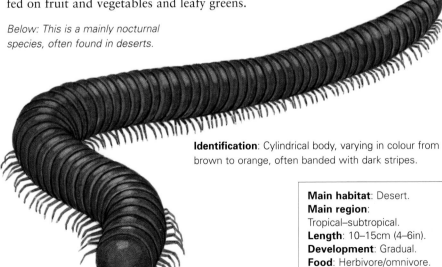

Identification: Cylindrical body, varying in colour from brown to orange, often banded with dark stripes.

Main habitat: Desert.
Main region: Tropical–subtropical.
Length: 10–15cm (4–6in).
Development: Gradual.
Food: Herbivore/omnivore.

SPIDERS

Spiders are well adapted to their predatory, carnivorous lifestyle, and they help to control many insect pests. Some, like the black widow spider, are dangerous, but most spiders are harmless to people. Although not all spiders actually build webs, all 35,000 known species can produce silk. The giant hairy tarantulas are some of the largest spiders in the American region.

Goliath Bird-eating Spider

Goliath Tarantula, *Theraphosa leblondi*

Often referred to as the 'king of tarantulas', this large spider from the forests of northern South America (notably Venezuela, Guyana, French Guiana and Brazil) can weigh as much as 160g (5.5oz), with a legspan of 25cm (10in) or more. Despite the common name, birds are not its usual prey. Although these spiders have indeed been seen to kill and eat birds, they normally prey on large insects, frogs, lizards and mice. They have venomous fangs, but bites are not fatal to people, being equivalent to a wasp sting and just causing local swelling and soreness. Many bites are venom-free. They have a further method of defence in being able to release irritating stinging hairs, and they can also make rather startling hissing sounds when disturbed. Goliath bird-eating spiders inhabit burrows in the moist soil of rainforests, either excavating these themselves or taking over an abandoned rodent burrow. Males live for between three and six years, but females are reported to live for as long as 25 years. The females can lay up to 400 eggs at a time, which hatch into miniature spiders.

Main habitat: Wet forests.
Main region: Tropical.
Length: 12cm (4.5in).
Legspan: 25cm (10in).
Development: Gradual.
Food: Carnivore.

Above: The goliath bird-eating spider has a good claim to be the world's largest spider.

Identification: Large and hairy, ranging in colour from light to dark greyish-brown. The legs are sturdy and covered in hairs.

Mexican Red-kneed Tarantula

Brachypelma smithi

This showy tarantula lives mainly in dry scrubland along the Pacific coast of Mexico and also in South America, although it is also found in forests. It spends most of the day in a silk-lined burrow, emerging at night to catch prey such as large insects, other spiders and small vertebrates. Like most tarantulas Mexican red-kneed tarantulas have a venomous bite and can release stinging hairs from their hind legs. This species is quite a popular (if slightly dangerous) pet, being large, boldly marked and furry, and also fairly easy to keep. These spiders are relatively placid and can be kept in a vivarium, but the pet trade is endangering the species in the wild, as is the destruction of its habitat. They breed at two years old and lay up to 1,000 eggs.

Above: The Mexican red-kneed tarantula only bites humans if provoked.

Main habitat: Scrub and desert; forest.
Main region: Tropical.
Length: 90mm (3.54in).
Legspan: 15–18cm (6–7in).
Development: Gradual.
Food: Carnivore.

Identification: A large, hairy spider with a dark brown body with reddish-orange and yellow patches on its leg joints.

Black Widow Spider

Latrodectus mactans

Main habitat: Varied.
Main region: Temperate, tropical.
Length: 6.5–10mm (0.26–0.39in).
Legspan: 20mm (0.79in).
Development: Gradual.
Food: Carnivore.

Identification: Mainly black and shiny with a rounded abdomen (female) and long, slender legs, with red spots on the back and a red patch on the underside. The male is smaller and has a longer abdomen, with red and white stripes along the edges.

Right: The name 'widow' comes from the fact that the female sometimes kills her partner after mating.

A notoriously venomous species of spider found across much of the Americas, from southern Canada through the USA to Mexico, the West Indies and South America. In North America it is most common in the south, especially in the desert regions of the south-west. It is not an aggressive species – most bites are the result of inadvertently stepping on or handling a spider, as for example when one finds its way into a shoe or glove, causing the spider to bite in self-defence. While the bite is painful and debilitating, it is rarely fatal. This is an adaptable spider, able to thrive in crevices of all kinds, including in sheds and garages as well as on wasteland, in wood stores and the like. The web is untidy and funnel-shaped and the female often hangs from it upside down.

OTHER SPECIES OF NOTE

Colombian Giant Red-legged Tarantula
Megaphobema robustum
A large tarantula from the rainforests of Colombia and Brazil, measuring up to about 20cm (8in) across. It has the typical tarantula hairy body and thick legs. In this species the legs are red, with black sections toward the base. The hind pair of legs are the hairiest and, as in many tarantulas, are used for defence. When threatened, this spider bobs up and down and flails its back legs, brushing their sharp irritant hairs against the attacker.

Pinkfoot Goliath Tarantula
Theraphosa apophysis
There is some argument as to whether this or the Goliath bird-eating spider is the world's largest tarantula. Although less well known, this pinkish-brown species from Venezuela can be massive, reputedly reaching a legspan of about 33cm (13in). In the wild it lives in deep burrows in the rainforest soil, emerging to hunt for large insects and small vertebrates.

Rosehair Tarantula *Grammostola rosea*
One of the most commonly available pet tarantulas, this species hails from the Atacama Desert of northern Chile and is also found in Argentina. It is adapted to desert and dry scrubland. Relatively small, with a legspan of about 12.5cm (5in), it is also rather docile and slow. Its hairy body and legs are a beautiful pinkish colour (hence the name), with a distinctly coppery sheen. It adapts quite well to captivity, and is easy to keep on a diet of insects.

Brazilian Wandering Spider

Phoneutria fera and *P. nigriventer*

Main habitat: Forest.
Main region: Tropical.
Length: 30–40mm (1.18–1.57in).
Legspan: 10cm (4in).
Development: Gradual.
Food: Carnivore.

Identification: Fairly large spider, and usually brown with rows of paler markings along the back. The legs are sturdy and long and the head bears very large and obvious fangs.

These venomous spiders belong to a family known as wandering spiders (Ctenidae). They are fast-moving, aggressive hunters, mainly found in the tropics where they often wander about on the forest floor. What makes them particularly dangerous is that not only do they have a toxic venom, they are also rather fearless and sometimes attack people. Bites from these spiders are very painful and sometimes fatal, and they are regarded as one of the most deadly of all spiders. They live mainly in northern South America, from northern Argentina to Brazil and Venezuela. They sometimes occur in and around houses, which is where most bites happen. They lurk in dark crevices and occasionally hide inside bunches of bananas, only to be imported along with them into other countries. For this reason it is sometimes called the 'banana spider'.

Above: Aggressive and highly venomous, the Brazilian wandering spider is very dangerous.

Golden Silk Spider

Banana Spider, *Nephila clavipes*

Not counting the very big tarantulas, this spider is one of the largest in North America, with a legspan of 12.5cm (5in) or more. It has a wide range, from the southern USA, through Central America and the West Indies, south to Argentina. Harmless web-weavers, golden silk spiders are welcome in gardens, where they catch many insects, some of which are harmful to garden plants, or a nuisance to people. The common name comes from the fact that the silk used to build their prominent, large webs is yellow, making the webs easy to spot. The webs can appear overnight as if by magic, sometimes right across a footpath. The alternative common name refers to the yellow colour of the spider.

Main habitat: Trees and bushes.
Main region: Warm temperate, tropical.
Length: 24–40mm (0.94–1.57in) (female); 6mm (0.24in) (male).
Legspan: 12.5cm (5in) (female).
Development: Gradual.
Food: Carnivore.

Identification: The female is much larger then the male. She has a silver cephalothorax, and an orange abdomen with white spots. The legs are banded brown and orange.

Above and right: Large and brightly coloured, the golden silk spider is easy to spot.

Social Spider

Anelosimus eximius

This small spider is found in the rainforests of South America, where it lives in colonies ranging from hundreds to many thousands of individuals. Together, the spiders co-operate to construct a sagging web, suspended hammock-like from long anchor threads high in the forest canopy, or sometimes over shrubs nearer to the ground. Such webs can be long-lasting and are constantly repaired by the colony members. These spiders also work together to trap and subdue prey, often many times bigger than each individual. Large numbers of spiders, each producing silk, can immobilize their prey most effectively. The individuals in each group are very closely related and work together for the good of the colony.

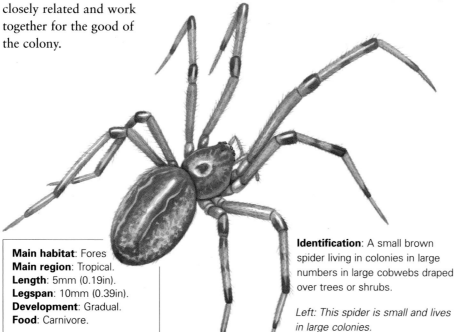

Main habitat: Fores
Main region: Tropical.
Length: 5mm (0.19in).
Legspan: 10mm (0.39in).
Development: Gradual.
Food: Carnivore.

Identification: A small brown spider living in colonies in large numbers in large cobwebs draped over trees or shrubs.

Left: This spider is small and lives in large colonies.

OTHER SPECIES OF NOTE

Banded Argiope Spider
Argiope trifasciata
A large web-builder that has an oval, striped abdomen and long, striped legs. Although big, it can be hard to spot in dappled shade as its striped markings disrupt its outline, rather like a tiger in the jungle.

Bold Jumping Spider
Phidippus audax
This is one of the most common of the jumping spiders in North America. It has four large eyes on the face and four smaller eyes on top of its head. These give it good vision for active hunting. They can also jump suddenly, covering a long distance. A special feature of this species is its large chelicerae (mouthparts), which are a bright iridescent green, and used in a threat display.

Crab Spider *Misumenoides formocipes*
A common American crab or flower spider, this species is usually found in the flower-heads of umbellifers such as Queen Anne's lace. It sits among the flowers waiting for insect prey to alight. The abdomen is yellow-orange and the cephalothorax and rear two pairs of legs are greenish-yellow. Only the long, crab-like front two pairs of legs are obvious, being dark brown.

MITES AND TICKS

Mites and ticks inhabit an impressive variety of locations worldwide (alpine, desert, terrestrial, freshwater, marine and underground), and are the most diverse group of all the arachnids. There are 35,000 species, 850 of which are ticks. Mites are tiny creatures, ranging in length from 0.08–16mm (0.03–0.63in). The slightly larger ticks range from 2–30mm (0.08–1.18in) in length.

Hard Tick

Ixodes spp., *Dermacentor* spp.

There are several hard ticks in these two genera and some of these are of medical significance as they can transmit diseases to people. Among the American species are the western black-legged tick (*Ixodes pacificus*) and the Rocky Mountain wood tick (*Dermacentor andersoni*), which can cause Lyme disease and Rocky Mountain spotted fever respectively. Female ticks lay their eggs in the soil or leaf litter. When the six-legged larvae hatch they seek out a small mammal, feed on the blood and then drop off before changing into the eight-legged nymph. The nymph also attaches to a mammal for a meal before moulting into the adult stage. It then drops off. The host species varies with the species of tick, some preferring deer, while others will select birds or small mammals. Adults of the species select large mammals including people. They wait on grass stems and shrubs and quickly attach themselves to a suitable passing host.

Identification: Small and rather hard-bodied with claw-like legs and well-developed mouthparts. *Ixodes* is blackish; *Dermacentor* is silver-grey and red-brown.

Main habitat: Scrub and grassland.
Main region: Temperate.
Length: 3–6mm (0.12–0.24in).
Development: Gradual.
Food: Carnivore.

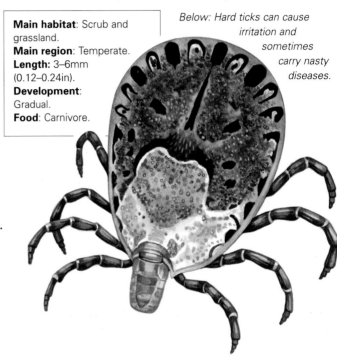

Below: Hard ticks can cause irritation and sometimes carry nasty diseases.

Harvest Mite

Trombicula spp.

Also known as 'scrub-itch mites' but perhaps best known in their larval form as 'chiggers', these mites live mainly in grassland and open forests. In North America they are found mainly in the Mid-west and southern states. The tiny larvae cause problems by feeding on the skin of mammals, including people, especially in hot and humid weather. They break down the skin cells by injecting enzymes, and then they feed on the skin, causing itching and swelling. After feeding, the larvae drop off to develop into nymphs and then adults with the full complement of eight legs. These are not parasitic, but feed on plant material.

Identification: Tiny and mainly orange-coloured. The larva is extremely small with six legs and chewing mouthparts.

Main habitat: Grassland and scrub.
Main region: Temperate, subtropical.
Length: 0.4mm (0.02in).
Development: Gradual.
Food: Carnivore (larva); herbivore (nymph and adult).

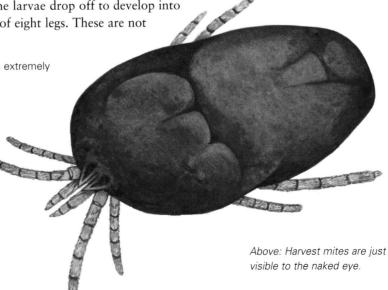

OTHER SPECIES OF NOTE
American Dog Tick *Dermacentor variabilis*
This is one of the most commonly encountered hard ticks. The larvae of this species feed on small mammals such as mice and also on birds, while the adults feed on dogs, raccoons and other large mammals, including humans. This is one of the species that can transmit Rocky Mountain spotted fever.

Above: Harvest mites are just visible to the naked eye.

SCORPIONS AND PSEUDOSCORPIONS

Scorpions live in a variety of habitats in tropical and warm temperate climates, and range in size from 1–20cm (0.5–8in). They are often found in deserts, but also in rainforests. They have pincer-like claws and an arching tail with a poisonous sting at its tip. As a general rule, scorpions with small claws and a large sting are dangerously venomous while those with large claws and a small sting are less so.

Durango Scorpion

Centruroides suffusus

Named after the Mexican state of Durango, this scorpion is found in dry habitats, where it lurks beneath rocks. Similar species are found in Texas and other southern states. It is a dangerous species, with strong venom that occasionally kills people, even healthy adults. It feeds mainly on insects and other arachnids. Like most scorpions, the Durango scorpion is mainly nocturnal. It can live for about 10 years. Although it can be kept in captivity, it is a dangerous, aggressive species and is therefore quite unsuitable as a pet.

Right: This dangerous scorpion is definitely one to be avoided.

Identification: Yellow or light brown in basic colour, with two dark stripes along the upper side of the abdomen. The pincers are slender and the tail is long, with a fairly large sting at the tip.

Main habitat: Desert.
Main region: Tropical.
Length: 80mm (3.15in).
Development: Gradual.
Food: Carnivore.

Giant Hairy Desert Scorpion

Hadrurus arizonensis

The largest of all North American scorpions, this is also an aggressive species, but although it will sting, its venom is not as toxic as some others and does not usually cause problems other than local discomfort, except when there is an allergic response. It lives mainly in the deserts of Arizona and California, and in southern Utah and Nevada. Being large it can include lizards in its diet as well as the usual range of invertebrates. The hairs on its body probably help it to detect vibrations caused by the movements of other animals, including its prey. In dry summer weather they retreat into burrows as deep as 2.5m (8ft) to escape the heat and find moisture.

Main habitat: Desert, semi-desert.
Main region: Tropical, warm temperate.
Length: 14–15cm (5.5–6in).
Development: Gradual.
Food: Carnivore.

Identification: Large, with a brown body and yellowish pincers and legs. The body has a covering of brown hairs. The tail is long.

Left and right: A large scorpion, capable of catching prey as big as lizards.

Dune Scorpion

Giant Sand Scorpion, *Smeringurus mesaensis*

Main habitat: Desert.
Main region: Tropical, subtropical.
Length: 7.6–10cm (3–4in).
Development: Gradual.
Food: Carnivore.

Identification: A large, rather slim scorpion, with a ghostly, colourless body and slender pincers and tail. The head and pincers are often pale yellow, as are the back and tail, while the sides of the body are translucent white.

Right: The pale dune scorpion blends well with its sandy habitat.

This scorpion is found in the sandy deserts of California, Arizona and Mexico where it lives in burrows. Its coloration gives it good camouflage in the sand, but it also makes this species attractive to collectors, and it is often sold embalmed inside plastic paperweights and ornaments. Rather aggressive, it will sting quite readily but is not usually dangerous, the sting being equivalent to that of a wasp. This is fast-moving scorpion that uses rapid bursts of speed to capture its prey. This scorpion is fairly popular as a pet, but is apparently not very easy to breed, so the practice puts more pressure on the wild populations.

OTHER SPECIES OF NOTE

Bark Scorpion *Centruroides exilicauda*
This is a dangerous scorpion from Mexico and the south-west USA. It has a powerful sting and can be deadly. Unfortunately it is sometimes found in houses. It rests in a head-down position, often above the ground on the bark of a tree or on the side of a rock. Like many dangerously venomous scorpions, its pincers are rather slender. It grows to a length of about 25–75mm (1–2.95in).

Devil (Stripe-tailed) Scorpion
Vaejovis spinigerus
A scorpion of dry, rocky sites and deserts in the south-west USA and Mexico. This species is not as dangerous as the bark scorpion, having a less powerful sting. It lives in the Sonoran desert and other deserts, and also nearby forests and grassland. The tail has a number of narrow stripes. It grows to between 50–70mm (2–2.76in).

Lesser Stripe-tailed Scorpion *Vaejovis coahuilae*
Closely related to the previous species, this small scorpion has a similar range. It lives in many different habitats, from pine forests to grassland and deserts. Its sting is painful and causes local swelling, but no long-lasting symptoms. It reaches a length of about 45mm (1.77in).

Didymocentrus krausi
A small scorpion from Central America, living in small burrows underneath stones and fallen tree trunks, mainly in Pacific forests. It grows to about 40mm (1.57in) and is light red-brown with a shiny body.

House Pseudoscorpion

Chelifer cancroides

Pseudoscorpions are more common than one might imagine, but are rarely seen as they are very small and hide away in crevices. Their favoured habitats are moss, leaf litter and underneath stones, logs or bark. They are active hunters, preying on other small arthropods such as bark-lice, thrips and springtails. Like scorpions, they are sensitive to vibrations and this helps them track down their prey, paralysing them with venom secreted from their sharp pincers.
Like spiders, they can produce silk, but from glands on their chelicerae, and use this to fashion protective cocoons in which to overwinter or protect their eggs. This species is often found inside houses or barns, in damp sites. The mating behaviour is interesting. The male marks out a mating territory with a special pheromone. When a female enters, he dances, vibrating his body and waving his pincers. He then deposits a sac of sperm on the ground and guides the female over it. The whole process may take ten minutes to an hour.

Main habitat: Soil, leaf litter.
Main region: Temperate.
Length: 3–4mm (0.12–0.16in).
Development: Gradual.
Food: Carnivore.

Identification: Tiny, with four pairs of legs and large pincers (pedipalps); the latter are more than twice as long as the legs. The body is rich brown and the abdomen clearly segmented.

Right: Tiny pseudoscorpions have no sting and are harmless.

INSECTS OF AUSTRALIA AND ASIA

Asia covers a huge area and many climate zones and includes almost every major habitat type, from the vast, dark coniferous forests of the north, through deserts, grasslands and temperate forests to the rich rainforests of the south. Australia too has its rainforests, but much of its interior is extremely arid. The large, colourful birdwing butterflies here are some of the most impressive insects, and also rather special to the region are the extraordinary leaf insects with their leaf-like flanges of cuticle that resemble foliage. Many of Australasia's 86,000 insect species are adapted to extreme conditions, such as the midges whose larvae can withstand the drying out of desert waterholes and temperatures of up to 58°C (136°F). The long isolation of New Zealand has produced some of the world's strangest insects, including the slow-moving and flightless wetas.

Above from left: A green carab beetle, a curved spiny spider and an orchid mantis.

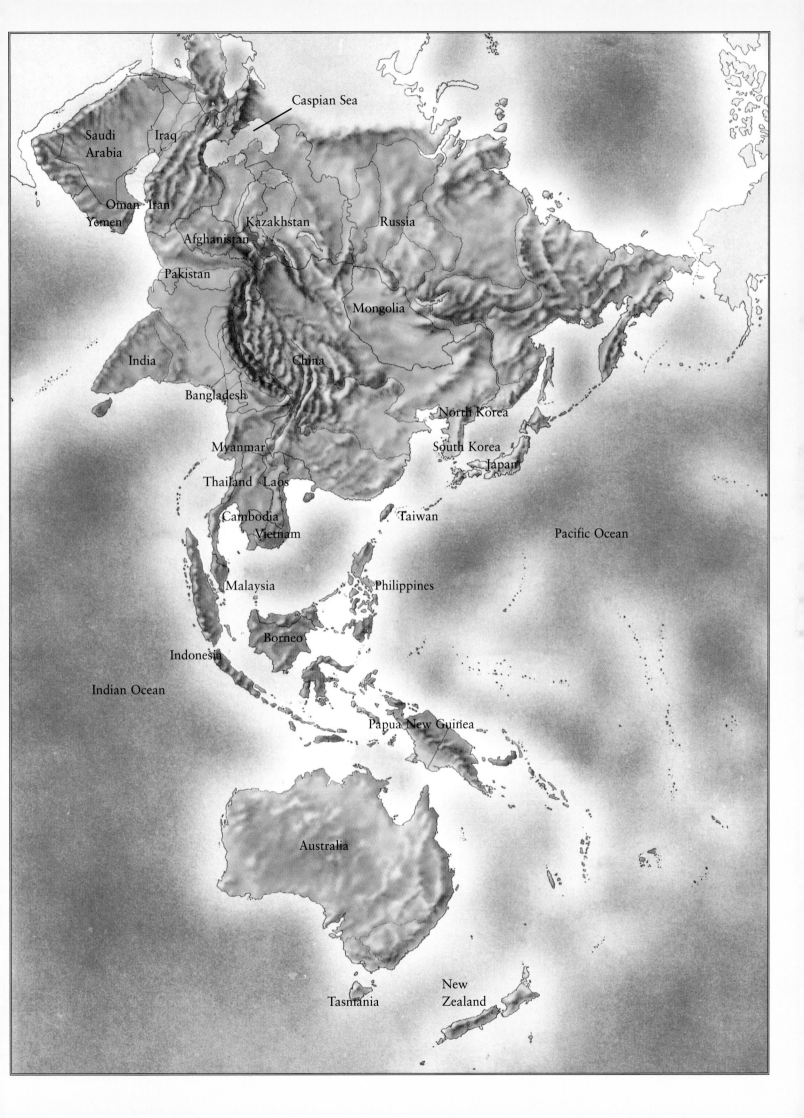

MAYFLIES AND STONEFLIES

Mayflies (Ephemeroptera) and Stoneflies (Plecoptera) spend most of their lives as aquatic larvae. Mayfly larvae have triple 'tails' or cerci protruding from the rear of the abdomen, while stonefly larvae have two cerci. Adult mayflies live only for a day or two, stoneflies for a few weeks. In Australia and New Zealand these insects are mainly found in or near cold streams and rivers.

Mayfly

Oniscigaster distans

This mayfly is on the wing from November to January. The main habitat for the larvae are clear forest streams, gravel-bottomed pools and soft-bottomed mountain lakes (cut off from rivers), usually with plenty of vegetation on the banks.

Right: Clear streams are the favoured habitat of this mayfly.

Main habitat: Streams and ponds.
Main region: Temperate – New Zealand.
Length: 28mm (1.1in).
Wingspan: 55mm (2.17in).
Development: Incomplete metamorphosis.
Food: Larva omnivorous; adult does not feed.

Identification: The adults are dark brown with a few lighter markings. The legs are lighter brown with dark brown ends. The sub-imago (sub-adult) is greyer than the adult, and it moults to the adult stage after three days. The larvae come in a variety of colours from brown through green to white but are always very well camouflaged and hard to find. They have feathery gills on the first six abdominal segments and a minute pair on the seventh.

Mayfly

Atalophlebia spp.

Common in Australia, these mayflies are found in slow-flowing streams and lakes, billabongs of the dry interior or in subalpine meltwater pools. The majority of species prefer cool, fast-flowing water. The nymphs feed on algae, rotting plant debris and minute organisms, though some species are predators. Adults do not feed. Mating occurs in swarms above the watery habitats they prefer, and eggs are laid on the surface and scatter as they sink to the bottom. Nymphs live on the bottom under stones or burrow in the substrate. Development takes up to one year with as many as 50 moults in some species.

Identification: Adults have two pairs of membranous wings – the hind pair much smaller than the forewings – short fine antennae and three long filamentous cerci. Nymphs look much like the adults but have no wings, the cerci are shorter and they have leaf-like gills along the sides of the abdomen. In some species the males have two very long white-tipped cerci, which may be longer than the rest of the body.

Left: These mayflies prefer still pools in which to breed.

Main habitat: Streams and lakes.
Main region: Australia.
Length: 10–25mm (0.39–1in).
Wingspan: 50mm (2in).
Development: Incomplete metamorphosis.
Food: Larva omnivorous; adult does not feed.

Tasmanian Giant Stonefly

Eusthenia spectabilis

Main habitat: Small streams and stony lake shores.
Main region: Tasmania.
Length: 40mm (1.57in).
Wingspan: 65mm (2.56in).
Development: Incomplete metamorphosis.
Food: Larva omnivorous; adult does not feed.

This insect is a giant, as often occurs when animals are isolated on islands. It is often found crawling around rocky shorelines of streams and rivers. This species sometimes flies toward lights at night, ending up in people's gardens.

Identification: Large, dark brown insect with lighter markings on the forewings, long antennae and cerci, and large hind legs. Most stoneflies are dull in colour but those of this genus have brightly coloured under-surfaces to their wings. Notable for its size, the largest in the region and one of the world's biggest.

Below: The Tasmanian giant stonefly is unusually large for a stonefly.

OTHER SPECIES OF NOTE

Striped Mayfly *Ameletopsis perscitus*
This is one of the most common mayflies in New Zealand, where it lives in clear waters. Related species are also found in Australia. It is unusual in having carnivorous nymphs, feeding mainly on other aquatic insect larvae. The adult is bronze-coloured with dark stripes on the wings.

Double-gilled Mayfly *Zephlebia nebulosa*
A medium-sized mayfly found in shady streams in the North Island of New Zealand. The adult has a dark body and long black front legs. The long wings are transparent and have the typical mayfly network of fine dark veins. The nymphs, which have very long cerci, cling to the surfaces of stones in the stream bed, grazing on algae.

Black Stonefly *Austroperla cyrene*
This stonefly family (Austroperlidae) is found mainly in Australia, New Zealand and also South America. The nymphs feed on decaying aquatic vegetation in flowing streams. They are quite easy to spot as they are marked in black, yellow and white. The adults are about 16mm (0.63in) long and shiny black, with long antennae.

Large Green Stonefly *Stenoperla prasina*
A large species, reaching 30mm (1.18in), with a dark green body. The nymph is a predator, feeding on other insect larvae, such as midges and mayflies. After one to three years the nymphs hatch into adults, but the latter live for only about a week. This genus is mainly found in eastern Australia and New Zealand.

Stonefly

Dinotoperla, Dinoperla spp.

These insects inhabit different freshwater environments from most stoneflies – cool settlement ponds or dammed pools on farms. The slow-moving nymphs feed on detritus and algae. The feathery gills of the nymphs are large, enabling them to obtain sufficient oxygen from the oxygen-poor still or slow warm waters in which they live. These habitats experience a great deal of temperature variation, and tend to lose oxygen.

Main habitat: Still water.
Main region: Southern Australia.
Length: 16mm (0.63in).
Wingspan: 25mm (1in).
Development: Incomplete metamorphosis.
Food: Larva omnivorous; adults do not feed.

Identification: The adults are mottled brown. The nymphs have a tuft of gills between the cerci – unusual in stoneflies.

Left: There are 35 species of this genus of stonefly in Australia.

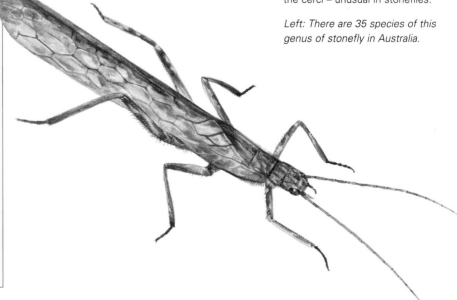

DRAGONFLIES

The order Odonata includes damselflies and dragonflies. One of the most ancient groups of insects, there are 5,000 species in 29 families. Australia has about 300 species. They are found all around the world, but are especially abundant in tropical regions and the Far East where they are sometimes used as ornaments. In China dragonfly nymphs are eaten in some areas as a delicacy.

Slender Skimmer

Orthetrum sabina

Skimmer dragonflies belong to the family Libellulidae. This is a common species all across southern Asia. It is found in cultivated areas such as parks and gardens and wasteland, wherever there is fresh water, including man-made watercourses such as drainage channels and artificial ponds. Adults fly all through the year.

Identification: Both males and females have pale yellow-green and black zebra stripes on the thorax. The front part of the abdomen (segments one to three) is swollen and striped like the thorax. The middle part (segments four to six) is thin and white, marked with black diamonds. The end (segments seven to ten) is fatter and black. When it is not flying it sits with its wings held forward.

> **Main habitat**: Watercourses, ponds and drains.
> **Main region**: Thailand, China, Malaysia, Indonesia and the Philippines.
> **Length**: 50–60mm (2–2.36in).
> **Wingspan**: 75mm (2.95in).
> **Development**: Incomplete metamorphosis.
> **Food**: Carnivore.

Right and far right: The slender skimmer has a very wide range across southern Asia.

Blue-spotted Hawker

Aeshna (Adversaeschna) brevistyla

Hawker dragonflies belong to the family Aeshnidae. The males and females are similar apart from the shape of the wings, those of the female being rounder than those of the male. They are strong fliers, patrolling still or slow-moving water and breeding in any permanent body of water. Females lay eggs in water plant tissue or mud.

Identification: This insect is a large dragonfly with a brown thorax and abdomen with cream zebra stripes, forming a 'cat's face' on the back near the base of the wings. There are also two bright blue stripes on the sides of the thorax. The leading edges of all the wings are pale orange. Like other dragonflies, blue-spotted hawkers have enormous compound eyes which cover most of the head, giving them excellent vision, and two pairs of large membranous wings, which they hold permanently out to the sides.

Above right: Like all hawkers, this dragonfly is strong and active in flight.

> **Main habitat**: Still or slow-moving permanent water.
> **Main region**: Australia.
> **Length**: 65mm (2.56in).
> **Wingspan**: 90mm (3.54in).
> **Development**: Incomplete metamorphosis.
> **Food**: Carnivore.

Australian Emerald Dragonfly

Hemicordulia australiae

Emerald dragonflies belong to the family Corduliidae. This species is sometimes called the 'sentry dragonfly', referring to its habit of hovering. Strong fliers, they never rest during the day. The males patrol about 50cm (20in) above small pools of water, which are cut off from the main watercourse by dense vegetation. They are quite shy and always fly up toward the sky if disturbed. However, if an intruder remains still they will eventually come to investigate. They rest overnight at the tops of trees, and like all members of their family they hang vertically from their perch, remaining motionless until the morning sun warms their flight muscles. When the young adults emerge from the nymph case they hunt far from the water while their bodies mature. The most noticeable feature of these young adult dragonflies is their brown eyes which only turn blue when they are mature enough to return to the water to look for mates. After the characteristic mating routine of wheel and tandem flying the female lays her eggs by zigzagging over the surface of the water, repeatedly dipping the tip of her abdomen into the water. This species has also colonized New Zealand where it is now quite common.

Main habitat: Still water with abundant vegetation.
Main region: Australia and New Zealand.
Length: 50mm (2in).
Wingspan: 70mm (2.76in).
Development: Incomplete metamorphosis.
Food: Carnivore.

Left: This dragonfly is quite common in Australia and New Zealand.

Identification: A medium-sized, long and slender insect with a narrow abdomen and thickened thorax. Both the males and females are pale yellow or green with iridescent blue-black markings. The head, with large eyes (green in male, brown in female) and mouthparts, is emerald green.

OTHER SPECIES OF NOTE

Bush Giant Dragonfly *Uropetala carovei*
This very large dragonfly is found only in New Zealand. Its body is yellow and black and may be 86mm (3.39in) long, with a wingspan of 13cm (5in). In flight the bush giant dragonfly is powerful but rather slow. It includes cicadas in its diet. The nymphs are unusual in that they live in wet tunnels in stream banks and emerge to hunt at night.

Red Percher Dragonfly *Diplacodes bipunctata*
A pretty, rather small species from the skimmer family with a wide range in Australia, New Zealand and some of the Pacific islands. The male is bright red and the female brownish-yellow. It often rests with its wings angled forward. The larvae live in warm, shallow water and the adults often rest on rocks.

Blue Skimmer *Orthetrum caledonicum*
This dragonfly is found in most of Australia, including Tasmania, as well as in New Guinea and New Caledonia. It is pale blue (male) or brownish-grey (female) and about 45mm (1.77in) long with a wingspan of 70mm (2.76in).

Giant Petaltail *Petalura ingentissima*
This very large Australian species is the world's bulkiest and its wingspan is up to 16cm (6in). The nymphs are unusual in living in soil, not in water. They spend years underground, feeding on other invertebrates, before eventually emerging as gigantic adults.

Yellow Emperor Dragonfly

Hemianax papuensis

Also known as the 'baron dragonfly', this is another member of the hawker dragonflies of the family Aeshnidae. It is a very large and common dragonfly, seen flying over water in the summer months. Strongly territorial, it will chase away any other flying insect found on its patch. It spends most of its time flying, defending its territory or hunting. It rarely sits still on sunny days and patrols on a regular route about 1m (3ft) above the water. Its territory can be up to 50m (164ft) along a watercourse. Females lay eggs under the surface of still or slow-moving water with dense vegetation. Sometimes they lay still while in tandem with a male.

Main habitat: Still or slow-moving fresh water.
Main region: Eastern Australia, New Zealand.
Length: 65mm (2.56in).
Wingspan: 10.5cm (4in).
Development: Incomplete metamorphosis.
Food: Carnivore.

Below: This common species is boldly marked in yellow.

Identification: Both males and females are a pale yellow colour with dark patterning along the narrow abdomen. The front edges of the wings, which are held straight out to the sides, are also pale yellow.

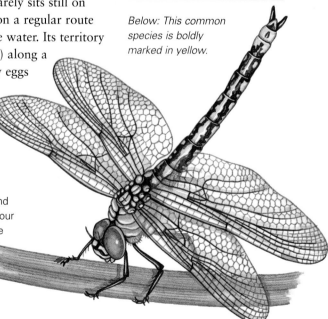

DAMSELFLIES

Damselflies (suborder Zygoptera) are generally smaller, more delicate and weaker fliers than dragonflies. Both their pairs of wings are similar in shape and size. The larvae, or nymphs, swim by wriggling their bodies using the three leaf-like gills at the end of the abdomen as paddles. Adult damselflies are usually seen near water.

Painted Waxtail Damselfly

Ceriagrion cerinorubellum

This insect is a very common and colourful damselfly. It has a wide distribution including Bangladesh, China, Indonesia, India, Sri Lanka, Myanmar, Malaysia, the Philippines, Singapore, Thailand and Vietnam. The adults are adept aerial hunters, ambushing flying insects, often twice their own size. Adults form mating wheels with males clasping females behind the neck and females curling their abdomens forward to touch the male's genitalia on the ventral surface of the abdomen. This position is held for ten minutes, during which time the male's genitalia removes sperm that has been deposited by other males during previous matings. The eggs are laid in water and the larvae feed on small aquatic invertebrates and moult up to 15 times before they emerge.

Main habitat: Gardens, marshland, ponds, lakes.
Main region: Tropical.
Length: 35mm (1.38in).
Wingspan: 45mm (1.77in).
Development: Incomplete metamorphosis.
Food: Carnivore.

Identification: The female has a blue-green thorax and eyes, her slim abdomen is black with yellow-orange at the front and rear. The male is similarly, though much more vividly, coloured, with bright red patches at the base and tip of the abdomen and bright blue eyes and thorax.

Left: The common name comes from the red tail tip of the male.

Yellow Featherlegs Damselfly

Copera marginipes

This damselfly is common in southern Asia, especially Thailand, and may be seen throughout the year. After mating in the characteristic mating wheel position the male clasps the female behind her head while she deposits her eggs to ensure no other male mates with her and removes his sperm from her body. The eggs are laid on submerged vegetation in a pool.

Main habitat: Ponds near trees.
Main region: Thailand, Malaysia, southern China, Taiwan.
Length: 35mm (1.38in).
Wingspan: 42mm (1.65in).
Development: Incomplete metamorphosis.
Food: Carnivore.

Identification:
This damselfly has a very slender abdomen, variable in colour (sometimes white) and banded, with a white tip. The rather short wings are transparent. Note that the wings are roughly equal in size and shape – a feature that distinguishes most damselflies from dragonflies. The female has pale brown markings; the male has yellow markings on his thorax, and yellow legs. The legs have feather-like hairs, hence the common name.

Left and below: This dainty damselfly is common in parts of South-east Asia.

Ochre Threadtail Damselfly

Nososticta solida

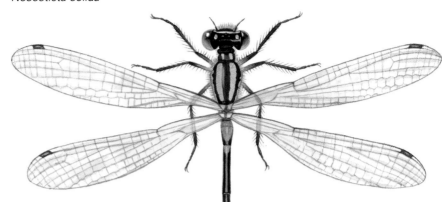

This insect is one of the most common damselflies around the Brisbane area of Queensland, Australia. It can be found resting in groups on plants in semi-shaded areas near running water in summer. When at rest it holds its wings tightly folded together vertically over the back of the abdomen. It is an excellent flier, able to hover for long spells and can fly backward, forward and sideways with equal precision.

Above and right: Like many damselflies, this species lives close to running water.

Identification: The male of this delicate insect has a yellow thorax with black patterning and a brown-yellow patch at the base of the wings. The narrow abdomen is black with yellow stripes. The female has similar black patterning but is a drabber pale-brown colour.

Main habitat: Shady plants near running water.
Main region: Temperate, tropical.
Length: 35mm (1.38in).
Wingspan: 40mm (1.57in).
Development: Incomplete metamorphosis.
Food: Carnivore.

Big Red Damselfly

Ceriagrion aeruginosum

This species belongs to the family Coenagrionidae, known as pond damselflies. The larvae live in still water. These are shy creatures, not easily seen as they hide among thick grass near water. When they do fly they fly quite slowly and do not seem unduly bothered by intruders. These insects rely on camouflage rather than speed of movement to defend themselves from predators. This species is found in China, Indonesia and Australia. Note that the wings taper markedly toward the base – another feature that distinguishes damselflies from dragonflies.

Main habitat: Thick waterside grass.
Main region: Tropical, subtropical.
Length: 40mm (1.57in).
Wingspan: 20mm (0.79in).
Development: Incomplete metamorphosis.
Food: Carnivore.

Identification: A beautiful damselfly with an orange-red abdomen and yellow-green thorax and head. There are blue bands toward the base of the abdomen. The abdomen is very long and thin and the wings rather short – only half the length of the whole insect.

Left: The bright red abdomen makes this damselfly easily visible in flight.

COCKROACHES

Cockroaches (Blattodea) were one of the first groups of winged insects to evolve. Their fossils have been found in rocks that are 300 million years old. They have broad, flattened oval bodies; the thorax partly covers the head and they have long thread-like antennae. There are about 4,000 species worldwide, of which 430 are found in Australia. Many more are found in Asia.

Pill Cockroach

Perisphaerus spp.

This cockroach lives in Malaysia, South-east Asia and Australasia. It is unusual in exhibiting remarkable parental behaviour. The nymphs are born white and blind, and they have unique tube-like mouthparts. The female has four holes at the bases of her mid and rear legs into which the tubes fit and the young suck liquid food from these holes. The nymphs cling to their mother's underside for two instars, and only at the third moult do they develop normal eyes and body colour.

Right: The pill cockroach resembles a giant woodlouse at first sight.

Main habitat: Soil.
Main region: Tropical.
Length: 30mm (1.18in).
Development: Incomplete metamorphosis.
Food: Omnivore.

Identification: A shiny black cockroach with a domed carapace, looking rather like a giant woodlouse, although being an insect it has only three pairs of legs. Most cockroaches are more elongated in shape so this species is very distinctive. It is wingless and can roll into a ball for defence, the tough cuticle providing very effective armour against attack.

Giant Burrowing Cockroach

Macropanesthia rhinoceros

Also known as the 'rhinoceros cockroach' or 'litter bug', this insect is the largest cockroach in the world and can weigh up to 50g (2oz). It lives in family groups, mainly in rotting timber, and gives birth to live young. The adults excavate tunnels in the ground to a depth of 1m (3ft) where the family lives permanently, feeding mainly on dead leaves. The tough, smooth body is well adapted to its burrowing lifestyle. When disturbed this cockroach emits a hissing sound. This wingless cockroach is found mainly in northern Queensland, Australia. The females produce between five and 30 live young once a year in early summer. The nymphs stay with their mother for about nine months, and moult about 12 times before reaching adulthood at three to four years. They can live for ten years.

Main habitat: Soil.
Main region: Tropical.
Length: 80mm (3.15in).
Development: Incomplete metamorphosis.
Food: Herbivore.

Identification: This is a heavily armoured, flightless cockroach with hairy legs. The body is dark brown or black and shiny, and the head is covered by the pronotum of the first thoracic segment. The antennae are rather short and fine.

Left: Long-lived and sturdy, this is a remarkable cockroach.

Australian Cockroach

Periplanata australasiae

This insect is native to Asia, and is also found in Australia, often inside homes. It hides away in cracks by day and flies to new territory by night. Its mobility and powers of dispersal make it very hard to eradicate. This species breeds throughout the year; after mating, the female produces an ootheca containing several eggs, which she carries around for some time. This cockroach feeds on most household scraps.

Identification: This insect is smaller than the very common and familiar American cockroach, to which it is closely related. Reddish-brown, with a yellow band across the base of the pronotum and yellow slashes along the base of the forewings. Very long antennae and long spiny legs. The forewings are thickened protective structures; the hindwings are folding membranous flight wings with a characteristic fan-like rear lobe.

Right: This is one of several species of cockroach that have taken to living inside houses.

Main habitat: Houses and other buildings.
Main region: Tropical and warm temperate.
Length: 30mm (1.18in).
Development: Incomplete metamorphosis.
Food: Omnivore.

Thorax porcellana
This cockroach from India shows very unusual behaviour. The female is flightless, with reduced hindwings. When the young hatch they crawl into the space below her hard forewings and stay there for their first two instars. Here they feed on liquid produced from pores in the female's abdomen. The young also pierce her cuticle with sharp mouthparts and drink her blood.

Archiblatta hoeveni
This is a Malayan species that grows to about 70mm (2.76in). It is reddish-brown with a flexible abdomen and very long legs that are black at the base and red lower down. The antennae are very long and have alternating black and white patches.

Wingless Cockroach *Cosmozosteria* spp.
The cockroaches in this genus have broad, flat bodies, mostly brownish, with bright yellow markings at the outer edges of the abdominal or thoracic segments. Wingless cockroaches may be seen in the daytime and often clamber among twigs in bushes. There are about 16 species in Australia.

Gisborne Cockroach *Drymaplaneta semivitta*
This is a small, about 25mm (1in) but attractively marked species found in western Australia and also in New Zealand (introduced). Mainly shiny and brownish-black, the main feature of this species is the broad creamy-yellow bands along the margins of the thorax.

Bush Cockroach

Polyzosteria limbata

The bush cockroach is also called the diurnal cockroach or Botany Bay cockroach. While most cockroaches hide by day and emerge at night, this species is unusual in that it is diurnal. It likes to bask on tree trunks in the sunshine, more in the manner of some true bugs, feeding on leaves, bark and pollen. To protect itself from predators this slow-moving cockroach produces a menacing hissing sound and also gives off a pungent smelling liquid. The bush cockroach is found in south-east Australia, in trees, bushes and heathland.

Main habitat: Scrub.
Main region: Tropical and subtropical.
Length: 45mm (1.77in).
Development: Incomplete metamorphosis.
Food: Herbivore.

Identification: The insects of this genus are large, often with metallic colours. This cockroach is wingless and shiny black and has iridescent bronze edges to its segments. Between 12 and 40 eggs are carried around by the mother in an ootheca, which protrudes from the back of the abdomen.

Left: The bush cockroach sits on the bark of tree trunks.

TERMITES

Termites (Isoptera) are highly social insects living in colonies. There are about 2,500 species, mainly found in tropical regions. About 350 species occur in Australia. Termites have 'castes' such as queens, workers and soldiers and these do different jobs in the colony. Only the reproductives are winged and their wings are shed soon after they have flown away to mate and found a new colony.

Giant Northern Termite

Mastotermes darwinensis

This insect is considered the most primitive of the termites – the only living species of the first termites descended from cockroaches. It nests in soil, trees or logs, eating wood and digesting the cellulose with the help of flagellate protozoans that live in its gut. It attacks wood in contact with the ground and can damage houses and sheds. It is found mainly in northern Australia and also New Guinea. Winged adults break out of a hole in the top of the mound and fly out in huge numbers in certain seasons. They land up to a kilometre from their birthplace and shed their wings. Males and females pair up and they begin to explore the area, female first. She settles into a suitable crevice and they mate, after which she produces a packet of eggs. The hatchlings consume secretions from the queen to take in the micro-organisms they will need. The young termites never become fertile so long as the queen feeds them – her saliva contains a hormone that inhibits sexual development. They care for the eggs the queen continues to lay. When the king or queen dies some of the young become sexually mature and take over the nest.

Main habitat: Soil.
Main region: Tropical Australia.
Length: 12mm (0.47in).
Development: Incomplete metamorphosis.
Food: Herbivore.

Identification: Large termites, with pale, unpigmented soft bodies, apart from the red heads and blackened jaws of the soldier caste.

Left: This large termite is quite common in tropical Australia.

Magnetic Termite

Amitermes meridionalis

These termites can respond to the Earth's magnetic field and create tall mounds that are aligned roughly north to south. They feed on dead plant material they find while burrowing in the ground. They have very soft cuticles and live in darkness, chewing the soil and mixing it with saliva to make a rock-hard cement with which to build their mound. The mated king and queen crawl into a crevice and begin to burrow downward, feeding, and the female laying eggs as they go. The queen continues to lay eggs all her life, the eggs becoming workers and soldiers. Workers dig burrows beneath ground and build a tower above ground. Temperature is crucial – if it gets too hot or cold the pheromone system of communication breaks down, stored food will rot or fungus gardens will not grow and the king and queen will dry out or become cold and die. Magnetic termites solve this problem by building a tall, 3m- (10ft-) wide mound which is narrowly wedge-shaped and aligned so that the thin crest lies north–south. In the mornings, the sun rising in the east shines on the flat side of the mound and warms it up.

Main habitat: Open country.
Main region: Northern central Australia.
Length: 4mm (0.16in).
Development: Incomplete metamorphosis.
Food: Herbivore.

Identification: Rather a small species of termite, best known for its habit of creating tall mounds. The mounds created by this species are remarkable structures, always aligned north–south and about 2.5–4m (8–13ft) tall.

Left: The magnetic termite builds tall mounds that take advantage of the sun.

Processional Termite

Hospitalitermes spp.

Main habitat: Open country.
Main region: South-east Asia.
Length: 4mm (0.16in).
Development: Incomplete metamorphosis.
Food: Herbivore.

Identification: The workers are rather ant-like with a pale rounded abdomen and a round head. The members of the soldier caste (shown) have a larger head with a nose-like forward projection.

Far right: Processional termites are named for their habit of foraging in lines.

This termite is a widespread species over much of South-east Asia. It takes its name from the fact that when the worker termites go out foraging they tend to follow each other in lines. Such processions may consist of up to 500,000 individuals, guarded on the outsides by individuals of the soldier caste. These termites may often be seen out in the open, searching for edible material, notably lichens gleaned from the bark of forest trees. Many are found in rainforests where they are very important in the overall ecological balance. Worker termites turn the food into convenient round balls, which they carry carefully back to the colony. Such food has been shown to be more nutritious than the wood diet of many other species and this may make it worth the risk of gathering in the open where there are many predators.

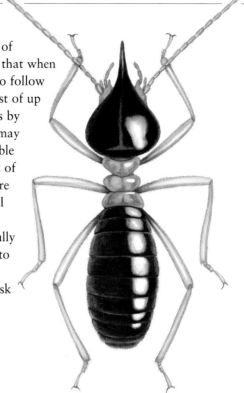

OTHER SPECIES OF NOTE

Spinifex Termite *Nasutitermes triodiae*
This termite is one of many whose soldiers have long snouts, used for squirting a sticky, noxious fluid, which is used in defence. The spinifex termite lives mainly in north Queensland, Australia, where it builds reddish mounds up to 7m (23ft) tall, towering above the grassland. A related species, *N. walkeri*, is common in the bush in Australia, mainly in New South Wales. It builds rounded nests in trees, and has been known to damage damp floor timbers.

Coptotermes acinaciformis
This termite is notorious as Australia's most destructive species. It lives in tree stumps, but also inside the tissues of living trees, where colonies often feed on the insides of branches, making them hollow. The famous Aboriginal didgeridoo is often made using a stem hollowed out naturally by these termites. They also nest underneath buildings and boats and attack any wood in them from below, using a network of underground tunnels.

Globitermes sulphureus
This termite is found mainly in South-east Asia. It is particularly interesting for its defensive strategy. When the termites encounter a threat, such as attacking ants, the soldier termites either attack using their mandibles, or they may even explode, covering the attacker with sticky fluid and killing themselves in the process, but reducing the threat to the colony. Because of this behaviour, they have been described as 'walking chemical bombs'.

Subterranean Black Termite

Odontotermes formosanus

Many species of termites of this genus are found in Asia. The subterranean black termite, for example, has a wide distribution from India and south China to Japan. This termite is quite a serious pest of crops and forest trees, and also affects man-made wooden structures. In south China, it causes considerable damage to buildings and also to wooden defences along reservoir dams and riverbanks. It lives underground and cultivates fungi as food. In some regions these subterranean black termites are eaten by people, and being rich in protein they are very nutritious. In southern India, for example, this species is eaten especially by children and pregnant women and is also used in traditional medicine. The closely related Indian mound-building termite, *O. redemanni*, constructs mounds from November to March during the cooler dry season. It emerges from the mounds to forage at night, especially from March to September.

Identification: The workers of this termite have a pale oval abdomen and a rounded orange head. The members of the soldier caste (shown) have a larger head with long pincer-like jaws.

Main habitat: Open country.
Main region: South-east Asia.
Length: 4mm (0.16in).
Development: Incomplete metamorphosis.
Food: Herbivore.

Below: Seldom seen, the subterranean black termite lives mostly below ground.

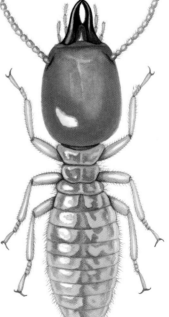

MANTIDS AND EARWIGS

Mantids are carnivorous insects with claw-like grasping front legs used for catching their prey. They look unlike any other insects. There are about 2,000 species found throughout the warmer regions, most of which live in Asia. About 160 species are found in Australia. Earwigs include both carnivorous and omnivorous species. They are more widespread than mantids.

Indian Flower Mantis

Creobroter meleagris

This flower mantis hails from India and other tropical parts of Asia. Its colour and shape enable it to hide in flowers using amazing camouflage, though it is not quite as gaudy as the orchid mantises. In some flower mantises the nymph stages resemble ants for protection, only becoming flower-like at their third instar. This species has impressive brightly coloured underwings, which it flashes to startle potential predators. Both males and females have long wings and are good fliers.

Identification: This is a green flower mantis with a comma on the top and a white eye-spot midway along each forewing. The legs are striped pale green or brown and cream. Each segment of the wide abdomen is marked with white.

Left: This is one of several mantids that hides in flowers to ambush its prey.

Main habitat: Flowers and shrubs.
Main region: Tropical.
Length: 35mm (1.38in) male 40mm (1.57in) female.
Development: Incomplete metamorphosis.
Food: Carnivore.

Chinese Praying Mantis

Tenodera aridifolia

This mantis, native to China, has been widely introduced elsewhere, including to Australia where it is well established, and North America. It has been found to be useful in pest control as it consumes the larvae and adults of many pest species. It is even marketed as a biological control agent, and also as a pet. It reaches adulthood in late summer, when it mates. The female lays up to 400 eggs in a foamy case (ootheca), stuck to a stem. The young hatch in spring and begin eating immediately. They feed mainly on other insects, spiders and the occasional small vertebrate.

Identification: A long, thin mantis – looks a bit like a stick insect. It is well camouflaged in its normal grassy habitat. It is coloured a variety of shades of green or brown with a green stripe down the leading edge of the wingcase.

Below: Females of the species die after the eggs are produced in the autumn. In any case, they cannot survive the first frosts.

Main habitat: Grassland and scrub.
Main region: Subtropical.
Length: 15cm (6in).
Development: Incomplete metamorphosis.
Food: Carnivore.

Common Green Mantis

Garden Mantis *Orthodera ministralis*

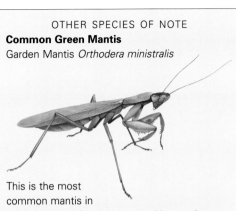

This is the most common mantis in Australia. It is often seen around homes in suburbs. It is small to medium-sized (30–50mm/ 1.18–2in) with a blue spot on the inside of its forelegs. The forewings cover the whole of the abdomen, and it has a broad, straight and rather flat thorax. It is most common in the north and north-east of Australia, but is also found else-where. It feeds on small insects such as moths.

New Zealand Mantis

Orthodera novaezealandiae

Closely related to the previous species, this is the only mantis native to New Zealand. It grows to about 40mm (1.57in) long and is almost completely pale green, blending well with leaves. This species may catch up to 25 flies each day. The ootheca contains several hundred eggs that hatch into miniature adults.

Giant Asian Mantis

Hierodula membranacea

This insect from tropical South-east Asia is large and heavily built. In fact, it is one of the largest of all mantids with a body length of 10cm (4in). In colour it varies from yellow-green to red-brown. The giant Asian mantis lives among trees and shrubs, where it spends much of its time waiting motionless, ready to ambush prey. Like many mantids it is a popular pet and can be bred in captivity. A wider range of other insects are taken as food.

Giant Earwig *Titanolabis colossea*

This is one of the world's largest earwigs, reaching a length of up to 55mm (2.17in). Wingless, it lives in the damp soil and rotting timber of the rainforests strung along the east coast of Australia. It has also been recorded in New Caledonia and Vanuatu. Its abdomen is quite long and broadest in the middle. The pincers are also rather broad.

Common Brown Earwig

Australian Earwig, *Labidura truncata*

This is a useful insect as it attacks and kills many caterpillars, including those of the codling moth, which is a pest of fruit trees. It uses its large pincers to grasp the soft-bodied larvae while feeding. It sometimes eats other insects, including other earwigs. Members of the earwig family Labiduridae are common over much of Australia, and this is the most common of them all. It favours sandy soils and like most earwigs it is nocturnal, hiding by day under logs, stones or bark.

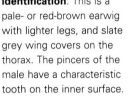

Main habitat: Soil surface.
Main region: Temperate, subtropical.
Length: 30mm (1.18in).
Development: Incomplete metamorphosis.
Food: Omnivore.

Identification: This is a pale- or red-brown earwig with lighter legs, and slate grey wing covers on the thorax. The pincers of the male have a characteristic tooth on the inner surface.

Right: This earwig is useful in pest control.

Seashore Earwig

Anisolabis littorea

This insect lives on the beach, typically under piles of damp seaweed. It is fairly common in eastern Australia and in New Zealand. Males and females, as well as nymphs of all sizes, may be found throughout the year. In the winter, however, they generally move to warmer parts of the beach. This species is also found in gardens and wasteland, under stones and logs. It runs away and hides in the narrowest of crevices if disturbed, but if it cannot get away or if the crevice is too small it will open its forceps in a threat display. It can inflict a painful and effective nip on a human finger. It is a predator capturing other creatures that live in the flotsam by running them down and grabbing them with its forceps. The victim's body case is ruptured by the forceps, and the earwig then devours the soft parts within. The female lays about 50 eggs in the sand away from other earwigs. She stands guard over them, and cleans them until they hatch at about 20 days, after which she continues to guard them until their second moult. The nymphs moult three more times before reaching adulthood.

Main habitat: Seashore: under moist debris above high tide mark.
Main region: Temperate, subtropical.
Length: 30mm (1.18in).
Development: Incomplete metamorphosis.
Food: Carnivore.

Identification: Deep, shiny blackish-brown colour, with orange legs and antennae. The male has slightly asymmetrical forceps.

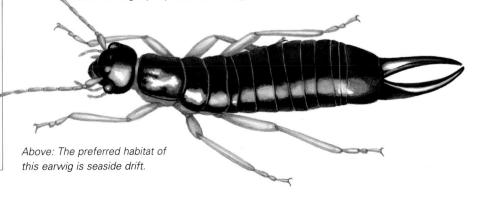

Above: The preferred habitat of this earwig is seaside drift.

CRICKETS

Crickets (Orthoptera) are mainly very active insects with powerful hind legs and known for their often loud 'songs' produced by rubbing movements of either their legs or wings. There are about 22,000 species, showing many variations on the basic body theme. Australia has more than 2,800 species and an even larger number live in Asia.

Cooloola Monster

Cooloola propator

Main habitat: Sandy soil.
Main region: Tropical, subtropical.
Length: 30mm (1.18in).
Wingspan: None.
Development: Incomplete metamorphosis.
Food: Omnivore.

This insect is so unlike any other cricket that when it was discovered in 1976 it was placed in its own family, the Cooloolidae. Virtually blind, it lives almost its entire life beneath the ground, where it preys mainly on soil invertebrates. It only comes above ground at night after heavy rainfall and so is rarely spotted. Cooloola monsters are found mainly in underground burrows, in sandy moist soil of the coastal rainforests of Queensland, Australia. Two further species have more recently been discovered from this highly unusual genus.

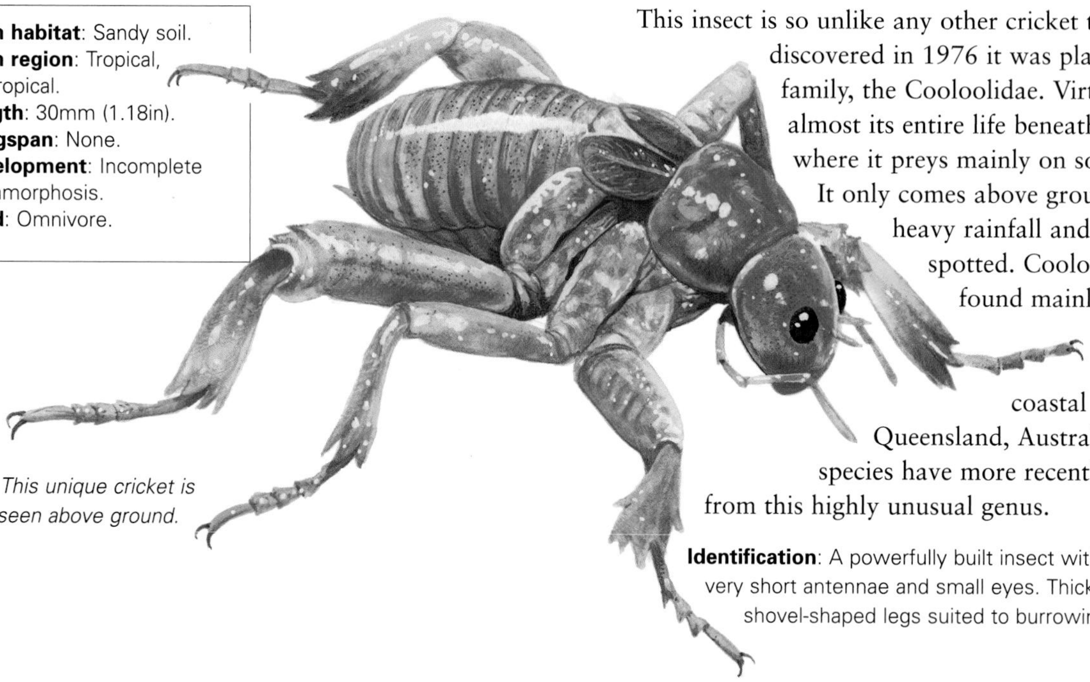

Right: This unique cricket is rarely seen above ground.

Identification: A powerfully built insect with a pale-coloured body, very short antennae and small eyes. Thickset hunched body and shovel-shaped legs suited to burrowing.

Giant Weta

Wetapunga, *Deinacrida heteracantha*

This insect is possibly the heaviest insect in the world, growing up to 90mm (3.54in) long and weighing as much as 70g (2.5oz) (heavier than a small bird). There are a number of species of weta in New Zealand, living in almost every type of habitat. Giant wetas have a limited range, being found on certain protected islands or high in the mountains out of reach of introduced predators such as cats, rats, ferrets and stoats. They have also suffered from habitat destruction. Nocturnal and flightless, the giant weta feeds mainly on leaves, flowers and fruit, unlike other wetas, which are mainly predators. All wetas have a fearsome defence display, waving their strong, spiny hind legs above their backs and stridulating loudly. They can inflict a painful bite but prefer to run away and hide. The female selects a mate and shares his territory for about six months then lays 100–300 eggs in the soil. These hatch one to four months later depending on soil temperature. The young mature at about 18 months.

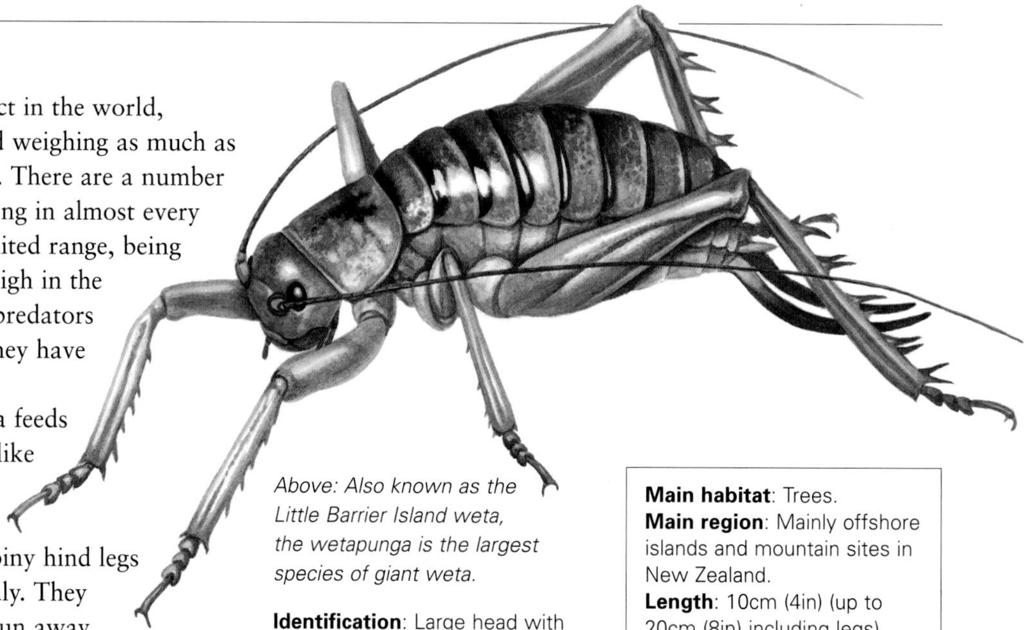

Above: Also known as the Little Barrier Island weta, the wetapunga is the largest species of giant weta.

Identification: Large head with strong jaws, and antennae twice as long as the body. The legs are also very long and used for clambering among twigs and branches.

Main habitat: Trees.
Main region: Mainly offshore islands and mountain sites in New Zealand.
Length: 10cm (4in) (up to 20cm (8in) including legs).
Wingspan: None.
Development: Incomplete metamorphosis.
Food: Omnivore.

Mountain Katydid

Mountain Grasshopper, *Acripeza reticulata*

Like many katydids, this Australian species has a bulky body, but its long legs are very thin and spidery. Its most remarkable feature is its brightly coloured abdomen, vividly striped with red, blue and black. This is revealed when it suddenly opens its wing cases as a defensive display when threatened. It is a lumbering creature which clambers rather than jumps, and its markings probably indicate it is distasteful to its potential predators. Its camouflaged wings help to hide it among fallen dead gum leaves. It feeds on grass, leaves and seeds.

Identification: The adult has a short, chubby body with mottled black/grey domed wing cases with two white stripes. Its thin legs are striped with pale grey-and-black bands. It has long antennae.

Left: This flightless grasshopper uses its wing cases for a defensive display.

Main habitat: Rainforest and wet eucalyptus forest.
Main region: Temperate, subtropical.
Length: 30mm (1.18in).
Wingspan: 30mm (1.18in) (female wingless).
Development: Incomplete metamorphosis.
Food: Herbivore.

OTHER SPECIES OF NOTE

Gum Leaf Katydid
Caedicia spp.
This is a common katydid in Australia, where it lives in gum trees, using its leaf-like camouflage to render it almost invisible. The forewings have the shape, colour and texture of eucalyptus leaves – even mimicking the central leaf vein. However, it can be heard, as the males produce a lot of noise at night by rubbing their wings together.

Spider Cricket *Endacusta* spp.
This small Australian cricket, about 20mm (0.79in) lives among rocks and under rotten wood. Active at night, it looks rather like a spider, with its long, thin legs (hence its common name). The body and legs are light grey and spotted with darker spots and blotches. The antennae are very long and thin.

Japanese Bell Cricket *Homeogryllus japonicus*
This cricket, native to Japan, is famed for its singing abilities, so much so that it is often caught and kept in small containers to entertain by its singing. It has inspired a great deal of art and music in Japan, and there is even an orchestra named after it!

Cave Weta *Gymnoplectron longipes*
This weta is found in caves in New Zealand's North Island. It has remarkably long legs and antennae, extending to 40cm (16in) from the tip of the antennae to the ends of the legs, though the body is only about 30mm (1.18in). Cave wetas feed on a wide range of food scavenged from the cave floor.

Gymnogryllus elegans

The male cricket of this tropical species produces a piercing territorial song so intense it may even hurt human ears. He amplifies this song by means of a 'megaphone' in the form of a little horseshoe-shaped mud wall around the entrance to his burrow. At night he stands with his rear end protruding through the gap in his wall as he sings. This cricket is quite common in the tropical forests of Malaysia.

Identification: Typical cricket shape, with long hind legs. The males have black and white markings.

Above: This long-legged cricket uses its high-pitched 'song' as a form of defence.

Main habitat: Rainforest.
Main region: Tropical.
Length: 30mm (1.18in).
Wingspan: 40mm (1.57in).
Development: Incomplete metamorphosis.
Food: Herbivore.

GRASSHOPPERS

Grasshoppers comprise the suborder Caelifera. They have much shorter antennae than crickets and usually a thinner, straighter body. Females do not possess an ovipositor. Male grasshoppers sing by rubbing a thickened vein which runs along each forewing against the inner surface of the long hind legs. The female's ears are thin round membranes on each side of the first abdominal segment.

Gum Leaf Grasshopper

Dead Leaf Grasshopper, *Goniaea australasiae*

All stages of this grasshopper rest on gum tree stems or on the ground around gum trees among the fallen leaves. The nymphs resemble dead leaves perfectly, both in shape and colour, having two forms – a greyish-brown form and an orange-brown form. They feed on leaves on the forest floor. The adults are less convincing leaf mimics as their wings give them away.

Identification: Adults are uniformly mottled reddish-brown with a prominent arched crest on the thorax. They have short antennae and large muscular hind legs. When disturbed the adults jump and fly a few metres away exposing their orange abdomen and hindwings. Both males and females have wings.

Main habitat: Dry eucalyptus woodland.
Main region: Australia.
Length: 50mm (2in).
Wingspan: 70mm (2.76in).
Development: Incomplete metamorphosis.
Food: Herbivore.

Left: The gum leaf grasshopper is cryptic at rest but colourful in flight.

Matchstick Grasshopper

Moraba spp.

The grasshoppers in the Eumastacidae family are well named, as they look like small twigs or matchsticks. Matchstick grasshoppers are found in tropical Asia and Australia, as well as many other parts of the world. The matchstick grasshopper takes its common name from its shape – its short, slim body is usually brownish. It has unusually large eyes, helping it to see in low light and at night.

Identification: Most species are wingless and sit tight against stems where they are hard to spot. They are about 30–35mm (1.18–1.38in) in length and the head is long and pointed.

Main habitat: Woodland and scrub.
Main region: Tropical, subtropical
Length: 35mm (1.38in).
Wingspan: None.
Development: Incomplete metamorphosis.
Food: Herbivore.

Right: This thin-bodied grasshopper, the colour of pale wood, is aptly named.

Oriental Migratory Locust

Locusta migratoria manilensis

Main habitat: Open country and crops.
Main region: Subtropical.
Length: 40–60mm (1.57–2.36in) (solitary locusts are bigger than gregarious).
Wingspan: 80mm (3.15in).
Development: Incomplete metamorphosis.
Food: Herbivore.

This is an eastern Asian subspecies of *Locusta migratoria* – the most widespread locust species in the world. The migratory locust has two phases – the solitary phase and the gregarious phase. When conditions are favourable, population numbers start to rise and the density of insects increases. Successive generations gradually change from having solitary habits to being gregarious. As numbers increase, the local food supply runs out and they swarm, flying off to find new food sources. A swarm of locusts may cover several hundred square kilometres with 40–80 million locusts per square kilometre. One locust can eat its own weight of food every day. However, they can be controlled biologically by the use of parasitic wasps, flies, nematodes and fungi, as well as by chemicals. Indonesia, in particular, experiences occasional outbreaks.

Identification: The colouring depends on the age and phase of the individual: gregarious nymphs are yellow-orange with black spots whereas the solitary nymphs are a less conspicuous green or brown. Gregarious adults are brown with yellow markings, the solitary phase being brown with green markings, as here.

Above: Swarms of locusts can be devastating to vegetation and crops.

OTHER SPECIES OF NOTE

Giant Green Slantface *Acrida conica* (Above)
This bright green grasshopper reaches 70mm (2.76in) long as an adult. The head is long and rather flattened and the legs are long and thin. It is well camouflaged unless disturbed, when it opens its wings to reveal its bright pink abdomen. Although it has wings, it rarely flies.

Vegetable Grasshopper
Atractomorpha australis
This is a common species of eastern Australia. It is small, about 30mm (1.18in) and bright green and hides among grasses. Although it feeds on the leaves of various plants it does not cause much damage. It has a conical pointed head and rather short antennae.

Froggatt's Buzzer *Froggattina australis*
A common green-and-brown grasshopper found in the Australian bush and grassy country. This small species takes its name from the loud buzzing sound it produces as it flies away.

Giant Grasshopper

Hedge grasshopper, *Valanga irregularis*

This insect is Australia's largest grasshopper and is found across much of that country. It can be a pest of commercial trees and shrubs such as coffee and citrus trees. In October–November up to 150 eggs are laid by each female at the bottom of a 90mm (3.54in) deep hole with a frothy plug. The eggs hatch during the summer and by April the insects are all mature. They feed on herbaceous plants, shrubs and trees.

Identification: The colouring on this insect is very variable. The nymphs are brightly striped green and black. The adults are creamy-brown to grey with small darker spots. The hind legs are fringed with orange-red spines tipped with black, which are used to attack predators such as birds. The young insect goes through several moults, with the wings appearing fully at the final moult. The smallest moult is 5mm (0.25in). The nymphs look identical to the adult except for colour and their lack of wings.

Main habitat: Varied.
Main region: Tropical.
Length: 60–90mm (2.36–3.54in).
Wingspan: 12cm (4.5in).
Development: Incomplete metamorphosis.
Food: Herbivore.

Below: Australia's largest, the giant grasshopper, is an impressive insect. The generation usually lives for one year.

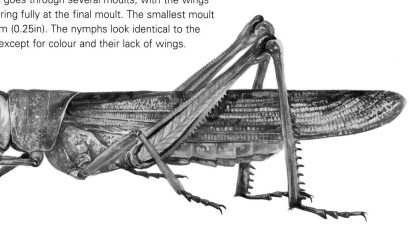

STICK AND LEAF INSECTS

Stick and leaf insects (Phasmida) are among the world's most remarkable insects. Most of the 2,500 or so species live in warm or tropical regions. There are about 150 species in Australia and 20 in New Zealand. Stick insects have long, thin twig-like bodies, though some have leaf-like projections as well. True leaf insects live mainly in South-east Asia and Papua New Guinea.

Spiny Stick Insect

Bristly Stick Insect, *Argosarchus horridus*

This is the largest of New Zealand's 20 stick insect species, and is widespread, usually feeding on the leaves of native trees. By day spiny stick insects settle on those parts of trees where their twiggy camouflage works best. Like all New Zealand species they are wingless and rely on camouflage for protection. Their defensive posture is to line their legs up alongside their bodies and freeze next to a twig. If this fails they let go of the twig and fall 'lifeless' to the ground. They can remain like this for hours, disappearing among the dead leaves and twigs of the forest floor. Many types of stick insect reproduce parthenogenetically, but this is rare in spiny stick insects.

Identification: Usually brown, although green forms do occur. Young stick insects are usually green, and as their name suggests they have spines all over their bodies and legs.

Main habitat: Forests of native trees.
Main region: Temperate.
Length: 15cm (6in) long.
Wingspan: None.
Development: Incomplete metamorphosis.
Food: Herbivore.

Above: This stick insect has protective spines on its legs and body.

Tessellated Phasmid

Ctenomorphodes tessulatus

This insect is a medium-sized stick insect of Australia. In autumn, the males, which are less than half the size of the females, fly around to find mates. After mating the females lay hundreds of eggs, one at a time, dropping them in leaf litter where they remain until they hatch in the spring. The sound of female stick insects dropping their eggs sometimes sounds like rain in the forest. The young hatch, looking much like tiny wingless adults, and make for the nearest tree, which they climb. They can reproduce parthenogenetically. Normally, unfertilized eggs only produce female young, but in the tessellated phasmid, males are also produced. These stick insects can reach plague proportions, forming swarms that can defoliate whole trees in eucalyptus forests. The insects begin to lose their brown-green coloration and become more brightly coloured as their density increases.

Main habitat: Woodland.
Main region: Temperate, subtropical.
Length: 14cm (5.5in) (female); 80mm (3.15in) (male).
Wingspan: 60–80mm (2.36–3.15in).
Development: Incomplete metamorphosis.
Food: Herbivore.

Identification: This species has a spiny thorax and legs. Although the male is smaller, it has longer wings (with a marbled pattern) and longer antennae. The female has large compound eyes.

Above: Unlike many stick insects, this species flies quite frequently.

Spiny Devil

Eurycantha horrida

Main habitat: Forest.
Main region: Tropical.
Length: 70mm (2.76in)
(male); 11cm (4.5in) (female).
Wingspan: None.
Development: Incomplete
metamorphosis.
Food: Herbivore.

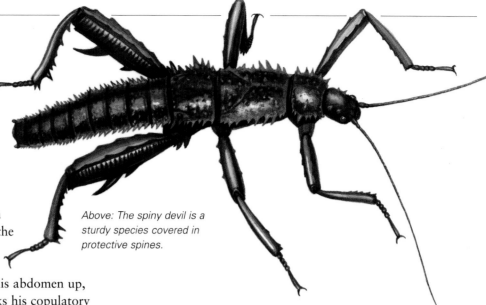

Above: The spiny devil is a sturdy species covered in protective spines.

This is a dark brown stick insect from Papua New Guinea. When disturbed the male will clasp the intruder (or his or her finger) between his spiny hind legs in a vice-like grip. He also curls his abdomen up, revealing bright white stripes and flicks his copulatory organ, (which resembles teeth-lined jaws) in and out through the tip of his abdomen. This threat display looks a little like a snake rearing in warning fashion and is made all the more repellent by the insect releasing a foul odour.

Identification: The body is large, stout and wingless, and bristling with sharp prickles. The spines along the abdomen are short and dense, while those on the upper hind legs are longer, curved and sharply pointed.

OTHER SPECIES OF NOTE

Titan Stick Insect *Acrophylla titan*
This is Australia's, and probably the world's, longest stick insect, at 25cm (10in) (female). The adult female is dark brown with grey or pink spots, and the hindwings have a brownish mottled pattern. The thorax and legs are spiny. The male is similar but much smaller, only about half the size.

Goliath Stick Insect *Eurycnema goliath*
This Australian species is only slightly smaller than the titan stick insect. It is not easily seen, however, as it tends to hide among the twigs at the tops of trees. It has quite a broad green body with white rings. The male is smaller, about 18cm (7in), and much slimmer, with longer antennae. When disturbed it displays a warning red colour under the wings.

Grass Stick Insect *Parasipyloidea* spp.
These extremely thin-bodied stick insects live among fine grass stems in the dry interior of Australia. They look exactly like a thin grass stem as they sit among the vegetation. The most common species is pale yellow-brown with darker lines along the body. It grows to a length of about 80mm (3.15in).

Giant Prickly Stick Insect *Extatosoma tiaratum*
This is a large stick insect, found in Papua New Guinea and Australia. It grows to about 15cm (6in) (female) and its rather plump body has leaf-like flanges, most obvious on the legs. The male is smaller and winged, and the first-stage nymph looks rather like an ant.

Javan Leaf Insect

Phyllium bioculatum

This leaf insect hangs from a branch and swings gently as if in a breeze. Only the male uses his hindwings to fly; in the female the wings have no function. The males are short-lived. Eggs are scattered on the ground and hatch in the spring, resembling various kinds of seeds to prevent them being eaten. When they hatch the young resemble their parents except that they have a more reddish colour, turning green soon after they begin feeding on leaves. They take several months to reach adulthood. This insect is found in South-east Asia, notably in Borneo, China, India, Java, Malaysia, Papua New Guinea, Singapore and Sumatra.

Main habitat: Forest.
Main region: Tropical.
Length: 46–68mm
(1.81–2.68in) (male);
67–94mm (2.64–3.7in)
(female).
Wingspan: 65mm (2.56in)
(male).
Development: Incomplete
metamorphosis.
Food: Herbivore.

Identification: This insect bears an uncanny resemblance to a dead or dying leaf. Its body is yellow-green. The abdomen is wide and extremely flattened, with an irregular shape. The forewings are also irregular and marked with a thickened 'vein' running along the leading edge and branching out to the crinkly edges. The 'leaf' is marked with random patches of red, brown and black, resembling patches of decay setting in, just as they would on an ageing leaf. The legs continue this disguise, the front pair being held out in front of the head so that the stubby head becomes the stalk of the leaf.

Above: Leaf insects are some of the best camouflaged of all insects.

BUGS

The insect order Hemiptera comprises the true bugs, although the word 'bug' is often used loosely to refer to any insect. True bugs are a large group, with more than 82,000 species found worldwide. They make up the most varied and numerous order of insects with incomplete metamorphosis. Australia has about 5,650 species. True bugs include cicadas, aphids and stink bugs among other groups.

Colossus Water Bug

Lethocera grandis

The huge aquatic bugs in this family (Belostomatidae) are the largest of all bugs and some of the most impressive of all insects. They live in static or slow-moving bodies of water with plenty of waterweed. The colossus water bug feeds on a wide range of other invertebrates, and sometimes catches frogs and fish. It flies readily from pond to pond and is often attracted to lights at night. The end of the abdomen carries a pair of organs that together form a breathing tube. The bug visits the surface from time to time to replenish air supplies. This bug is eaten as a delicacy in parts of Asia. The female glues her eggs to the male's back and he broods them until they hatch.

Main habitat: Still or slow-moving water.
Main region: Tropical.
Length: 90mm (3.54in) long.
Wingspan: 10cm (4in).
Development: Incomplete metamorphosis.
Food: Carnivore.

Identification: Brown body, broad, oval and flattened with front legs adapted to form sharp clawed pincers for seizing prey. The hind legs are broad and flat for swimming. The antennae fold into special grooves on the sides of the head and the eyes are large.

Left: This large bug is a fierce predator.

Water Strider

Limnometra cursitans

Water striders and pond skaters belong to the family Gerridae. This Australian species is a brown water strider. It walks on its middle and rear legs which are twice as long as its body, its shorter front legs being used for grabbing and holding prey. Predators and scavengers, water striders are sensitive to the ripples caused by unfortunate creatures that fall on to the surface of the water and cannot get airborne again. They run toward such creatures and suck out their juices. When threatened they disappear under the water, the velvety covering of their bodies trapping air to breathe. When danger has passed they pop back up again. Although they have wings, they rarely fly.

Main habitat: Surface of fresh water.
Main region: Tropical.
Length: 25mm (1in) long.
Wingspan: 45mm (1.77in).
Development: Incomplete metamorphosis.
Food: Carnivore.

Left: Water striders hunt by walking along the surface film.

Identification: These are delicate insects with very long legs tipped with water-repellent hairs, enabling them to run about on the surface of bodies of water without breaking the surface tension.

Cotton Harlequin Bug

Tectocoris diophthalmus

Main habitat: Gardens and
cotton plantations.
Main region: Tropical and
subtropical.
Length: 16mm (0.63in).
Wingspan: 30mm (1.18in).
Development: Incomplete
metamorphosis.
Food: Herbivore.

This bug from eastern Australia belongs to a family (Scutelleridae) known as
jewel bugs – so-named because they are often bright metallic colours. The
female stands guard over her clutch of eggs glued to a plant stem, to
protect them from predators and parasitoids.
This bug feeds by piercing the tissues and
sucking the sap of plants such
as hibiscus and cotton. It
can cause the transmission of fungus
to these plants.

Identification: The scutellum (the triangle below the pronotum,
between the wings) is so large that it covers the wings, giving this bug
a beetle-like appearance. The adult female is bright orange with black
blotches, black legs and antennae. The male is metallic blue with red
patterning. The young nymphs are metallic blue, purple and orange and
cluster together on their food plant.

Right: This colourful bug is an occasional pest of cotton plants.

OTHER SPECIES OF NOTE

Green Jewel Bug *Lampromicra senator*
This is an attractive Australian jewel bug,
sometimes spotted in gardens. It is small,
about 10mm (0.39in), and rather beetle-like
in shape. Its body is bright and shiny, with a
greenish-golden upperside featuring an
orange central patch, and bright orange
upper legs.

Dry Nut Stink Bug *Spermatodes* spp.
This small bug grows to about 10mm (0.39in).
It bears a remarkable resemblance to a dry
nut, hence the common name. Domed and
brown, its surface is covered in pinprick-like
dots. By keeping still on a branch or twig it
escapes notice from potential predators.

Spined Citrus Bug *Biprorulus bibax*
This common shield bug feeds on citrus
fruits. It is about 20mm (0.79in) long,
green, and has thorn-like, black-tipped
spines projecting forward at each side of
the thorax. The nymphs are green and
orange with black spots. This bug can
damage citrus crops, especially lemons
and mandarins.

Shield Bug *Poecilometis* spp.
The shield bugs of this genus are the
largest shield bugs in Australia. They feed
on gum and wattle trees and are common
in forest and scrub. They have rather long
wings and flat bodies with a delicate pattern
in grey and white, although some species
are reddish. Also known as stink bugs, they
protect themselves by producing a nasty-
smelling liquid.

Lychee Stink Bug

Tessaratoma papillosa

Giant shield bugs (Tessaratomidae) are found
almost everywhere but are especially numerous
in tropical and subtropical regions. They are
named for the large shield-shaped scutellum
which covers part of the abdomen. They are
herbivores, sucking juices from plants. While
some are masters of camouflage, others are
brightly coloured and use other ways to protect
themselves from predators, earning the name
stink bugs. Both adults and nymphs produce a
foul-smelling liquid when molested. This species
is a serious pest of lychees and longan in China,
Vietnam, Thailand, Myanmar, the Philippines
and India. It has one generation per year, and
adults aggregate and overwinter on lychee
bushes. In spring the
females move to plants
with new flowers and
shoots in order to mate and
lay their eggs on the backs of
leaves from spring through to late
summer. Nymphs appear in summer,
mature and overwinter
before mating.
They feed on the
terminal shoots
and flowers and
may damage up to 30 per cent of
fruits of these commercial crops.

Right: The nymph is a very colourful insect.

Main habitat: Cultivated fruit
farms.
Main region: China, Vietnam,
Thailand, Myanmar,
Philippines and India.
Length: 27mm (1.06in).
Wingspan: 50mm (2in).
Development: Incomplete
metamorphosis.
Food: Herbivore.

Identification: The adult is a
large brown shield bug, with a
heavy darker scutellum reaching
a point more than halfway along
the abdomen. Its oval body is
serrated around the sides and
rear end, resembling a dead leaf.
Underneath it is silvery. The
nymph (shown) is more oblong in
shape with a reddish back edged
with orange
and black.

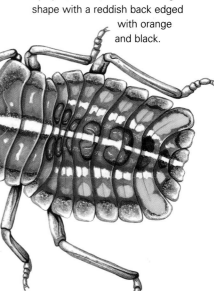

Assassin Bug

Velinus malayus

Assassin bugs (family Reduviidae) are a large group of tissue-sucking predators. They have a long, thick, menacing beak used for killing and consuming their victims. They kill other insects by stabbing them through a vulnerable part of their cuticle and then injecting saliva, which both paralyses the body and digests the tissues. Then the victim is sucked dry. In some species the nymphs have the macabre habit of carrying around the sucked-out bodies of their prey as a form of camouflage. This species, from the Malaysian rainforests, specializes in killing bees. It first gathers sweet odorous resins from plants, spreads them on its front legs, and then lies in wait. The smell attracts the bees toward it.

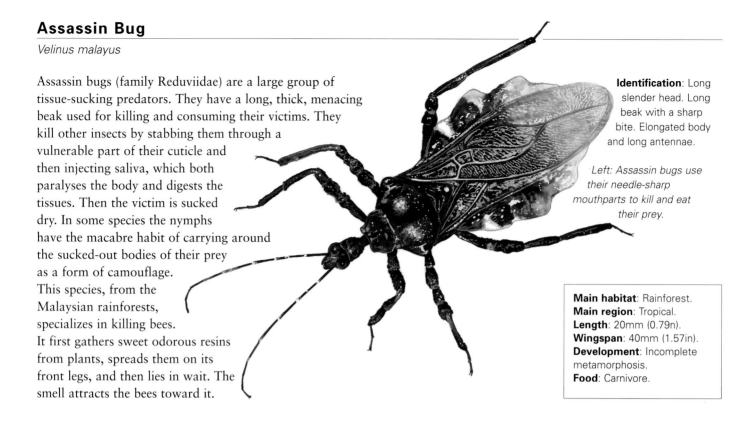

Identification: Long slender head. Long beak with a sharp bite. Elongated body and long antennae.

Left: Assassin bugs use their needle-sharp mouthparts to kill and eat their prey.

Main habitat: Rainforest.
Main region: Tropical.
Length: 20mm (0.79n).
Wingspan: 40mm (1.57in).
Development: Incomplete metamorphosis.
Food: Carnivore.

Crape-myrtle Aphid

Sarucallis kahawaluokalani

This tiny aphid is native to South-east Asia. In some regions where it has been accidentally introduced (including the USA) it has become a serious pest of crape myrtle (*Lagerstroemia*). It sucks sap from the plant and utilizes only the nutrients it needs, the sugar being ejected as honeydew and covering any surface beneath it with a sticky coating. The eggs hatch early in spring, as the leaf buds break open on the food plants. All the hatchlings are female. They moult four times and can reach adulthood in five days when conditions are optimum. Then they give birth to live young by parthenogenesis. Several generations of unfertilized females are born throughout the summer. Each nymph is born with the next generation of nymphs already developing inside her and she can give birth within 24 hours. As the days get shorter and the temperature drops males are produced and mating takes place. The eggs are placed in crevices on the crape myrtle and stay there over winter, hatching in the spring.

Main habitat: Crape myrtle plants.
Main region: India, China, Korea, Japan; introduced to USA.
Length: 1mm (0.04in).
Wingspan: 2mm (0.08in).
Development: Incomplete metamorphosis.
Food: Herbivore.

Identification: Nymphs are pale to bright yellow with black spikes on the abdomen. Adults are also bright yellow but have black spots and two black tubercles on the back of the abdomen. All adults bear wings which are held like a roof over the body and have black markings.

Left: The nymph of the crape-myrtle aphid has short black projections on its body. Adults differ slightly in colour and may produce two mottled wings that are held back over the body.

Frog Cicada

Venustria superba

Main habitat: Rainforest.
Main region: Tropical.
Length: 35mm (1.38in).
Wingspan: 85mm (3.35in).
Development: Incomplete metamorphosis.
Food: Herbivore.

Identification: A medium-sized cicada with a brown body and black tip to the abdomen (male). The wings are long and slightly grey (male).

Right: Frog cicadas take their name from their croaking calls.

Cicadas are found throughout the tropical and subtropical regions of Asia and Australasia. Their most obvious feature is their ability to produce sound. Eggs are laid in incisions on stems and branches of trees and when they hatch the young fall to the ground, burrowing into the soil with their huge mole-like forelegs. The larvae live underground sucking sap from plant roots. They often take many years to grow to maturity. When they do reach adulthood they wait until night, climb the tree and shed their skin, leaving it attached to the tree. The soft, pale insects that emerge soon become hardened and much darker to blend in with their surrounding. Adult cicadas feed on sap from stems and leaves. This species, from north-east Queensland, Australia, produces a characteristic low-pitched, frog-like call late in the afternoon. The call is produced by vibrating two membranes on each side of the abdomen, the sounds being amplified by large cavities.

OTHER SPECIES OF NOTE

Common Assassin Bug *Pristhesancus* spp.
A common species in Australia where it is widespread. Also known as the 'bee killer', it grows to about 20mm (0.79in) and is an active hunter of other invertebrates. It has an orange body with black legs and long antennae. It should not be handled as it can inflict a painful bite.

Double Drummer Cicada *Thopha saccata*
This is one of Australia's largest and loudest cicadas. It grows to 45mm (1.77in) long and is a brightly coloured orange and black. The song of this species is a loud whine, varying in pitch; it sings through the day and into the evening. It is often found in open dry forests.

Emperor Cicada *Pomponia imperatoria*
This is the world's largest cicada, living in the highlands of South-east Asia, notably peninsular Malaysia. It is about 11cm (4.5in) long with a wingspan of 22cm (8.5in). Its body is dark and the wings are transparent with prominent veins. The song of this giant cicada is said to be deafening.

Bladder Cicada *Cystosoma saundersii*
This unusual cicada is bright green and the male has an inflated, bladder-like abdomen, which acts to amplify its song. This Australian cicada grows to about 50mm (2in) long and is hard to spot among green foliage. The forewings are shaped and patterned like leaves to give it protection.

Passion-vine Hopper

Scolypopa australis

Leaf-hoppers are generally small, slender insects with large heads and a tapering body. Some are brightly coloured but others have camouflage colours of browns, greens and blues. They have long, spiny hind legs that enable them to jump as their name suggests, and they are good fliers. All are sap-suckers and many live on only one type of plant. Passion-vine hoppers are native to eastern Australia and have been introduced to New Zealand. They feed on many plants including citrus, kiwi, hydrangea, jasmine, wisteria and passion-vine, sucking sap with their backward pointing rostrum and excreting large amounts of honeydew. If the hoppers have been feeding on a poisonous plant and their honeydew is used by bees to make honey it can cause severe poisoning and even death in humans that eat it. Eggs are inserted into soft plant stems where they overwinter.

Right: This hopper has large mottled wings.

Main habitat: Gardens, parks and fruit plantations.
Main region: Temperate, subtropical.
Length: 5–6mm (0.19–0.24in).
Wingspan: 15mm (0.59in).
Development: Incomplete metamorphosis.
Food: Herbivore.

Identification: The passion-vine hopper is brown with transparent forewings mottled with black-and-white markings and held like a tent over the abdomen. Nymphs are pale with brown markings and have a tuft of white waxy filaments on the abdomen, which they can fan out like a peacock.

CADDIS FLIES, LACEWINGS AND RELATIVES, ANTLIONS AND SCORPION FLIES

These insects belong to a diverse range of groups. Caddis flies have aquatic larvae, most of which build protective tubes, but antlions and scorpion flies lay their eggs on land. The adults of all of these insects are winged. Lacewings and relatives have net-veined wings. Scorpion flies have a beak-like face.

Marine Caddis Fly

Philanisus plebeius

Most caddis fly larvae live in fresh water, but this species is unusual in living in tidal seawater pools in New Zealand and south-eastern Australia. The female lays her eggs inside the coelomic cavity of a starfish in the spring and autumn. She has a noticeably long ovipositor. The larvae hatch and escape out into the rock pool. There they feed on seaweed fragments. They construct their cases from sand grains and bits of seaweed and are very hard to spot. The adults fly mainly at night, and are very moth-like.

Identification: The larval case is highly camouflaged, with pieces of seaweed, and fragments of shells and corals. The adult is pinkish-beige in colour with long wings and long antennae.

Main habitat: Rock pools.
Main region: Temperate, subtropical.
Length: 18mm (0.71in) (adult); 9mm (0.35in) (larva).
Wingspan: 30mm (1.18in).
Development: Complete metamorphosis.
Food: Herbivore (larva).

Above: This caddis fly is unusual in that its larvae live in salt water.

Snowflake Caddis

Asmicridea grisea

Most adult caddis flies fly at night and are attracted to light. This species from Tasmania, however, is one of the few day-flying caddises. It is abundant in many parts of the island and when the adults emerge from the water they attract large numbers of fish, especially brown trout, which come to feed on both the pupae and adults. This in turn attracts anglers who are therefore very aware of caddis fly emergence periods. Dams on rivers such as the Shannon have affected the abundance of these insects and it has become necessary to adapt the pattern of the release of the water from them to ensure that these caddis flies can complete their life cycle.

Identification: The adult is a pretty insect, rather moth-like and with striking pure white wings, hence the common name. A mass emergence of these caddis flies can indeed resemble a snowstorm.

Main habitat: Rivers.
Main region: Temperate.
Length: 20mm (0.79in).
Wingspan: 35mm (1.38in).
Development: Complete metamorphosis.
Food: Omnivore (larva).

Below: The fluttering white adults look a little like snowflakes.

Antlion

Acanthaclisis fundata

Main habitat: Dry sandy areas.
Main region: Temperate, subtropical.
Length: 30mm (1.18in) (adult); 25mm (1in) (larva).
Wingspan: 60mm (2.36in).
Development: Complete metamorphosis.
Food: Carnivore.

Identification: The larvae are unlike the delicate adults. They have elongated, flattened bodies with hairs that face forward. The hind legs are large and have forward-pointing claws used for moving through the sand. Their large heads bear strong jaws.

Right: The adult antlion looks a bit like a dragonfly.

Antlion eggs are laid singly in dry soil. These insects live in dry areas or sheltered undercliffs and overhangs. Each larva digs a deep pit by moving in a circle backward and flicking sand outward. The larva then digs itself into the bottom of the pit with its jaws just below the deepest part. Any insect that finds its way into the pit starts an avalanche of sand and tumbles down toward the antlion's jaws. The antlion increases the chaos by tossing sand at its victim. Once caught, the prey is sucked dry and the shell flicked out of the pit before the trap is set again. The larva undergoes about three or four moults and then spins itself a silk cocoon in which it pupates. The pupa works its way to the surface before the adult emerges. There are usually several generations per year.

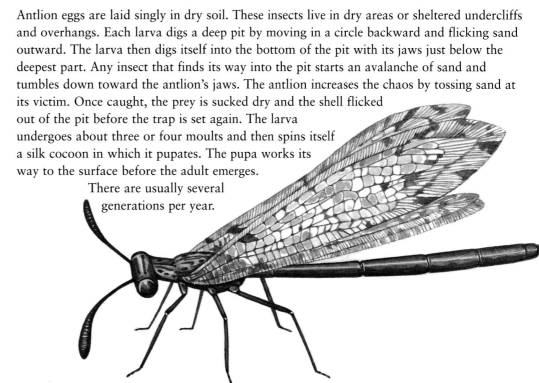

OTHER SPECIES OF NOTE

Longhorned Caddis *Triplectides* spp.
The larva of this species lives inside a hollow piece of stick or reed in slow-flowing streams. The adult is about 15mm (0.59in) long, and has extremely long antennae, about five times as long as its body. It has a pale brown body and wings and darker eyes.

Tasmanian Lacewing *Micromus tasmaniae*
This small light brown lacewing is found in Tasmania and New Zealand. Its wings have a covering of fine hairs. Like most lacewings, both the larvae and adults eat smaller insects, especially greenfly and other aphids.

Mantisfly *Campion* spp.
Mantisflies are related to lacewings and antlions. They take their name from the front legs which are hinged for catching prey, like those of mantids. Otherwise they resemble lacewings with their clear wings. This Australian species grows to about 20mm (0.79in) long. It has yellow-and-black wasp-like markings along its back.

Giant Antlion *Heoclisis fulva*
This large antlion is very hairy and dark grey as an adult, with long net-veined wings. Members of its genus are found mainly in Australia, but also in China and Malaysia. This species is common in much of Australia. It is often attracted to lighted windows.

Hanging Scorpion Fly

Harpobittacus tillyardi

Main habitat: Eucalyptus forest.
Main region: Temperate, subtropical.
Length: 30mm (1.18in).
Wingspan: 60mm (2.36in).
Development: Complete metamorphosis.
Food: Carnivore.

Scorpion flies have a beak-like rostrum with mandibles at the tip. They have two equally sized membranous wings with strong venation. They have long legs and in some species the males have upturned abdomens which make them look a little like scorpions. The larvae are caterpillar-like with strong mandibles. The males often offer nuptial meals to their prospective mates. The eggs are laid in or near the ground and the larvae live in moist leaf litter, moss or similar environments, scavenging on dead insects. They pupate in the soil. This species is one of a group known as hanging scorpion flies from their habit of catching their prey by hanging from branches and dangling their long back legs which have claws.

Identification: This Australian species is one of the most common and the largest in its family (Bittacidae). It is reddish-brown and looks a little like a cranefly with beak-like mouthparts, large claws on their hind legs and two pairs of wings. The adults are slow fliers.

Right: This species uses its powerful hind legs to capture its prey.

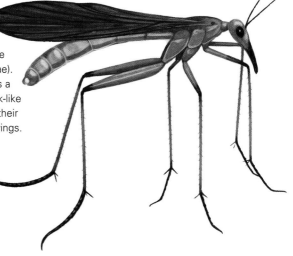

BEETLES

*With more than 300,000 known species, beetles (Coleoptera) are by far the largest of the insect orders.
In Australia there are estimated to be more than 28,000 species, and in Asia there are many more. The
beetles are found in all sorts of habitats and occur in a wide variety of shapes and sizes. Amazingly,
about a third of all animal species are beetles.*

Ground Beetle

Mecynognathus dameli

This species is one of Australia's largest ground beetles. It is a
flightless predator and runs down live prey, catching its victims in
its large sharp jaws. Although the larvae resemble caterpillars in
appearance, they are also predators and capable of moving quite
fast. This species is found in the Cape York Peninsula, Australia.
All ground beetles are carnivorous, eating other invertebrates as
well as scavenging. They rarely fly, as their name suggests – indeed
many are flightless. Nevertheless, they are active creatures, with
long slender legs which enable them to run very fast after their
prey. Their formidable jaws make them efficient predators. They
are stout flattened beetles; the elytra or wing cases have grooves
or striations along their length and form a perfect oval shape
over the abdomen. Many ground beetles are active both by
night and during the day.

Identification: Large and shiny black, with a
large head and huge jaws. The thorax is
also large.

*Above left and right: Sleek and
shiny, this ground beetle is a
speedy predator.*

Main habitat: Ground.
Main region: Tropical.
Length: 75mm (2.95in).
Wingspan: None.
Development: Complete
metamorphosis.
Food: Carnivore.

Green Carab Beetle

Stink Beetle, *Calosoma schayeri*

This large, active ground beetle feeds on caterpillars
and other slow-moving prey. Found in most
habitats, it hides under stones and logs by day
but comes out to hunt at night. It is sometimes a
nuisance as it is often attracted to lights, for
example in shopping centres. It squirts an
unpleasant smelly substance from the tail as a
defence when handled, but it is also a useful
beetle as it preys on pests of crops such as army
and cutworms, which are pests of cotton. Like
the adults, the larvae are also predacious.

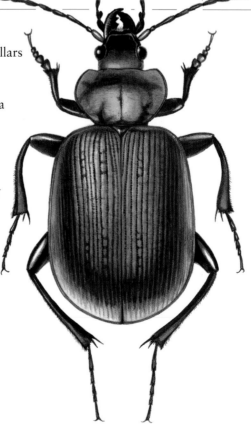

*Above and right: This beetle is an attractive metallic
green with a relatively large wingspan.*

Main habitat: Ground.
Main region: Temperate,
subtropical.
Length: 24mm (0.94in).
Wingspan: 48mm (1.89in).
Development: Complete
metamorphosis.
Food: Carnivore.

Identification: A beautiful,
plump, iridescent green beetle
with ridged elytra. It has a
rounded thorax and a rather small
head, and the long legs typical of
members of this family. As in
most beetles, the tough
forewings form a protective cover
for the membranous hindwings,
which are used for flight and
folded beneath the elytra, when
not in use.

Violin Beetle

Mormolyce phyllodes

> **Main habitat**: Tropical forests.
> **Main region**: Tropical.
> **Length**: 90mm (3.54in).
> **Development**: Complete metamorphosis.
> **Food**: Carnivore.

Identification: This large brown beetle has an extraordinarily shaped body, like a flat violin. The 'neck' of the violin is the very long, thin head and thorax. The 'body' of the violin is the ridged wing cases which are drawn out into wide flanges on each side. This beetle also has very long, thread-like antennae.

This is one of the world's strangest beetles. Also called the 'fiddle beetle', the common names refer to its violin-like shape. This unusual beetle lives in forests in Malaysia and Indonesia. Its remarkable shape and colouring enable it to hide in the soil, under bark or between layers of bracket fungus, where it lays its eggs. It eats other beetles, insect larvae and snails. It ejects butyric acid in self-defence, which can paralyse human fingers for up to 24 hours. Mainly because of its extraordinary shape and large size, it is much sought after by collectors and it is now protected by Malaysian law.

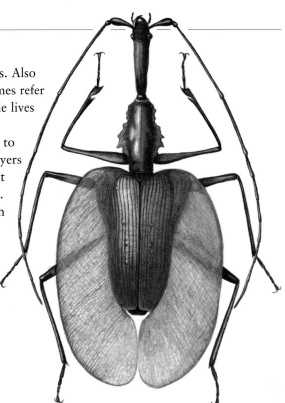

Right: The thin flanges extending from the wing cases are a unique feature.

> OTHER SPECIES OF NOTE
>
> **Japanese Tiger Beetle** *Cicindela japonica*
> A beautiful tiger beetle found in Japan, and also in Korea, China and Vietnam. It is about 25mm (1in) long and is prettily patterned in shiny green-blue, red, black and white. Even its legs have these glinting metallic colours.
>
> **Tamamushi Beetle** *Chrysochroa fulgidissima*
> This jewel beetle from the family Buprestidae is a very large metallic green beetle with reddish longitudinal stripes. Found in forests in Japan, it lays its eggs on dead tree trunks and the larvae burrow into the wood. A 7th-century Buddhist shrine exists in Nara, Japan, decorated with 9,000 wing cases of this beetle. Jewel beetles are flattened, elongate beetles with large eyes and short antennae, and legs that can be folded against the body. They are usually brightly coloured, often metallic. The most spectacular come from South-east Asia.
>
> ***Chrysochroa buqueti***
> Another jewel beetle, this species is about 50mm (2in) long, with an iridescent red-gold head and thorax, cream and shiny blue-black banded wing cases, and brown membranous wings. The abdomen beneath the wings is metallic blue. The adults of this species fly in April and feed on the flowers of trees. This beetle lives in tropical forests in Vietnam, Thailand and Malaysia.
>
> ***Megaloxantha daleni***
> A large, 55mm (2.17in) metallic green jewel beetle from Thailand and Malaysia with irregular creamy-white patches on the lower surfaces of the elytra. Male and female are similar but the female is larger.

Blue-spotted Tiger Beetle

Cicindela aurulenta

Tiger beetles belong to the same family as ground beetles. They have downward-turned heads and powerful curved jaws. Their larvae live hidden in burrows with their jaws level with the surface, ready to grab any insect that wanders too close. The blue-spotted tiger beetle is found in Asia, notably in Borneo, China, Indonesia, Malaysia and Thailand. Like all tiger beetles it is a fast-running predator feeding on a range of other invertebrates.

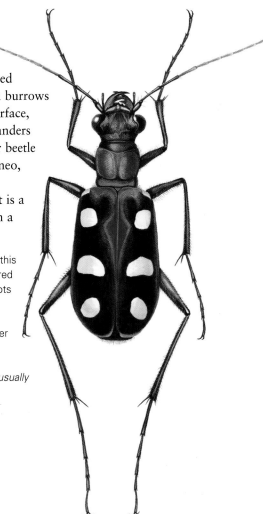

Identification: The body and head of this beetle are iridescent blue-green with red edges and bright yellow or orange spots on the elytra. It has large prominent compound eyes, long, fine black antennae and very long legs. The upper legs are covered in stiff hairs.

Right: The legs of this species are unusually long, even for a tiger beetle.

> **Main habitat**: Ground.
> **Main region**: Tropical.
> **Length**: 20mm (0.79in).
> **Wingspan**: 30mm (1.18in).
> **Development**: Complete metamorphosis.

Jewelled Frog Beetle

Sagra buqueti

The family Chrysomelidae, to which this beetle belongs, is a large one. Most species are small and oval or rounded, and many are brightly coloured. Both adults and larvae are vegetarians – some feed on pollen or seeds but most feed on foliage. Many therefore are serious pests, though a few have been used for the biological control of weeds. The jewelled frog beetle comes from peninsular Malaysia. It lays its eggs in groups on plants and seems to particularly favour jungle vines.

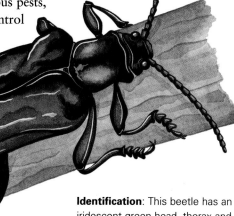

Main habitat: Rainforest.
Main region: Tropical.
Length: 27mm (1.06in).
Wingspan: 54mm (2.13in).
Development: Complete metamorphosis.
Food: Herbivore.

Left and below: The hind legs of the male jewelled frog beetle are disproportionately large.

Identification: This beetle has an iridescent green head, thorax and legs and its glossy elytra shimmer with metallic shades of blue, green and gold. The name frog beetle comes from the large, swollen hind legs, which are used by the males when they battle to win females.

Oil Beetle

Blister Beetle, *Mylabris tiflensis*

This species belongs to the oil or blister beetle family (Meloidae), named for the oily secretions its members produce. When molested the beetle exudes this liquid from its leg joints – it is both a poison and an irritant, and may cause blisters. Female blister beetles are very choosy about their mates and the males have to indulge in complex courtship rituals to entice them. In this species from India, sexual activity is restricted to a few brief minutes just before dusk when many individuals in a locality will mate. Courtship involves the male facing the female and touching her head with his rapidly vibrating front legs. Eggs are laid in the soil and the long-legged, active larvae that hatch set out to find food. Adult oil beetles are slow-moving plant eaters but the larvae are parasites of bees or grasshoppers. Some species lie in wait for bees, hitching a lift back to the nest; others look for grasshopper egg cases in the soil. Once they find a suitable food supply they moult into a more maggot-like form and devour the eggs. In China, some species of this genus are used in traditional medicine.

Main habitat: Dry sites.
Main region: Tropical.
Length: 25mm (1in).
Wingspan: 45mm (1.77in).
Development: Complete metamorphosis.
Food: Herbivore (adult).

Identification: Large beetles with distinct 'necks' between their heads and thoraxes, usually shiny black. They have brightly coloured elytra patterned with warning red, yellow or black stripes and are less well armoured than other beetles. *M. tiflensis* has bright red and black stripes across its wing covers.

Left: This oil beetle has bright warning coloration.

Eucalyptus Longhorn

Phoracantha semipunctata

Main habitat: Eucalyptus trees.
Main region: Subtropical.
Length: 14–30mm (0.55–1.18in).
Wingspan: 45mm (1.77in).
Development: Complete metamorphosis.
Food: Herbivore.

Identification: A long brown-black beetle with large eyes and stout antennae that are longer than the insect itself. The wing cases are pitted and have cream and black patches over the front half. The near half is smoother and tipped with cream spots. It has large biting jaws and long, strong legs which end in obvious claws.

This beetle belongs to the longhorn family (Cerambycidae). It is a serious pest as it feeds on and kills eucalyptus trees. It has spread from its native Australia to many other countries through imported timber. Its life cycle takes a year and the adults fly in early summer. The females are attracted to trees that are stressed by dry conditions, and lay eggs on or under the bark. The larvae that hatch either bore straight into the tree and form very extensive galleries in the cambium beneath the bark or they burrow just under the surface of the bark, leaving a dark line. The full-grown larvae excavate a chamber, plugging the entrance with sawdust.

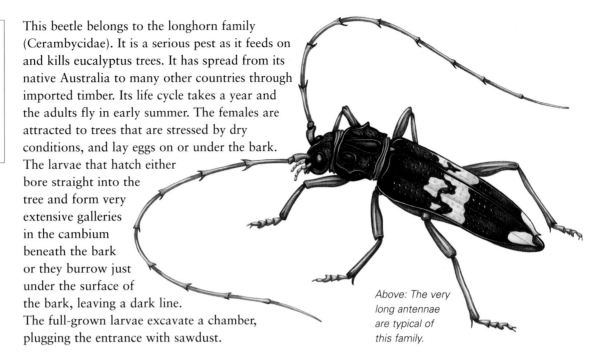

Above: The very long antennae are typical of this family.

OTHER SPECIES OF NOTE

Leaf Beetle *Paropsis obsoleta*
This beetle is a member of the Chrysomelidae, a very large family with more than 35,000 species. Its genus is common in Australia. This species is creamy-yellow, with a network of darker lines and several red-brown spots.

Eucalyptus Tortoise Beetle *Paropsis charybdis*
Another leaf beetle, this species from Australia and New Zealand has a hard, domed carapace, giving it a tortoise-like appearance. It is found on a range of gum trees, feeding on the leaves, and hibernates underneath the bark in winter. The adult is about 14mm (0.55in) long and light brown with darker speckles.

Xixuthrus microcerus
This large, brown longhorn beetle is a giant, growing to more than 75mm (2.95in) long. It has sharp claws and very large toothed jaws. It lives in forests in Australia, Borneo and Papua New Guinea. The grubs are edible and form part of the traditional Aboriginal diet in northern Queensland, Australia.

Wallace's Longhorn *Batocera wallacei*
Australia's largest beetle grows to 80mm (3.15in). It lives in the Cape York Peninsula, and also in Papua New Guinea. Its antennae are longer than its body and curve strongly backwards toward the tips. It feeds on the sap of trees, including the introduced breadfruit tree, and is sometimes called the 'breadfruit beetle'. It can make a loud noise by rubbing its thorax against its abdomen.

Huhu Beetle

Haircutter, *Prionoplus reticularis*

This longhorn is the largest beetle in New Zealand, reaching 45mm (1.77in). The long, white, fat larvae were once widely eaten by the Maori. The males fly at night with a loud whirring noise and are attracted to lights and therefore often bang into windows loudly or fly into houses and crash about. The long legs and antennae are covered with sharp hooks and if they land in hair they can be hard to disentangle – hence the name 'haircutter'. Their powerful jaws can give a painful nip. The 3mm- (0.12in-) long cigar-shaped eggs are laid in batches of ten or so in crevices on dead wood and the larvae hatch after about four weeks. The larvae bore into the wood, consuming it as they go and getting bigger over the next two to three years, growing to about 50mm (2in) long.

Main habitat: Trees.
Main region: Temperate, subtropical.
Length: 45mm (1.77in).
Wingspan: 75mm (2.95in).
Development: Complete metamorphosis.
Food: Herbivore.

Identification: This beetle is oval in shape, with antennae about half as long as the body. The brown wing cases are marked with an attractive and delicately veined pattern in cream.

Left: Adult huhu beetles emerge from decaying wood in midsummer.

Giant Stag Beetle

Dorcus hopei

The giant stag beetle lives in Japan and Korea, mainly on Japanese oak and yew. It takes one to two years to complete its life cycle but the adults can overwinter and live for two to three years. The eggs are laid in September and the grubs develop over the next two years or so before pupating. The adult emerges one month later. Like many stag beetles, the jaw shape of the males can vary. The adults appear from June to September and feed on nectar, flowing tree sap or ripe fruit. They are nocturnal. The jaws of the male stag beetle are like the antlers of a deer, in both shape and function. They are used mainly in jousting contests between rival males competing for access to a female.

Main habitat: Trees.
Main region: Temperate.
Length: 70mm (2.76in).
Wingspan: 90mm (3.54in).
Development: Complete metamorphosis.
Food: Herbivore.

Identification: A wide, flat, mainly black beetle. The male has a wide head with huge gaping jaws with one 'tine' or branch on each. The female has a narrow head with quite normal jaws. The larvae are large and grub-like and feed on rotting wood.

Left: The 'antlers' of the giant stag beetle are not as dangerous as they appear.

King Stag Beetle

Rainbow Stag Beetle, *Phalacrognathus mulleri*

This rather rare stag beetle from northern Australia is regarded by many as one of the most beautiful of all beetles and is therefore much sought after by collectors. Australia's largest stag beetle, it breeds in rotting wood on fallen or standing trees, living or dead. Both adults and larvae feed on moist rotten wood, and the adults also eat plant sap and nectar. Eggs are laid at any time of the year in groups of up to 30. The male attends the female while she is laying. Larvae hatch after ten to 14 days. When fully grown the larva constructs a pupal cell inside which the adult develops, taking about a week to gain its metallic colours. It can stay within the pupal cell for up to eight months. The adults dig their way out using their large jaws and powerful legs. They can live for 18 months in captivity. The males wrestle by trying to get their jaws beneath an opponent to flip him over.

Identification: Large, shiny, metallic green-red beetle. The colours vary with the light and angle. The horns stick out straight in front and curve upward, rather like scimitar blades.

Main habitat: Rainforest.
Main region: Tropical.
Length: 24mm (0.94in) (females), 70mm (2.76in) (males).
Wingspan: 50mm (2in).
Development: Complete metamorphosis.
Food: Herbivore.

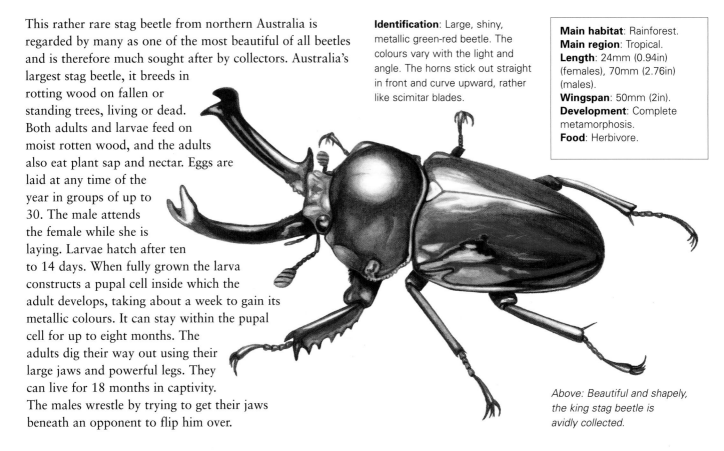

Above: Beautiful and shapely, the king stag beetle is avidly collected.

Rhinoceros Beetle

Elephant Beetle, *Xylotrupes gideon*

Main habitat: Trees.
Main region: Warm
temperate–tropical.
Length: 45–60mm
(1.77–2.36in).
Wingspan: 65mm (2.56in).
Development: Complete
metamorphosis.
Food: Herbivore.

This is a common and widespread scarab beetle from mainland South-east Asia, Indonesia and Australia. The males use their horns to try and knock a rival male off a branch when a female is in the vicinity. The females give off pheromones when they are looking for a mate. These beetles sometimes gather in large numbers. Females lay about 50 eggs in decaying vegetable matter on which the larvae feed, hatching in about three weeks. The adult emerges after about one month and lives for two to four months. Adults are seen in the summer in Australia but in the tropical parts of their region they are present all year round. Though they are harmless they can make an alarming hissing squeak if molested by rubbing the abdomen against their wing cases. They also have large and very sharp claws on the ends of their legs.

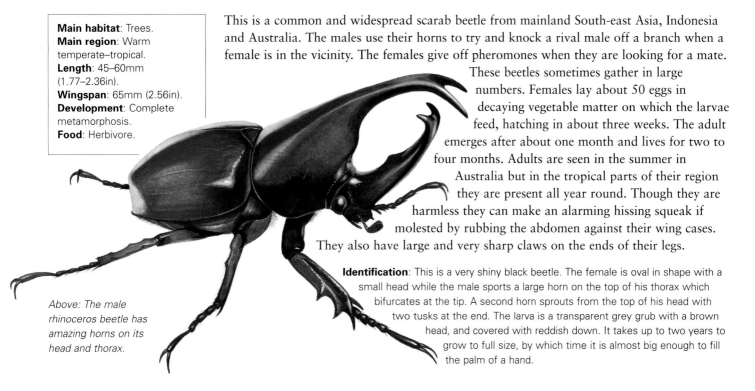

Identification: This is a very shiny black beetle. The female is oval in shape with a small head while the male sports a large horn on the top of his thorax which bifurcates at the tip. A second horn sprouts from the top of his head with two tusks at the end. The larva is a transparent grey grub with a brown head, and covered with reddish down. It takes up to two years to grow to full size, by which time it is almost big enough to fill the palm of a hand.

Above: The male rhinoceros beetle has amazing horns on its head and thorax.

OTHER SPECIES OF NOTE

Large Green Chafer
Tanguru Chafer, *Stethaspis suturalis*
This shiny, bright leaf-green beetle lives in forests in New Zealand. The larvae feed on tree roots and the adults feed on leaves. The adults emerge from their pupae en masse in the summer and are often seen flying in large numbers at the edges of forests at dusk, making a buzzing sound.

Spotted Flower Scarab *Polystigma punctata*
This is a common flower scarab in Australia, especially in the eastern coastal region. It is about 15mm (0.59in) long and has a shiny yellowish body with several dark spots. Flower scarabs feed on nectar as adults and fly well from flower to flower.

Brown Stag Beetle *Syndesus cornutus*
Native to Tasmania, this small, dark brown stag beetle grows to about 25mm (1in). It is also found in New Zealand. By day it lurks underneath the bark of trees from which it emerges at night in order to feed on sap. The larvae live and feed in rotting timber.

Helms' Stag Beetle *Geodorcus helmsi*
This is a hefty stag beetle from New Zealand, where it lives on logs and trees. It is a rare species found mainly in the wet regions of the west of South Island. The adults can reach more than 40mm (1.57in) in length and they have a heavy domed carapace. They are flightless and so are easy prey for introduced rats and mice. They emerge at night to forage, often feeding on sap that exudes from the trunks of trees.

Flower Beetle

Agestrata luzonica

A slow-moving, but fast-flying scarab beetle from Malaysia and South-east Asia. It feeds on fruit, pollen and nectar from flowers. The larvae live in and feed on decaying wood. The male and female are similar in appearance but not in size, the female being larger than the male.

Main habitat: Forest.
Main region: South-east
Asia.
Length: 40–45mm
(1.57–1.77in).
Wingspan: 50mm (2in).
Development: Complete
metamorphosis.
Food: Herbivore.

Identification: A large, green scarab beetle with both the appearance and feel of metal. The body is rather oblong and flattened. It has a long head with a shield-shaped plate covering its mouthparts, and prominent globular eyes. Its legs are long and rather spiny.

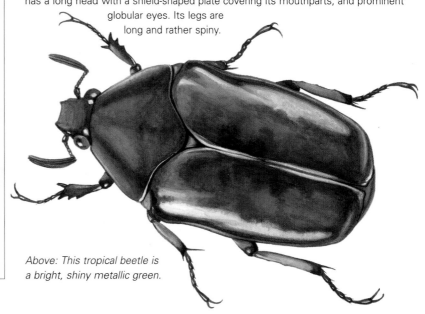

Above: This tropical beetle is a bright, shiny metallic green.

Transverse Ladybird

Coccinella transversalis

This is one of the most common ladybirds encountered around towns throughout Australia. Both adults and larvae feed on aphids on all sorts of plants. They are active during the day. The adults hibernate in communal roosts through the winter, emerging in the spring to mate and lay between 20 and 50 eggs, which are attached to the leaves of aphid-infested plants. The larvae can consume 350 aphids during the three weeks it takes them to become a pupa. The adult emerges after seven to ten days. Like other ladybirds this species can release a foul-smelling and tasting yellow fluid from its legs when molested, hence the bright warning coloration.

Main habitat: Urban areas, forest, heath and woodland.
Main region: Temperate and subtropical.
Length: 4–6mm (0.16–0.24in).
Wingspan: 12mm (0.47in).
Development: Complete metamorphosis.
Food: Carnivore.

Above: The transverse ladybird is one of Australia's most common ladybirds.

Identification: Circular and convex beetle with short, clubbed antennae and short legs. It is yellow, orange or orange-red with black irregular patches on the back, and a central dark stripe.

Harlequin Ladybird

Multicoloured Asian Ladybird, *Harmonia axyridis*

This is an example of biological control gone out of control. This multi-coloured ladybird comes from Japan, China and other parts of Asia, and was introduced into North America in the 1970s to control plant pests. It did extremely well but also drove out native species of pest-eating insects, including some native ladybirds. Harlequin ladybirds are also a nuisance as they overwinter in large numbers in crevices inside houses. They have recently been seen in Britain and it is thought they might out-compete and even kill native ladybirds there. They are large and adaptable and once the aphid populations start to fall in the summer they will turn on other insects.

Identification: The insect is very variable – sometimes red with black dots or black with white, yellow or red dots.

Main habitat: Trees, crops, gardens.
Main region: Temperate.
Length: 6mm (0.24in).
Wingspan: 12mm (0.47in).
Development: Complete metamorphosis.
Food: Carnivore.

Right: The harlequin ladybird has a reputation as a fierce competitor.

OTHER SPECIES OF NOTE

Firefly *Pteropteryx* spp.
In the rainforests and mangrove swamps of Borneo, this most spectacular of all the world's fireflies begins to flash as the sun sinks. Groups flash as they fly in arcs, and eventually whole trees are lit up with their yellow-green flashing lights. In time these flashes begin to synchronize and the tree itself seems to pulse with light. The males flash in synchrony to enhance their individual messages and send them far across the swamps to attract females.

Bennett's Painted Weevil
Eupholus bennetti
One of the largest and most colourful weevils, growing to about 26mm (1.02in), this species from Papua New Guinea has bold longitudinal stripes of pale and dark blue, and stout blue-coloured legs. It has a very tough, knobbly cuticle. The colour may be an indication to potential predators that this insect tastes unpleasant.

Giraffe Weevil *Lasiorhynchus barbicornus*
This is the longest weevil in the world at up to 80mm (3.15in), the snout making up almost half that length. Males have their antennae at the ends of their snouts while females have theirs halfway along their snouts. This is so that the females can bore holes in rotting wood deep enough to lay their eggs, which they do from October to March. The larvae feed on yeasts and other fungi within the wood for up to two years. When the adult emerges from the wood it cuts a hole with a perfectly square section. Captain Cook's crew apparently collected this extraordinary insect when they first visited New Zealand in 1769.

Botany Bay Diamond Weevil
Chrysolophus spectabilis
Black, with brilliant metallic blue or green scales, this weevil was first collected by Sir Joseph Banks at Botany Bay on Captain Cook's expedition of 1770. It is a very common species in eastern Australia. The larvae feed on the wood of certain acacia (wattle) trees, drilling holes in the trunks, branches and roots.

Blue Mountains Firefly

Atyphella lychnus

Main habitat: Rainforest.
Main region: Temperate, subtropical.
Length: 6–9mm (0.24–0.35in).
Wingspan: 12mm (0.47in).
Development: Complete metamorphosis.
Food: Carnivore (larva).

Identification: Soft, flattened body, with large eyes and short antennae. Above, it is golden brown with a blackish head and lower body. The luminescent region is toward the end of the abdomen on the underside.

Right: Though modest in appearance, fireflies can produce startling flashes of light.

Fireflies, or lightning bugs, use coded flashes of light to attract mates. These are produced by chemical reactions within special bioluminescent organs in their abdomens. Some also use trickery – flashing to lure members of a different species in order to eat them. Both males and females flash, but many females are flightless and resemble larvae. The active larvae feed on invertebrates such as snails. Male fireflies fly about flashing, and the females, which have wings but rarely fly, give an answering blink. The eggs, larvae and pupae are also luminous. This firefly species is found in eastern Australia, notably in the Blue Mountains, hence its name. It is thought that the adults do not feed.

Bamboo Shoot Weevil

Cyrtotrachelus longimalus

Weevils (family Curculionidae), comprise the largest of the beetle families, in fact one of the largest families in the animal kingdom. They have an elongated face called a rostrum with the chewing mouthparts at the end. Their antennae have a definite 'elbow'. Eggs are laid in plant tissue and the legless larvae tunnel and feed within the plant. Adults also feed on plants from the outside. Many weevils are serious pests of a very wide range of crops and stored foods, though a few are being used to tackle invasive plants. Both adults and larvae of this tropical Asian weevil feed on bamboo shoots, the larvae causing the most damage by their burrowing activities; they are serious pests, causing the death or deformation of bamboo culms. There is one generation per year; the adults overwinter in cocoons in the soil and emerge in the spring or early summer. Both adults and larvae can be found on bamboo shoots from May to October. The adults suck sap from the tops of the shoots. One to three eggs are laid in a feeding hole and the legless larvae tunnel straight down into the tissues, moulting five times. When they are fully fed they drop to the ground, burrow down 10–30cm (4–12in) and pupate a few weeks later. The adults emerge within 15 days.

Main habitat: Bamboo plantations.
Main region: Bangladesh, China, India, Myanmar, Sri Lanka.
Length: 18–38mm (0.71–1.49in).
Development: Complete metamorphosis.
Food: Herbivore.

Identification: This species varies in colour from light brown to dark reddish-brown. The larvae are cream-coloured, becoming darker as they reach maturity.

Left: This weevil is one of several insects that specialize in eating bamboo, feeding on nothing else.

FLIES

Flies form a very diverse order of insects, with more than 120,000 known species. They are perhaps the most commonly encountered of all insects, occurring in almost all land habitats. In Australia alone there are more than 7,750 species. In many temperate countries, flies make up about a quarter of all insects. Adults range in length from 5–50mm (0.19–2in), with a maximum wingspan of 80mm (3.15in).

Bush-fly

Musca vetustissima

This common Australian fly is a major pest which breeds in dung in pastures and flies to nearby rural and urban areas during the summer. Bush-flies are one of the main reasons for the traditional cork hat worn by farmers (and tourists) developed to keep flies off faces. Bush-flies are attracted to people and animals in large numbers, attempting to feed on sweat, mucus, blood or tears around mouths, noses, eyes or wounds. In so doing they spread diseases such as eye or gut infections from one person to the next. It is the female flies that are most persistent, as they need protein and minerals for their eggs to develop. Female bush-flies can lay up to five batches of up to 50 eggs each, usually deposited in animal dung. They find the body fluids and the dung they need by smell and appear in seconds at newly produced dung. These flies flourish in Australia because there are so many cows: a single cowpat can produce 2,000 flies! The time taken from egg to maggot to pupa to fly is between two and ten weeks, depending on the weather. In the cooler southern areas the flies die in the winter, but in the tropical north they keep breeding all year.

Main habitat: Farmland, especially near livestock.
Main region: Temperate, tropical.
Length: 10–12mm (0.39–0.47in).
Wingspan: 20mm (0.79in).
Development: Complete metamorphosis.
Food: Omnivore.

Above: Bush-flies can be a nuisance in the Australian outback.

Identification: Rather like a typical housefly in shape. The compound eyes are large and dark orange. The body is mainly grey-black with stripes on the back.

Japanese Blowfly

Calliphora nigribarbis

These blowflies are associated with birds, and in particular with poultry. The organism that causes avian flu has been found in the guts of these insects, among others, although it is not yet known if they are implicated in the spread of the disease. They have been found to migrate in large numbers across Asia from the north-west to the south-east in the autumn. In Japan, the adults tend to appear from November to June in towns and from October to July in mountain forests. The adults and developing stages are not generally found between the months of July and October, and it is thought that they aestivate during this time, then migrate from one area to the other.

Main habitat: Grassland.
Main region: Temperate.
Length: 19mm (0.75in).
Wingspan: 40mm (1.57in).
Development: Complete metamorphosis.
Food: Omnivore.

Identification: A hairy fly with a shiny blue-black abdomen, dark grey striped thorax and large dull red eyes. The wings are quite long and transparent, with obvious black veins.

Left: This blowfly may help to spread avian flu.

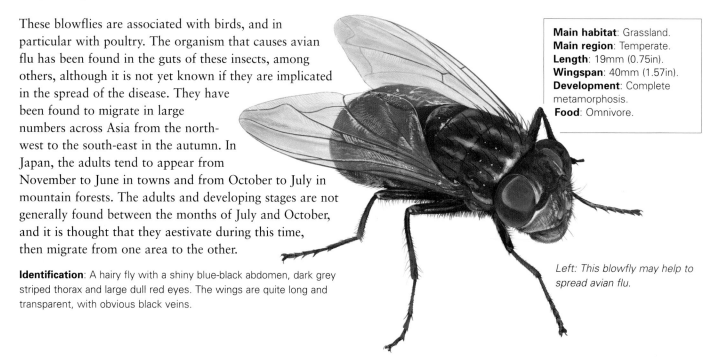

Drone Fly

Eristalinus aeneus

Main habitat: Flowers (adult); stagnant water (larva).
Main region: Temperate.
Length: 15mm (0.59in).
Wingspan: 25mm (1in).
Development: Complete metamorphosis.
Food: Herbivore.

Identification: This species resembles a honey bee, but has only one pair of wings. Brownish-black hairy body with a striped thorax. The compound eyes are very large, pale and spotted. The larvae are aquatic 'rat-tailed maggots', with a grub-like (though hard) body 20mm (0.79in) long with a 30mm (1.18in) long extensible breathing tube 'tail'.

Drone flies are from the same family as hoverflies (Syrphidae), and they share their behaviour – typically hovering frequently as they flit about, from flower to flower. The adults mimic various species of harmful wasps or bees. They feed on pollen and nectar, while the aquatic larvae feed on decaying material in stagnant water. Adult drone flies seem to be particularly attracted to yellow flowers, and also to the smell of decay. The female lays four or five white elongate oval eggs each less than 1mm (0.04in) long, close to stagnant water. The larvae, once they have grown sufficiently large, move into dry ground where they pupate within the larval skin. This species has a very wide global distribution.

Below: This drone fly has unusually large eyes.

Oriental Screw-worm *Chrysomya bezziana*
The larvae of this fly feed on the flesh of mammals. The oriental screw-worm is found across much of tropical Africa and Asia as far as Papua New Guinea. It belongs to the same family (Calliphoridae) as bluebottles and blow-flies. It lays eggs on live mammals, including humans, usually in a wound or orifice. The adult fly has a dark metallic green body, black legs and an orange-yellow face. The larvae are maggot-like with bands of spicules around their bodies. They burrow deep into the flesh, rasping through tissue with hooked mouthparts.

Blowfly *Amenia imperialis*
The shiny green body (variably spotted white) of this blowfly contrasts strongly with its yellow head and chocolate-brown eyes. It has long, clear wings and rather stout hairy legs and grows to a length of about 16mm (0.63in). This species is quite common, especially in coastal regions of Australia, where it can be seen feeding on native flowers. Its larvae are parasites of snails, hence its other common name of 'snail blowfly'.

Bee Fly *Exoprosopa* spp.
This Australian bee fly lives in dry inland sites, mainly in southern Australia. About 20mm (0.79in) long, it has long grey wings, with large darker patches, giving it a distinctly bee-like appearance. It has a hairy body and a long proboscis for probing into flowers to feed on nectar. Unlike those of bees, the legs are long and spindly. Bee fly larvae are parasites, feeding on the grubs of bees and their food stores.

New Zealand Sandfly

Austrosimulium ungulatum

Also called 'buffalo gnats' or 'turkey gnats', these tiny flies are usually black, orange-red or brown. They carry no diseases but are considered one of the worst nuisances in New Zealand because of the way they incessantly bite people, leaving itchy lumps on the skin. The males sip nectar or plant sap, while the females suck blood for the extra protein needed to produce their eggs. They only attack during daylight, as they locate prey by sight and smell. The female uses her saw-like mouthparts to puncture the skin so that the blood wells up. Her saliva contains an anticoagulant, which incidentally causes the itching. She can then lap up the blood with her proboscis. The eggs are laid in water and the larvae attach themselves to a substrate by a hook-like appendage or 'holdfast' and feed by filtering particles from the water with a bristle-like comb around the mouth. There are about a dozen species in New Zealand, but only two bite humans. This species is found in large numbers on the west coast of the South Island, the Fiordland region.

Main habitat: Open spaces near water.
Main region: Temperate.
Length: 2.5mm (0.09in).
Wingspan: 5mm (0.19in).
Development: Complete metamorphosis.
Food: Herbivore (male); carnivore (female).

Identification: Tiny, dull, velvety-black fly, with a hunched body.

Below: The bite of this sandfly can cause irritation to the skin.

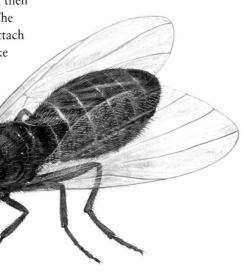

Hexham Grey Mosquito

Ochlerotatus (Aedes) alternans

This is the largest of Australia's 230 or so species of mosquito. Though the adult females are aggressive biters of humans and other animals, both by day and night, they are thought to be mainly a 'nuisance' pest. However, they are known to carry a disease called Ross River virus, which causes inflammation of joints. Ross River virus infects people when they are bitten by a mosquito carrying this virus. The mosquitoes probably pick up the virus from kangaroos and other marsupials, which are a natural host for the virus. The larvae feed on other mosquito larvae. Mosquitoes (family Culicidae) contain some of the most dangerous insect pests. They have long, narrow wings, long slender legs and syringe-like mouthparts. Female mosquitoes feed on the blood of vertebrates and in so doing spread serious diseases such as malaria and dengue fever. Males feed at flowers. Most live in tropical regions and tend to be host-specific.

Main habitat: Damp sites.
Main region: Temperate to tropical.
Length: 20mm (0.79in).
Wingspan: 35mm (1.38in).
Development: Complete metamorphosis.
Food: Carnivore (larvae and adult female).

Above: This long-legged mosquito is Australia's largest.

Identification: This species is large, with a hairy body, patterned in grey, brown and cream and very long mouthparts. The wings are mottled and have pale scales. The legs are very much longer than the body and striped grey and cream. Its larvae feed on the larvae of other mosquitoes.

Fungus Gnat

Glow-worm, *Arachnocampa* spp.

Most fungus gnats (family Mycetophilidae) are vegetarians, feeding as their name implies on the fruiting bodies of fungi. This species, however, is carnivorous. The larvae, known as glow-worms, live in large colonies under overhangs of all sorts, such as bridges, banks and hollow trees, but most famously in caves. Each larva secretes a long, transparent mucus hammock suspended by strings of silk. From this it dangles several dozen very fine double strands of silk, each threaded with beads of a glue-like substance. The larva baits its line by glowing steadily with an electric blue light. Caves with large colonies of these larvae can glow with many points of strange light. Night-flying insects are drawn to the light and become stuck, whereupon the larva senses the vibrations of the captive's struggles, leans out of its tube, and reels in its line by eating it, and devours the victim when it comes into reach. There are several species, found in Australia and New Zealand, with another reported from Fiji.

Identification: The adults are midge-like flies about 15mm (0.59in) long with long, narrow legs and rather short wings. The larvae are about 3–5mm (0.12–0.19in) at first, growing to about 30mm (1.18in).

Main habitat: Caves, overhangs, hollow trees and the like.
Main region: Temperate, tropical.
Length: 15mm (0.59in) (adult).
Wingspan: 15mm (0.59in) (adult).
Development: Complete metamorphosis.
Food: Carnivore (larva).

Above: This gnat is best known for its luminous larvae that attract other night-flying insects by luring them into their traps.

Cranefly

Leptotarsus imperatoria

Main habitat: Damp
habitats.
Main region:
Temperate–subtropical.
Length: 40mm (1.57in).
Wingspan: 75mm (2.95in).
Development: Complete
metamorphosis.
Food: Herbivore (larva).

This impressive yellow-and-black cranefly from Australia is one of the world's largest. The eggs are laid in damp soil, and the larvae feed underground on rotting plant parts. They take about a year to grow up to 25mm (1in) long, at which point they are ready to pupate. The pupa wriggles itself up close to the soil surface so the adult can emerge above ground. The males often gather where females are emerging from their pupae. The female will mate immediately and lay eggs, living as an adult for only a few hours. Males live for a few days, often flying into houses toward lights. Craneflies (family Tipulidae) are mainly found in damp habitats. They have long, slender bodies and wings and very long legs. They are slow and rather clumsy in flight and easily shed their legs if handled or attacked. Their larvae are long and slender with a tough cuticle and are known as 'leatherjackets'. They feed under the ground on plant roots. Most adult craneflies do not feed.

*Above: Long-legged and brightly
coloured, this cranefly is an
impressive insect.*

Identification: A giant yellow-and-black species, with a wingspan of 75mm (2.95in) and a body length of up to 40mm (1.57in).

OTHER SPECIES OF NOTE

Cranefly *Gynoplistia* sp.
This is a large genus of cranefly, with many species found in Australasia, including in New Zealand. Most are relatively small, growing to about 16mm (0.63in) and many have banded bodies, wings and legs, some striped yellow and black and looking a little wasp-like from a distance. This may afford them some protection from birds and other predators.

Long-legged Fly *Sciapus* spp.
The flies in this genus are small, only about 6mm (0.24in) long, and are quite often seen on vegetation, including in gardens. Seen close, they are pretty insects with bright metallic colours, shining green and bronze. Apart from the colours, their most obvious feature is their long legs. The larvae live in damp soil or under the bark of trees. Adult long-legged flies catch and eat smaller insects, such as aphids. These flies often gather in groups to display on horizontal leaves or tree trunks. Some species live near water, while others prefer humid forests.

Green Long-legged Fly *Parentia malitiosa*
This and the previous species belong to the family Dolichopodidae. This species is native to New Zealand. It is about 5mm (0.19in) long and has a bright shiny green thorax and abdomen with reddish eyes. There are many species in this genus, most rather colourful. They flit like tiny jewels among the foliage, hunting for insects, or visiting flowers to sip nectar.

March Fly *Plecia* spp.
These flies, in the family Bibionidae, are dark-coloured and sturdily built and tend to move rather slowly. They are quite common, especially in warm regions. Those in this genus have an orange thorax and rather short antennae. Some resemble wasps.

Tiger Cranefly

Nephrotoma australasiae

This cranefly from northern Australia is well named: its body is striped like a tiger. It also goes by the names of 'mosquito hawk', 'gallinippers', and 'Jimmy spinners'. It is found mainly in damp forests and the female lays her eggs in wet soil on the forest floor. This cranefly spends most of its time resting on leaves in shady areas.

Main habitat: Damp
woodland.
Main region: Tropical.
Length: 17mm (0.67in).
Wingspan: 30mm (1.18in).
Development: Complete
metamorphosis.
Food: Herbivore (larva).

Identification: Black body with yellow bands on abdomen and a V-shaped pattern on the thorax. The male is slightly smaller than the female. The legs are black and very long and slender, and the abdomen rather pointed. The thorax and head are bright yellow and the eyes are black. The wings have a greyish sheen.

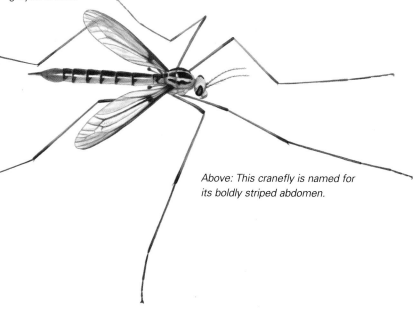

*Above: This cranefly is named for
its boldly striped abdomen.*

Antler Fly

Stalk-eyed Signal Fly, *Achias australis*

The antler fly is one of the most unusual of tropical flies. The males take part in jousting contests over females, rather like rutting deer. The males spend most of their time defending shaded courtship territories on the bark of rainforest trees. Competing males 'eye' each other up face-to-face in ritualized contests; the one with the shorter eye stalks usually backing off, while the winner gets to mate with the female. They can often be spotted resting on vegetation in damp sites. These flies belong to the family Platystomatidae. The common names come from the antler-like eye-stalks of the males and the fact that the wings are in almost constant motion, being used to signal to each other, in a kind of semaphore.

Left: This strange fly literally has its eyes out on stalks.

Main habitat: Rainforest.
Main region: Tropical.
Length: 10mm (0.39in).
Wingspan: 20mm (0.79in).
Development: Complete metamorphosis.
Food: Omnivore.

Identification: This is a rather small fly, only about 10mm (0.39in) long, with a body shape not unlike a fruit fly. However, its head is quite extraordinary, with the eyes widely separated at the ends of long antler-like stalks. In the female, the head is broad and shaped like a hammer, again with the eyes at each end. The wings are mottled and coloured.

Australian Robber Fly

Blepharotes splendidissimus

Main habitat: Open forest.
Main region: Temperate–subtropical.
Length: 40mm (1.57in).
Wingspan: 75mm (2.95in).
Development: Complete metamorphosis.
Food: Carnivore.

Identification: This Australian species is a large, brown, heavily built robber fly, with a flat abdomen with fringe-like tufts of hair around the segments. It has stout, bristly legs and a long sharp proboscis. Its wings are long and marked with cream and brown.

Right: Robber flies chase and capture their prey in flight.

Robber flies or assassin flies (family Asilidae) include the largest of all flies. Some are slender and some are stout, but they all have a humped thorax, long abdomen and a sharp beak-like proboscis, as well as a hairy head, body and legs. These flies are predatory as their name suggests, in both the larval and adult forms, though some larvae eat rotting material. They have large eyes and tend to sit in sunny spots watching for suitable insects or spiders to come along. They then turn toward the prey and take to the air. Victims are grabbed on the wing with the strong front legs and stabbed with the sharp proboscis, after which saliva is injected, which both paralyses and digests the prey. Then the body contents are sucked out. Eggs are laid in soil, rotting wood or inside plants. The spindle-shaped maggots live in rotting wood or soil. Australian robber flies make a buzzing noise in flight and may often be seen carrying their prey.

Horsefly

Erephopsis guttata

Main habitat: Near water or swamps.
Main region: Temperate, subtropical.
Length: 25mm (1in).
Wingspan: 50mm (2in).
Development: Complete metamorphosis.
Food: Herbivore (adult male); carnivore (larva, adult female).

Identification: A large, black-bodied horsefly species, with white spots on its abdomen. The mouthparts of the adult female are large and blade-like and can cut and pierce the skin. The eyes are large and prominent.

This horsefly lives mostly in damp sites or near woodland. The adults only live for three or four weeks. The bullet-shaped larvae live in water, mud or wet sand, and may take up to two years to develop. They have a breathing tube at the end of their bodies and are predators, feeding on other aquatic larvae and small vertebrates. This species is quite common in eastern Australia.

Left: Horseflies have particularly large eyes.

OTHER SPECIES OF NOTE

Bush Gadfly *Scaptia adrel*
A New Zealand member of the horsefly family (Tabanidae), this fly lives in forests where it can often be seen sunbathing in clearings or visiting flowers to find nectar. While many (female) members of the family bite and feed on blood, this is one of a number of harmless flower-feeders. The bush gadfly has a rather broad, hairy body and resembles a bee. It also looks and sounds bee-like in flight.

Robber Fly *Laphria* spp.
This is a common genus of robber fly, with most species growing to about 25mm (1in). It is quite stout-bodied (for a robber fly), with a dark grey or black abdomen, and hairy legs. Like all robber flies, these species have sharp, piercing mouth-parts and large eyes, giving them acute vision to track down live prey, often caught on the wing.

Spider-hunter *Leptogaster* spp.
This genus of robber fly specializes in catching stationary prey, rather than hunting flying insects. Some species pounce on spiders as they sit in their webs, thus turning the tables on the arachnid hunters. They have long legs and a long narrow abdomen.

Australian Soldier Fly *Inopus rubriceps*
Common in Australia and New Zealand, this species likes to sunbathe on leaves. The common name refers to the fact that some species have bright colours, not unlike some military uniforms. This species, however, is mainly dark, with a red head. The larva is a pest of pastures and some crops, such as maize. It is sometimes called the sugarcane soldier fly.

Horsefly

Tabanus striatus

Horseflies or marchflies (family Tabanidae) are chunky flies that can be a nuisance to livestock and to people, as the females bite mammals to feed on blood. The males are mostly nectar-feeders and are harmless. If attacked by a swarm of female marchflies, livestock can be weakened through loss of blood. After cutting a wound, the female fly just sits there lapping up the nutritious blood, which supplies her with the energy needed to develop her eggs. This species, which has a wide range including China, India, Pakistan, Russia and South-east Asia, is a carrier of surra, a disease of horses.

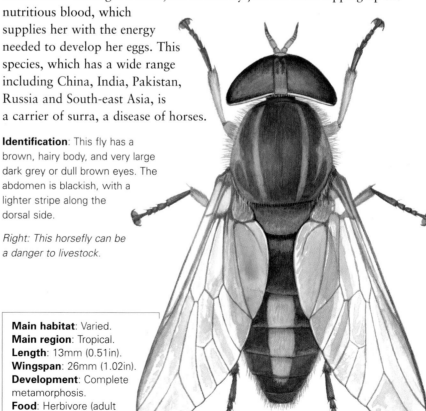

Identification: This fly has a brown, hairy body, and very large dark grey or dull brown eyes. The abdomen is blackish, with a lighter stripe along the dorsal side.

Right: This horsefly can be a danger to livestock.

Main habitat: Varied.
Main region: Tropical.
Length: 13mm (0.51in).
Wingspan: 26mm (1.02in).
Development: Complete metamorphosis.
Food: Herbivore (adult male); carnivore (larva, adult female).

BUTTERFLIES AND MOTHS

Butterflies and moths (order Lepidoptera) are found in nearly every habitat where there is vegetation and flowers. Butterflies are mostly day-flying, whereas moths are typically night-flying. Of the approximately 200,000 species. There are more than 20,800 species of butterflies and moths in Australia, and many times more than that number in Asia. Wingspans range from 3–32cm (1.18–13in).

Common Green Birdwing

Cairns Birdwing, *Ornithoptera priamus*

Main habitat: Tropical forest.
Main region: Tropical.
Length: 50mm (2in) (male); 63mm (2.48in) (female).
Wingspan 15cm (6in) (male); 19cm (7.5in) (female).
Development: Complete metamorphosis.
Food: Herbivore.

Identification: There are many subspecies and forms across the range but all the males are iridescent shades of green or blue with black markings in varying patterns, a yellow abdomen and black head and thorax, often with bright red side patches. The larger females are rather drably coloured – brown with cream or white markings.

This magnificent butterfly belongs to the swallowtail family (Papilionidae), and was the first ever birdwing species to be described, in 1717. Carl von Linné described it again in 1758. It is also one of the biggest and best-known birdwings because it has such a wide distribution and range of habitats. Among other regions it occurs in the Moluccas, Solomon Islands, Papua New Guinea and in Australia. Like many other birdwings it is often collected and sold as mounted specimen, which poses a threat in some areas.

Above: Birdwings are beautiful insects, with long colourful wings.

Asian (Chinese) Yellow Swallowtail

Xuthus Swallowtail, *Papilio xuthus*

This splendid butterfly is another member of the swallowtail family, Papilionidae. It inhabits deciduous and mixed forests and has a wide distribution including China, Japan and Korea, the far south-east of Russia (Amur), and also Hawaii. This insect has two generations per year, adults of the spring generation being much smaller than those of the summer generation. The adults fly from May to August, feeding from flowers. The caterpillars feed on plants of the rue family and overwinter as pupae.

Right: The hindwings of swallowtails end in streamer-like tails.

Main habitat: Deciduous and mixed forests.
Main region: Temperate, tropical.
Length: 25mm (1in).
Wingspan: 8–10cm (3.15–4in).
Development: Complete metamorphosis.
Food: Herbivore.

Identification: This is a typical swallowtail butterfly, each hindwing having a characteristic long tail extension. These butterflies are strikingly patterned with black, white and pale yellows over their wings and bodies, and the sexes are rather similar. The hindwings have a number of short blue stripes and red dots.

Golden Emperor Moth

Loepa katincka

Main habitat: Varied.
Main region: Tropical.
Length: 30mm (1.18in).
Wingspan: 9–10cm (3.54–4in).
Development: Complete metamorphosis.
Food: Herbivore (larva).

Identification: This moth is large and vivid yellow with brown and gold zigzag patterns and four eye-spots. The male has bushy feathered antennae to detect the pheromones emitted by unmated females. The caterpillars are blue-grey with diagonal stripes of black and luminous pale green triangles along the side, and brown hair tufts with metallic blue protrusions along the back.

Closely related to the silkworm moths, emperor moths belong to the family Saturniidae. They are all large insects and include the largest of the lepidopterans. Eye-spots are widely used as defensive tactics throughout the insect world. Large, widely spaced eyes normally belong to large, possibly predatory, animals, and birds treat them with respect. Usually these insects hide their eye-spots until their camouflage is discovered and they are disturbed by a predator. They then flash their eye-spots suddenly, gaining momentary hesitation on behalf of the predator and allowing the insect to get away. This species is found in India and across much of south and South-east Asia. The eggs hatch within a few days and the caterpillars feed on shrubs until autumn, shedding their skin four times.

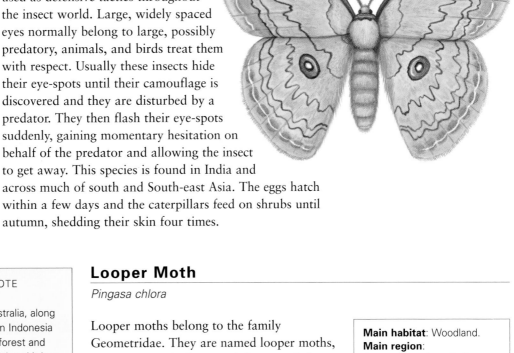

Below: This pretty moth is rather short-lived as an adult.

Ulysses Butterfly *Papilio ulysses*
This butterfly lives in north-east Australia, along the coast of Queensland, and also in Indonesia and Papua New Guinea, in tropical forest and gardens. Its wingspan is 14cm (5.5in) and it has long tails on the hindwings. It is hard to spot when perched, as the undersides of the wings are blackish-brown. In flight the metallic blue of the upperwings is striking.

Great Nawab *Polyura eudammipus*
A medium-sized species (wingspan about 90mm (3.54in)) found from eastern China to Malaysia, Myanmar and Nepal. It is creamy white above with dark tips to the forewings. Each hindwing has two tails and a row of white spots in a dark band. The underside is a bright shiny silver with chocolate-coloured stripes.

Large Silverstripe *Childrena childreni*
An orange species with many dark spots, especially on the forewings. The undersides of the hindwings are grey-green, with a network of silver bands. It is found mainly in northern India, Myanmar, Bangladesh and southern China.

Japanese Emperor *Sasakia charonda*
This species, related to the European purple emperor, lives in eastern China, Korea and Japan. Underneath it is silvery on the hindwings. The forewings are brown with silver spots and tips. The male is beautiful above – mainly brown with white spots, and with a glinting blue or purple iridescent sheen on the inner wings. Its wingspan reaches about 12cm (4.5in).

Looper Moth

Pingasa chlora

Looper moths belong to the family Geometridae. They are named looper moths, or sometimes 'inch worms', because their slender, well-camouflaged caterpillars move in a characteristic way, extending the front part of their bodies, attaching their true legs to a twig and then drawing up the hind part of their bodies and attaching with the rear 'prolegs' – so looping or measuring their way along the twig. When at rest they look very much like a twig. The caterpillars feed on leaves and are active during the day. They pupate on their food plant or in the soil, producing a flimsy silk cocoon.

This looper from Australia has highly cryptic coloration also as an adult and can be very hard to spot when resting on a tree.

Right: Looper moths are well camouflaged on eucalyptus tree bark.

Main habitat: Woodland.
Main region: Temperate–subtropical.
Length: 18mm (0.71in).
Wingspan: 40mm (1.57in).
Development: Complete metamorphosis.
Food: Herbivore.

Identification: The adults have broad, triangular wings, when at rest pressed flat against the substrate. Pale silvery grey with fine, black wavy lines.

WASPS, BEES, ANTS AND SAWFLIES

These hymenopterans are highly specialized insects, with chewing mouthparts and (mostly) four membranous wings (ants are usually wingless, except for reproductives at certain times). This is a huge and diverse order, with an estimated 280,000 species. In Australia there are about 15,000 species, and many times more than this in Asia, especially in the tropical regions.

Orange Caterpillar Parasitic Wasp

Netelia producta

This is an ichneumon that is very common in Australia, and is often attracted to the light of houses on summer evenings. The larvae are parasites of large caterpillars. The female paralyses a caterpillar by stinging it. She then lays her egg near the victim's head. When it hatches, the larva clings to the caterpillar's head until the latter is ready to pupate in the soil. Then the wasp larva kills and eats its host and eventually emerges as an adult. The adult female can deliver a painful sting.

Identification: This is a delicate insect with very long antennae, large black eyes, a slim reddish-orange body and legs, and clear wings, veined with black. The thin antennae are a little longer than the abdomen.

Main habitat: Cosmopolitan.
Main region: Australia.
Length: 20mm (0.79in).
Wingspan: 15mm (0.59in)
Development: Complete metamorphosis.
Food: Herbivore (adult); carnivore (larva).

Below: This ichneumon has extremely long antennae.

Figwort Sawfly

Tenthredo scrophulariae

This insect is a common sawfly of Europe and Asia, often found in damp sites. The adults hunt for small flies and other insects. The larvae are vegetarians and feed on the leaves of figwort.

Identification: Many of the species in this genus have black-and-yellow striped adults that mimic true wasps, and this species is a similar size to common wasps. It differs from a wasp, however, as it does not have a 'wasp waist'. This species has orange antennae. There are four yellow membranous wings, the rear ones being shorter than the front ones. The wings are orange along the leading edge, and they are always held outstretched when at rest. The legs are black at the top and yellow lower down. The larva is large and caterpillar-like, coloured white to light greenish-blue, with black dots.

Main habitat: Damp sites.
Main region: Asia and Europe.
Length: 12mm (0.47in).
Wingspan: 26mm (1.02in).
Development: Complete metamorphosis.
Food: Herbivore (larva); carnivore (adult).

Above: Like many sawflies, this species has a wasp-like appearance.

OTHER SPECIES OF NOTE

Paperbark Sawfly *Lophyrotoma zonalis*
Dark blue on the body and wings, with orange spots on the thorax and orange rings on the abdomen, this Australian sawfly is quite distinctive. Its larvae feed on the paperbark tree (*Melaleuca quinquenervia*), which has become invasive in some regions (for example Florida where it was introduced), and the paperbark sawfly is being investigated as an agent of biological control.

Black and White Ichneumon Wasp *Gotra* spp.
These handsome ichneumons, about 16mm (0.63in) long, have antennae that curve outward at the tip. The antennae are black with a white mid-section. The background colour is shiny black with a number of white transverse stripes and spots. The legs are long, and black with a central white patch and red at the base. These ichneumons often crawl over leaves looking for moth cocoons, into which they lay their eggs.

Bulldog Ant *Myrmecia* spp.
These large powerful ants, about 25mm (1in) are some of the best known in Australia. They construct large nest mounds often decorated with stones and pieces of vegetation, up to 1m (3ft) in diameter. They are active and aggressive and have big eyes and large, toothed jaws (shaped rather like the pincers of an earwig) with which they can catch a wide range of prey. They also have powerful stings. Although the adult ants catch small insects they mostly take these back to the nest to feed to the larvae. The adults themselves are omnivorous, and include nectar and gum trees in their diet.

Meat Ant *Iridomyrmex* spp.
These medium-sized ants, 8mm (0.31in) are common in dry regions in Australia. They are black with long legs and a large, red head. They are often seen foraging in large numbers in the open, for other insects and honeydew. Like most ants, they nest underground. On the surface above the nest is a mound of soil, usually covered in small stones and studded with tiny entrance holes. Meat ants are some of the most efficient scavengers to be found.

Weaver Ant

Green Tree-ant, *Oecophylla smaragdina*

Weaver ants are found across Asia, from India to Taiwan, and from South-east Asia to Australia. They are quite common in forests and plantations, where they feed on smaller invertebrates as well as nectar. These ants build very complex nests by folding living leaves and stitching these together (hence the term weaver). This provides perfect camouflage among the tree branches and protection from both predators and the weather. They fold each leaf by forming a chain of ants from one edge to the other then they shorten the chain by losing one ant at a time until the edges are brought together. Then a worker brings a larva in its mandibles and squeezes it so it produces silk which is used to bind the leaf edges together. The colony can be very large, consisting of several rolled leaf nests over many branches, or even trees. Some trees have evolved to attract weaver ants with nectar, as the presence of the ants deters herbivores that might otherwise browse on the foliage. Weaver ants, both pupae (called ants' 'eggs') and adults, are eaten by people; they are said to be a rich source of vitamin C. In China these ants are also used to protect fruit orchards from insect pests. They are encouraged to invade the orchard by the use of bamboo bridges between the trees.

Identification:
These are large orange-red ants with a green head and abdomen. Their legs are long and thin and also orange-red, as are the elbowed antennae.

Main habitat: Forest, plantations.	
Main region: Tropical.	
Length: 9mm (0.35in).	
Wingspan: None.	
Development: Complete metamorphosis.	
Food: Omnivore.	

Left and below: These ants have a fierce bite and can also sting.

Honeypot Ant

Warumpi, *Camponotus inflatus*

These small ants live on desert sands or in eucalyptus scrub in southern Australia. They forage during the cool night. Both the common name and the specific epithet refer to the extraordinary way in which they store honeydew and nectar. One type of worker, known as a 'replete', never leaves the nest, but hangs from the ceiling of domed galleries 2m (6ft) down in the earth. Workers forage for honeydew and nectar, obtained from flowers, galls, aphid secretions and special glands on eucalyptus leaves. They return and feed this nutritious mixture to the repletes whose abdomens swell to the size of a grape. During the dry season workers stroke the abdomens of the repletes until they regurgitate droplets of honey. When a replete has been emptied it resumes normal worker size and duties. The nest entrance is a tiny hole in the ground which descends straight down for up to 2m (6ft) and then branches into horizontal galleries where the repletes live. Other ants sometimes try to raid the nests and kidnap the repletes. Honeypot ants have long been used as a source of food by aboriginal people, who dig them up to eat.

Identification: Brown or black, with two or three yellow stripes across the abdomen. The repletes look amazing when full – the skin between the abdominal segments is stretched and transparent so that the 'honey' within is visible.

Main habitat: Dry desert or scrub.	
Main region: Subtropical.	
Length: 10mm (0.39in).	
Wingspan: None.	
Development: Complete metamorphosis.	
Food: Omnivore.	

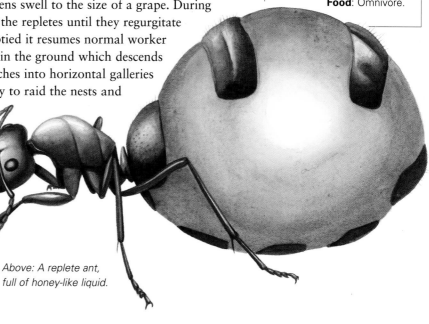

Above: A replete ant, full of honey-like liquid.

Giant Mason Bee

Megachile (Chalicodoma) pluto

This very large species is generally thought to be the largest of all bees. Also known as Wallace's giant bee, it was discovered by the famous naturalist Alfred Russell Wallace in 1859, in Indonesia. After having been thought to have become extinct, it was rediscovered in 1981 on Bacan Island and some other nearby islands. Giant mason bees, which belong to the leaf-cutter or mason bee family, nest inside termite colonies. Up to six female bees burrow into the termite nest and create cells where they lay their eggs and care for the larvae and pupae as they develop. It is not yet clear if the female bees live together socially and are dependent on each other, or whether they merely share a nest hole and rear their own young independently. The large jaws and associated mouthparts are adaptations for collecting tree resin. The females line their cells with a mixture of tree resin and wood chips. These cells may be more than 40mm (1.57in) long by 20mm (0.79in) wide and each is half-filled with pollen before an egg is laid in it.

Identification: A huge bee with a massive head and powerful, large, beetle-like jaws. The body is covered with black velvety hairs with some white on the head and first abdominal segment.

Main habitat: Forests.
Main region: Tropical.
Length: 23mm (0.91in) (male); 40mm (1.57in) (female).
Wingspan: 63mm (2.48in) (female).
Development: Complete metamorphosis.
Food: Herbivore.

Above: This is the world's largest species of bee.

Giant Oriental Honey Bee

Apis dorsata

Main habitat: Forest.
Main region: Tropical.
Length: 19mm (0.75in) (worker).
Wingspan: 30mm (1.18in).
Development: Complete metamorphosis.
Food: Herbivore.

Identification: Rather similar in overall shape to the common honey bee, but much larger. The head and thorax are dark, as is the tip of the abdomen. The abdomen has paler yellow or orange bands and may be very orange toward the base. The queens and drones are mainly brown.

Right: The giant oriental honey bee is the largest of all the honey bees.

This insect is another large bee, and the largest of the honey bee genus (*Apis*). It occurs across south and South-east Asia. Its sting delivers a virulent poison and is long enough to penetrate traditional bee-keeping protective garb. The giant honey bee builds its nest out in the open and has therefore evolved very effective defence. The colony, which may consist of several hundred thousand bees, builds a single huge honeycomb up to 3m (10ft) across, either high up on a cliff or 20–30m (65–98ft) up from a branch in the crown of a tall tree – usually one that emerges above the canopy. There are often several colonies in the same tree. The comb is always covered and protected by a mass of worker bees, while about a third of the colony are out foraging nectar and pollen. When a bee stings an intruder it emits a pheromone that almost instantly attracts more bees from its own colony and they all then home in to attack the intruder. However, as stinging kills the bees they also have a warning display, which may deter the intruder without costing so many lives. This display consists of the outer layer of bees performing a rippling 'Mexican wave' across the face of the comb. Bees leaving the colony generally fly upward, while those returning enter from underneath. Bees not yet old enough to forage leave the nest in relays every afternoon to leave their droppings well away from the nest, thus avoiding give-away clues building up at the bottom of the tree, and also familiarizing the bees with the landmarks they will need when they begin foraging. When the flowering season in their part of the forest is over the bees consume all the honey in the comb and together with the queen migrate to find a new location. The flight may take weeks with many stops on the way.

Giant Japanese Hornet

Sparrow Bee, *Vespa mandarina*

Main habitat: Varied.
Main region: Temperate, subtropical.
Length: 50–55mm (2–2.17in).
Wingspan: 76mm (3in).
Development: Complete metamorphosis.
Food: Carnivore (larva); Omnivore (adult).

Identification: Dark brown, shiny thorax, black-and-yellow striped abdomen, brown wings and legs, orange head with brown eyes and large purple-brown mandibles.

Right: Largest of wasps, the giant Japanese hornet is a fierce predator with a potent sting.

This impressive wasp is the world's largest. It is found from south-east Russia across to Korea, China, Japan, Indochina, Nepal and India. It has a sting 6mm (0.24in) long which injects a potent poison containing chemicals that cause extreme pain, damage tissues and attract other hornets to the victim. These wasps kill about 70 people a year in Japan. Their favourite prey, however, are honey bees. Colonies send out scout hornets to search for honey bee colonies or hives, and may fly up to 80.5km (50 miles) in a day. When a scout finds a suitable colony it marks the site with a pheromone. Other hornets then target the site en masse. Each giant hornet can kill honey bees at a rate of 40 per minute, and together they can wipe out a hive within hours. Like all wasps this hornet can sting repeatedly, but bees are usually killed by being decapitated by the hornet's huge mandibles.

OTHER SPECIES OF NOTE

Dwarf Honey Bee *Apis florea*
A small honey bee found from Oman and Iran through Pakistan and India and South-east Asia. It lives mainly in tropical forests and sometimes in adjacent farmland. The smallest of the honey bees, it nests in small colonies, normally slung from the branch of a tree. This bee has a black head and thorax, black-and-white bands on the lower abdomen and an orange base to the abdomen.

Blue-banded Bee *Amegilla cingulata*
An attractive medium-sized, 15mm (0.59in) bee, common in Australia, this species is related to the larger carpenter bees. Its name refers to the fact that the mainly black abdomen has narrow blue horizontal bands – rather unusual in a bee. This contrasts strongly with the bright orange furry thorax and head. This species uses a special technique known as 'buzz-pollination' when feeding at flowers. By buzzing in the flowers they release pollen that gathers on their fur-like hair.

Stingless Bee *Trigona carbonaria*
Also known as 'sweat bee' or 'sugarbag bee', these bees are mainly black and very small, only about 5mm (0.19in) long and look like miniature honey bees. They are harmless and are often seen visiting flowers for nectar and pollen. The typical nest site is inside a hollow branch or log. Although they cannot sting, the workers of this species will defend their nest by biting with their mandibles.

Spider Wasp

Cryptocheilus bicolor

Many wasps have a parasitic stage in their life cycle, their larvae being parasitic, usually on the larvae of insects or other invertebrates. Spider wasps, also known as 'digger wasps' provide their larvae with a living spider on which to feed. The female wasp digs a burrow in sand by scratching it away with her front legs and kicking it backward with her rear legs. Then the wasp sets out to hunt for a suitable spider, stings it to paralyse it and drags it back to the burrow. A single egg is then laid on the spider's abdomen and the burrow is sealed. The wasp arranges sand around the burrow to disguise its presence. The egg hatches and the larva feeds on the living spider.

Main habitat: Sandy soil.
Main region: Temperate–tropical.
Length: 35mm (1.38in).
Wingspan: 65mm (2.56in).
Development: Complete metamorphosis.
Food: Carnivore (larva); herbivore (adult).

Identification: This Australian species is a large, slender, black wasp with orange-gold legs, a bright yellow-orange face and large eyes. Its antennae are long, curving and yellow. The wings are translucent with a yellow sheen and there is an orange band around the abdomen.

Above: The spider wasp often hops and runs, flicking its wings.

MILLIPEDES AND CENTIPEDES

Millipedes and centipedes are distinguishable from insects by their many legs (at least nine pairs and often many more), and by their uniform bodies (not divided into head, thorax and abdomen). There are approximately 15,000 species, ranging in length from 0.05–30cm (0.02–12in). They are distributed widely, from the Arctic to the tropics, and even in some deserts.

Asian Forest Centipede

Scolopendra subspinipes

Also known as the Vietnamese centipede or orange-legged jungle centipede, this species from South-east Asia and other tropical regions is large and quite aggressive. Nevertheless it is a popular terrarium pet, although care must be taken to avoid getting bitten – its venom is quite powerful, causing pain and swelling. The Asian forest centipede catches many other invertebrates and paralyses them by injecting venom from its sharp fangs. It hides in damp places, for example in leaf litter, by day, and hunts at night.

Main habitat: Forest.
Main region: South-east Asia.
Length: 15–19cm (6–7.5in).
Development: Gradual.
Food: Carnivore.

Below: This centipede has a nasty, venomous bite.

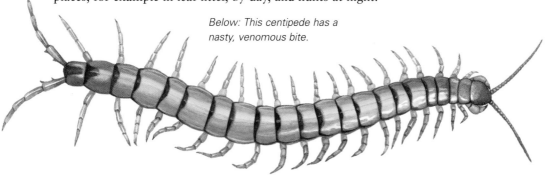

Identification: This centipede is generally brightly coloured, often with alternate orange, yellow or even blue and black segments. It has 21 segments with shiny brown plates along its back, and orange and blue legs, although there are many colour variations. One form has a greenish-grey shiny body, yellowish legs and a red head and antennae.

Giant Centipede

Ethmostigmus rubripes

This is one of the largest centipedes found in Asia, occurring from China to Indonesia. It is also found in Australia. It is a common, solitary, nocturnal predator and is found in both dry and moist regions hiding under stones, logs and leaf litter. This centipede can be found in a very wide range of habitats, including forest, woodland, heathland, rainforest, desert, and also some urban areas. It will bite if disturbed or handled, producing severe pain and some swelling.

Identification: It has a dark greenish-brown or yellow body with yellow legs and antennae. Sometimes shows bluish bands between the body segments.

Below: Another active predator with a painful bite.

Main habitat: Varied.
Main region: Tropical.
Length: 7.5–16cm (3–6in).
Development: Gradual.
Food: Carnivore.

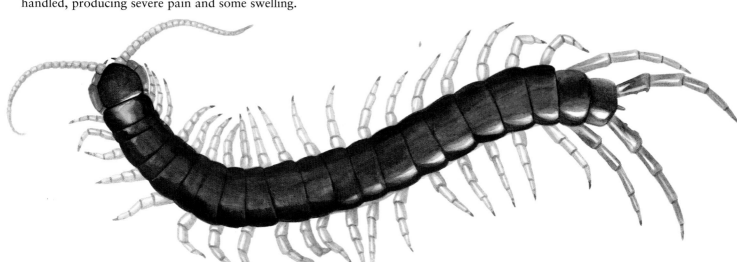

Vietnamese Rainbow Millipede

Aulacobolus rubropunctatus

Main habitat: Leaf litter.
Main region: Vietnam.
Length: 50mm (2in).
Development: Gradual.
Food: Herbivore.

Below: Colourful and unusual, this species is sometimes kept as a pet.

Identification: A long millipede with a shiny body. The segments have black and grey transverse stripes with a reddish central line along the back. The legs and antennae are bright red.

This large millipede from South-east Asia has become quite popular as a pet. It is fairly easy to keep and is happy on a diet of greens and fruit. It is an active species, and spends less time curled up than some other large millipedes. Many people now breed this species in captivity which it is hoped will take the pressure off collection from the wild. The Vietnamese rainbow millipede produces its eggs inside cases, which look rather like the millipede's droppings. They need damp conditions for successful hatching and the young emerge after about five weeks. The young are like miniature adults and grow slowly to mature size, gradually becoming more colourful.

OTHER SPECIES OF NOTE

Earth Centipede *Geophilida* spp.
These centipedes have a long, slender body and are usually found in damp places such as underneath rocks and logs and in leaf litter in wet forests. They have more pairs of legs than most other centipedes and the legs are also rather short. The head is long and rectangular in shape. Earth centipedes feed mainly on soft-bodied prey such as worms, and are not harmful to people.

House Centipede *Allothereua maculata*
This centipede is often found inside houses, where it consumes spiders and insects. It is small, to about 25mm (1in), but its legs are very long and slender, as are its antennae. It can run very fast on its flexible legs and seeks out dark places in which to hide during the day. The last pair of legs are so long and thin that they look like a second pair of antennae. It is common in southern Australia and is found in woodland as well as in houses.

Red Stone Centipede *Paralamyctes grayi*
Stone centipedes have a flattened body and head, and 15 pairs of legs. This species is found in forests and grows to about 32mm (1.26in). The female has special spurs between the last pair of legs. She uses these to roll her eggs in soil to give them some camouflage. The species in the *Paralamyctes* genus are common to Australia, New Zealand, India, South Africa and the southern tip of South America.

Greenhouse Millipede *Oxidus gracilis*
This is an Asian species that has become established in other parts of the world. It can be found in greenhouses, where it sometimes becomes a pest. It belongs to a group known as flat-backed millipedes due to the fact that its segments are distinctly flattened across the top.

Pill Millipede

Procyliosoma tuberculatum

This species is the largest millipede native to New Zealand. It is found mainly in the humus and leaf litter of native forests on the North and South Islands. Some millipedes are able to roll up completely into a ball to protect themselves, rather in the manner of pill woodlice.

Main habitat: Forest floor.
Main region: Temperate.
Length: 22mm (0.87in) (male); 40–50mm (1.57–2in) (female).
Development: Gradual.
Food: Herbivore.

Identification: Males are 11mm (0.43in) wide and 22mm (0.87in) long, females are 22mm (0.87in) wide and 40–50mm (1.57–2in) long but they tend only to reach their full size on offshore islands where there are no predators. They are black or brown-black with a chestnut-coloured head.

Below: The flexible pill millipede can roll up into a tight ball.

SPIDERS

Spiders are well adapted to their predatory, carnivorous lifestyle, and many also help to control troublesome insect pests. Some, like the Sydney funnel-web, are very dangerous, but in fact most spiders are harmless to people. Although not all spiders actually build webs, all known species can produce silk. Body lengths range from 4–90mm (0.16–3.54in) with the largest legspan reaching 26cm (10in).

Australian Tarantula

Selenocosmia spp.

These spiders belong to the tarantula or bird-eating spider family (Theraphosidae). They are also known as whistling or barking spiders. These latter names come from the fact that most species are capable of producing sounds by rubbing rows of spines on their palps (the front pair of sensory organs) against spines at the base of their jaws. They make these sounds when they sense danger, to deter predators. This genus is from south-western Australia. Australian tarantulas usually take insects, other invertebrates, lizards and frogs. They are long-lived, the males living for about five years, the females up to 12 years. Despite their scary appearance, they are not very aggressive, although they can deliver a painful venomous bite and should not be handled.

Main habitat: Varied, from rainforest to dry scrub.
Main region: Warm temperate, tropical.
Length: 60mm (2.36in)
Legspan: 16cm (6in).
Development: Gradual.
Food: Carnivore.

Identification: Large and heavy bodied, with powerful fangs. Colour ranges from pale to dark brown, with a silver sheen. Some species have hairy legs.

Left: Tarantulas are large, with powerful jaws.

Sydney Funnel-web Spider

Atrax robustus

Funnel-web spiders are found mainly in eastern Australia, often in tree bark, soil or in rotten wood of forested areas, or under rocks. There are about 40 species in the genera *Atrax* and *Hadronyche*. The Sydney funnel-web spider is the most famous, being notorious for its venomous bite. It is aggressive, responding to harassment by exuding a drop of venom from the tip of the fangs and then attacking. The bite of the male spider is both painful and potentially lethal to people. However, there have been no reported deaths since the introduction of effective antivenom. Its fangs strike downward, whereas more advanced spiders have fangs that close together sideways. The funnel-web spider, as its name suggests, makes a cone-shaped web that leads into a cool, damp crevice under a rock, in tree bark, soil or in rotten wood, where it hides. Trip-lines radiate out from the cone, warning the spider of approaching prey. These spiders are nocturnal and the females rarely stray from their burrows.

Main habitat: Under rocks and wood; sometimes in houses.
Main region: New South Wales, Australia.
Length: 15–35mm (0.59–1.38in).
Development: Gradual.
Food: Carnivore.

Identification: The Sydney funnel-web spider has a glossy, dark red-brown carapace, red legs and a light brown, black or plum-coloured abdomen. The long spinnerets are obvious at the end of the abdomen.

Left: This aggressive spider has a bite that can be lethal.

Woodlouse-eating Spider

Dysdera crocata

Main habitat: Under stones.
Main region: Temperate.
Length: 15mm (0.59in).
Development: Gradual.
Food: Carnivore.

Identification: This spider has only six eyes while most other spiders have eight. It has a brick-red carapace and legs and a pale grey abdomen.

This spider has a wide distribution around the world both naturally and through having been introduced. It builds a silk cell under a stone where it hides by day, but does not build a web to catch its prey. Instead it is an active nocturnal hunter, specializing in catching woodlice using its enormous fangs. It is unaffected by the unpleasant secretions with which woodlice defend themselves from most other predators. Its strong fangs are specially evolved for dealing with the tough armour of woodlice. The female lays about 40 eggs within a silk cell in the summer and remains with them until the young hatch and disperse in the autumn. They take up to 18 months to reach maturity and may live for another two or three years. This relatively large spider is capable of delivering a painful bite.

Above and far left: This spider uses its pincer-like jaws to catch woodlice.

OTHER SPECIES OF NOTE

Banded Huntsman Spider
Isopeda insignis
A flat-bodied, fast-moving spider, with some 40 related species in Australia. It is about 32–40mm (1.26–1.57in) long (female) and is sometimes spotted dashing across walls and ceilings, especially at night when it emerges to hunt. It kills its prey with a bite from its sharp, curved fangs. The cephalothorax is rounded and the abdomen oval and it is grey-brown in colour, with alternating light and dark bands on its legs. Huntsman spiders have eight eyes in two rows of four.

Red-headed Mouse Spider
Missulena occatoria
A chunky, powerful spider up to 25mm (1in) long living in burrows up to 30cm (12in) long, sometimes in gardens. The dark, shiny legs are crab-like and sturdy and the male has a pale blue abdomen and bright red head and fangs. This spider has toxic venom and can inflict a dangerous bite. It is found across much of mainland Australia.

Kimura-gumo *Heptathela kimurai*
This trapdoor burrowing spider from Japan exhibits a number of interesting primitive features, including signs of body segmentation and centrally placed spinnerets. It lives in a tunnel with a flap-like lid and sets out to hunt, creating a line of silk as it goes, thus facilitating its return journey. The family to which it belongs (Liphistiidae) is known only from China, Japan and South-east Asia.

Tasmanian Cave Spider

Hickmania troglodytes

This primitive spider, found only in Tasmania, is regarded as a living fossil, the last of an ancient lineage. Its nearest relatives are in South America. It lives throughout Tasmania and can be seen in large numbers near the entrances of caves and drains, and under bridges and logs. It weaves a huge sheet-web, which can be 1m (3ft) across and hangs beneath it by its long legs, trapping prey such as crickets, beetles, spiders or flies that blunder into the web. Tasmanian cave spiders mate from late winter to spring. The male plucks at the female's web and when he has her attention he taps her with his long front legs and retreats. The eggs are laid in a large complex double-walled structure that protects them, and which is often decorated with fragments of wood for disguise. The female guards the egg sac for up to ten months.

Main habitat: Dark, damp sites.
Main region: Tasmania.
Length: 13–20mm (0.51–0.79in).
Legspan: 18cm (7in)
Development: Gradual.
Food: Carnivore.

Identification: A large spider. Males have smaller bodies but longer legs than females and have a distinct kink near the end of each second leg. The carapace is reddish-brown and the abdomen is dark grey-brown.

Above: This long-legged spider lurks in damp caves.

Australian Wolf Spider

Lycosa godeffroyi

Wolf spiders (Lycosidae) are large spiders with the excellent eyesight of nocturnal hunters. This is one of two common Australian species, found in varied habitats, from wet forests to meadows and dry shrubland. It is also commonly seen in suburban gardens, running across lawns and in among leaf litter, tracking down its prey of smaller invertebrates. Some can also kill small toads, including the young of the invasive cane toad. Females can often be seen carrying a rounded silken egg sac. When the young hatch they crawl on to her back. The bite of a wolf spider can cause local swelling and pain or itchiness. This species lives in a burrow that leads down about 15cm (6in), and then runs horizontally for another 15cm (6in).

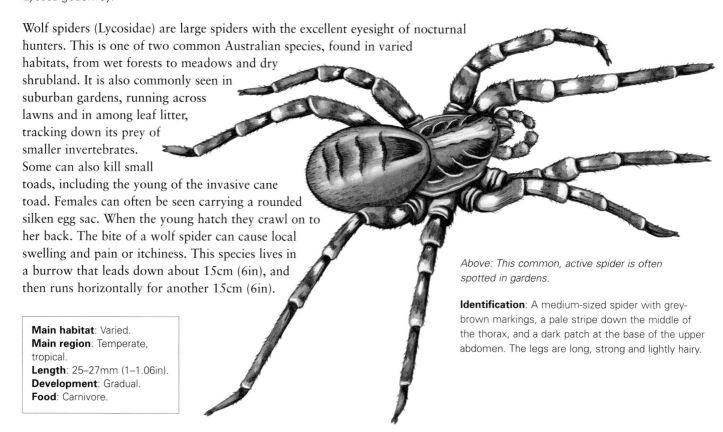

Above: This common, active spider is often spotted in gardens.

Main habitat: Varied.
Main region: Temperate, tropical.
Length: 25–27mm (1–1.06in).
Development: Gradual.
Food: Carnivore.

Identification: A medium-sized spider with grey-brown markings, a pale stripe down the middle of the thorax, and a dark patch at the base of the upper abdomen. The legs are long, strong and lightly hairy.

Curved Spiny Spider

Horned Spider, Wishbone Spider, *Gasteracantha arcuata*

The genus *Gasteracantha* contains some of the strangest of all the orb-web spiders (Araneidae). Orb-web spiders weave circular webs constructed by stringing radiating spokes from a central hub to surrounding structures and then creating a spiral thread which is glued to each spoke. Prey is caught in the sticky web, bound in silk and taken to a retreat to be consumed. The spiny spiders build vertical webs among tree branches and then sit at the centre in an opening. This species lives in trees and is found in India, Malaysia, Singapore, Indonesia, Thailand, Myanmar and Sri Lanka.

Main habitat: Forest.
Main region: Tropical.
Length: 10mm (0.39in).
Development: Gradual.
Food: Carnivore.

Identification: The female curved spiny spider has a hard, flat, crab-like body with a pair of long curved spines rising up like horns from the sides of her abdomen. The function of these is unclear, but they may help protect the spider from attack. The spider is black with the back of the abdomen bright orange-yellow with black dots. The male is much smaller and plainer without the projections.

Left and far left: The curved spiny spider has an almost crab-like appearance.

Ornamental Tree-trunk Spider

Herennia ornatissima

Main habitat: Woodland.
Main region: Tropical.
Length: 12–15mm
(0.47–0.59in) (female);
5–6mm (0.19–0.24in) (male).
Development: Gradual.
Food: Carnivore.

This species is one of the many spiders that camouflages itself against tree bark. Another orb-web spider, it spins a web a few millimetres above the lichen-covered bark of a tree and then sits boldly at its centre, camouflaged among the lichens and bark. The female disguises her nearby egg case with pieces of bark stuck into the white silk wrapping. The male is a tiny creature, which lurks near the females awaiting a chance to mate. This species has a wide range, including China, Singapore, India, Malaysia, Thailand, Myanmar, Vietnam, Papua New Guinea and Sri Lanka.

Identification: The female has a flattened, whitish abdomen, patterned with red and with fluted edges. The abdomen is red underneath, with a central black spot. The legs are long and rather thin.

Below: This is a tropical species, which spins a web just above the layer of tree bark in which it lives.

Far left: An ornamental tree-trunk spider with a second spider on its back.

OTHER SPECIES OF NOTE

Peacock Spider, Gliding Spider
Maratus (Saitis) volans
One of the most fascinating of jumping spiders (Salticidae), this species not only jumps, but the male can actually glide, using flaps on the sides of the abdomen, as far as 20cm (8in) – quite a distance considering the spider only measures some 4–5mm (0.16–0.19in) long. The flaps, edged in white, are also used by the male in a courtship dance along with the white-tipped legs. The male is beautifully patterned above in bright blue and red stripes, hence the alternative common name of peacock spider. Salticid spiders live mostly on vegetation. The rectangular thorax, stout body, rather short legs, their distinctive eye arrangement and their jumping capabilities make them one of the most easily recognizable of spider families.

Spitting Spider *Scytodes thoracica*
This spider has a wide distribution. Unlike most spiders, its cephalothorax is larger and broader than the abdomen. The jaws are small. The spitting spider is brown in colour with darker spots, and also has long, spotted legs. It takes its common name from the fact that it spits out poisoned silk threads in which to trap its prey. In addition to the usual spider silk glands at the tip of its abdomen, it has extra silk glands toward the front of its body and these are close to the poison fangs. When it gets near to its prey it spits out two poisoned threads in a zigzag, covering and immobilizing it, after which it can feed on its victim. The fluid congeals quickly into a gummy mass.

Japanese Scorpion Spider

Arachnura logio

Many spiders use mimicry to avoid predation and to deceive their prey. Some rely on the colours and markings on their bodies but others have also changed their shape and behaviour to improve their camouflage. Spiders of this genus are found throughout Asia and Australia. This species is from China and Japan. The female constructs an orb web with a sector missing, and hangs a string of brown egg masses down the spoke in the middle of the open sector. She then waits head-down in the middle of her web beneath the eggs, looking like a dead leaf caught in the web. This spider lives mainly in coniferous woodland. It can bite if provoked, leading to local swelling and pain. The males are smaller than the females and do not have the 'tail'.

Identification: This spider is a leaf-mimic, being brown in colour with an elongated abdomen tipped by blunt projections that resemble the stalk of a leaf, or the curved tail of a scorpion.

Right: A remarkable spider that mimics a leaf.

Main habitat: Woodland.
Main region: Temperate.
Length: 25–28mm (1–1.1in).
Development: Gradual.
Food: Carnivore.

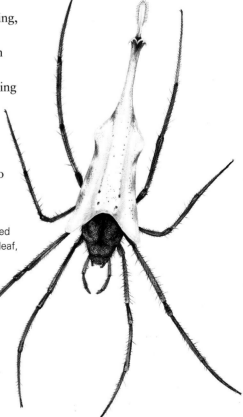

MITES, TICKS AND SCORPIONS

Like spiders, mites, ticks, pseudoscorpions and scorpions belong to the class Arachnida, and have four pairs of legs. Some mites and ticks carry diseases of animals and humans. Scorpions are most diverse in tropical regions, and are often found in deserts and rainforests. The much smaller pseudoscorpions (or false scorpions) range from 2.5–8mm (0.9–0.31in) long and are mostly found in leaf litter; they do not sting.

Hard Tick

Haemaphysalis concinna

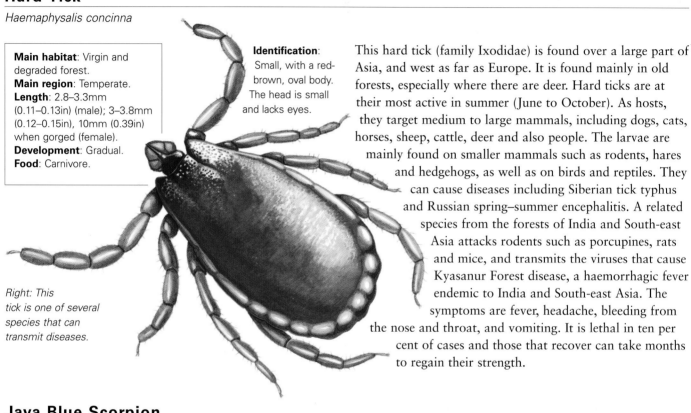

Main habitat: Virgin and degraded forest.
Main region: Temperate.
Length: 2.8–3.3mm (0.11–0.13in) (male); 3–3.8mm (0.12–0.15in), 10mm (0.39in) when gorged (female).
Development: Gradual.
Food: Carnivore.

Identification: Small, with a red-brown, oval body. The head is small and lacks eyes.

Right: This tick is one of several species that can transmit diseases.

This hard tick (family Ixodidae) is found over a large part of Asia, and west as far as Europe. It is found mainly in old forests, especially where there are deer. Hard ticks are at their most active in summer (June to October). As hosts, they target medium to large mammals, including dogs, cats, horses, sheep, cattle, deer and also people. The larvae are mainly found on smaller mammals such as rodents, hares and hedgehogs, as well as on birds and reptiles. They can cause diseases including Siberian tick typhus and Russian spring–summer encephalitis. A related species from the forests of India and South-east Asia attacks rodents such as porcupines, rats and mice, and transmits the viruses that cause Kyasanur Forest disease, a haemorrhagic fever endemic to India and South-east Asia. The symptoms are fever, headache, bleeding from the nose and throat, and vomiting. It is lethal in ten per cent of cases and those that recover can take months to regain their strength.

Java Blue Scorpion

Malaysian Blue Scorpion, *Heterometrus cyaneus*

Scorpions use their stings mainly in self-defence and the potency varies considerably between species, from lethal to innocuous. They detect ground vibrations with sensory organs on the underside of the abdomen. After the female has laid her eggs and they have hatched she forms ramps with her pincers for the babies to crawl up on to her back where they remain until they have moulted. This species of scorpion from South-east Asia is found mainly in forests, where it normally lurks underneath logs or leaf litter.

Identification: This is a large, black scorpion with a blue sheen on its back and claws. Like all scorpions it has four pairs of legs, two fang-like chelicerae and two large pincers. The abdomen ends in a curved tail with a poison sac and sting on the end.

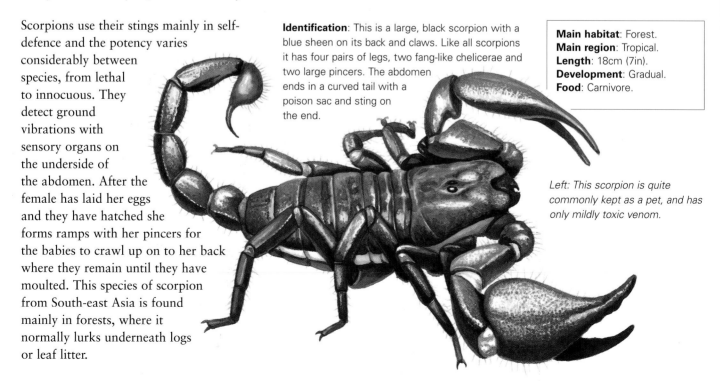

Main habitat: Forest.
Main region: Tropical.
Length: 18cm (7in).
Development: Gradual.
Food: Carnivore.

Left: This scorpion is quite commonly kept as a pet, and has only mildly toxic venom.

Desert Scorpion

Urodacus yaschenkoi

Main habitat: Desert.
Main region: Tropical.
Length: 7–12cm (2.5–4.5in).
Development: Gradual.
Food: Carnivore.

Identification: Typical scorpion shape, with medium-sized claws and sting. The body is greyish and the claws, legs and tail are reddish-orange, blending well with the sandy desert habitat.

Right: This scorpion is capable of spraying its venom if molested.

This scorpion is found in the deserts of the interior of Australia, roughly from the Murray River in South Australia and north-west Victoria across to the north-west of Western Australia. It lives for up to ten years, reaching maturity at about four years. It mates and moults in early summer and about 20 young are born in late summer or early autumn. This scorpion is remarkable in being able to thrive in such a hostile environment, where the annual rainfall may be less than 25cm (10in). It has a nasty sting, but is not lethal. The sting causes local pain that can last several hours. It can also spray venom when molested. This scorpion digs spiral burrows up to 1m (3ft) deep where it rests in relative cool.

OTHER SPECIES OF NOTE

Scrub Typhus Mite *Trombicula* spp.
Mites of this genus can transmit the disease known as scrub typhus or tsutsugamushi disease, which occurs in many parts of South-east Asia and results in fever, headache and rash. The second name is Japanese and means 'small, dangerous creature'. These mites are sometimes known as 'chiggers'. They are found across a wide area, including much of south and east Asia, including China, Japan, Korea and eastern Russia, in varied habitats from rice paddies to plantations, dry semi-deserts and even beaches. They are also found in Australia. The mite larvae attach themselves to people in places, such as forest clearings and scrub, where intermediate hosts such as rodents are also found. This mite has a very complex life cycle, involving six stages: egg, larva, three types of nymph, and finally adult. The larvae feed on animal hosts where they may remain for ten days or longer, feeding mainly on lymph and dead skin. One of the nymph stages and the adult mites feed mainly on other arthropods.

Black Rock Scorpion *Urodacus manicatus*
A dark blackish-brown scorpion widespread in southern and eastern Australia, and found under rocks and logs. It feeds on a range of prey including beetles, cockroaches, centipedes, millipedes, spiders and worms, waiting at or near its burrow until prey comes into range of its large and powerful claws. Its sting is painful but not deadly. The black rock scorpion grows to a length of about 70mm (2.76in).

Pseudoscorpion

Geogarypus taylori

Pseudoscorpions, or false scorpions, resemble scorpions but they are tiny and have no tail or sting, although some produce venom through their claws. They can be found in most terrestrial habitats under rocks, leaf litter, bark, and in soil. They feed on other small invertebrates. They sometimes disperse by clinging with their pincers to the legs of flies or beetles. They produce silk from their chelicerae to make nests for overwintering and egg protection and have complex mating behaviour involving pedipalp waving, leg tapping and abdomen quivering, followed by a dance with locked pincers. The mother carries the eggs and hatchlings in a silk sac that hangs from her abdomen, and she feeds the babies on a milky secretion.

Main habitat: Soil.
Main region: Temperate, subtropical.
Length: 1–2mm (0.04–0.08in).
Development: Gradual.
Food: Carnivore.

Identification: The head and chelicerae are dark, and the rather broad, rounded abdomen is striped in cream and black, with a series of parallel bands.

Right: This species is one of about 150 pseudoscorpions found in Australia.

GLOSSARY

Abdomen Body region behind the thorax of an arthropod.

Antenna Sensitive narrow projections from the head. Also known as 'feelers'.

Aposematic Coloration on the body, warning that the animal is distasteful.

Appendage Extension of the body such as leg, wing or antenna.

Arachnid Member of the class Arachnida, which includes spiders, scorpions, mites and relatives.

Arthropod Invertebrates with jointed limbs.

Carnivore Animal that feeds on the flesh or tissues of another animal.

Carrion Dead animal flesh or tissues.

Caste Individuals specialized for a set task in a colony of social insects.

Cephalothorax Body region comprising a fused head and thorax, typically found in arachnids.

Cerci Paired, filamentous outgrowths from the tip of the abdomen.

Chelicerae Pincer-like mouthparts, typically found in arachnids.

Chemoreceptor Sense organ sensitive to traces of chemicals in the environment, as taste or smells.

Chitin The complex compound in the arthropod cuticle that gives it strength and rigidity.

Chrysalis The pupa of lepidopterans.

Cocoon Silk covering produced by some mature larvae before pupation. Also the silken egg sac of spiders.

Comb Rows of wax or papery brood cells in nests of social wasps and bees.

Communal Insects that share a nest. Each female rears her own brood.

Below: A newly hatched Heliconius charithonia *caterpillar on a passionflower tendril.*

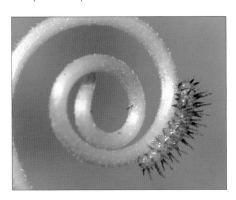

Compound eye Type of arthropod eye made up of many separate light-sensitive units.

Coxa The section of an arthropod leg that attaches to the body.

Crustacean Member of the class Crustacea.

Cryptic The patterning and colours of an animal that give it protection by allowing it to blend into a background.

Cuckoo-spit The foamy substance produced by some bugs to protect them from predators and from drying out.

Cuticle The outer layer of an arthropod's skin, forming the exoskeleton.

Diapause A resting period during development, triggered by specific conditions.

Diploid The normal state in which each cell nucleus contains two sets of chromosomes.

Diurnal Active mainly during the day.

Dorsal The back (usually the upper) surface of the body.

Drone Male honey bee.

Ecdysis Moulting of the old skin during arthropod growth.

Ectoparasite A parasite that lives and feeds on the surface of its host.

Elytra Modified hardened forewings that act as protective wing cases for the more delicate hindwings.

Endoparasite A parasite that lives and feeds inside the body of its host.

Endopterygote An insect in which wings develop inside its body. Involves a pupal stage and complete metamorphosis.

Exarate A pupa in which appendages are free of the rest of the body.

Exopterygote An insect in which the wings develop outside its body, during incomplete metamorphosis.

Exoskeleton The rigid outer body covering of an arthropod.

Femur The third joint of an insect leg, and usually the most powerful.

Gill Thin tissues that facilitate gaseous exchange during respiration.

Haemolymph The body fluid (blood) of an arthropod.

Hemimetabolous Insect with incomplete metamorphosis, with no pupal stage.

Above: A male polyphemus moth, Antheraea polyphemus, *displaying eye spots.*

Herbivore Feeding on plant material.

Hexapod A six-legged arthropod. Insects are the main hexapod class, the others being collembolans, diplurans and proturans.

Holometabolous Insect with complete metamorphosis, with a pupal stage.

Host The organism from which a parasite gains its nourishment.

Hymenopteran Member of the order Hymenoptera. Includes wasps, bees, ants, sawflies and wood wasps.

Imago The mature adult stage.

Instar Growth stage in the life of an immature arthropod, between moults.

Invertebrate Animals lacking a backbone.

Labium The lower lip in the mouthparts of an insect.

Larva The first juvenile stage in the life cycle of an animal that undergoes metamorphosis. The larva usually looks very different from the adult.

Lepidopteran Butterflies and moths, member of the order Lepidoptera.

Mandible Paired appendages near the mouth of an insect and some other arthropods, used for grasping and cutting food or for defence.

Maxilla Paired mouthparts behind the mandibles of an insect and some other arthropods, used for swallowing food. Sometimes the second pair of maxillae are fused to form the labium, or the maxillae may be modified into a proboscis.

Metamorphosis Development from egg to adult involving stages with different body shapes. May be

Above: A pill woodlouse, Armadillidium vulgare, *uncurling from a protective ball.*

complete where there is a pupal stage, or incomplete where the larva (or nymph) grows slowly to adult form.

Mimicry Where one species (the mimic) gains protection by resembling another species (the model). The mimic is often harmless, and the model may be distasteful or venomous.

Moulting The shedding its outer covering by a growing arthropod.

Myriapod Member of the superclass Myriapoda.

Nectar Sweet liquid produced by flowers to attract insect visitors. Nectar is a high-energy food exploited by many insects.

Nocturnal Active mainly at night.

Nymph The young larval stage of insects with incomplete metamorphosis.

Ocellus A simple eye with a single lens. Also used for the eye-like markings on the wings of some lepidopterans.

Ootheca Case containing the eggs of certain arthropods, especially orthopteran insects.

Ovipositor Egg-laying tube at the end of the abdomen of some female insects.

Paedogenesis Reproduction before the adult mature stage. Seen sometimes in insects where nymphs may be pregnant with the next generation.

Palp Appendage near the mouth of an arthropod, used for touching and tasting.

Parasite An organism that feeds on another (the host) without killing it.

Parasitoid An organism that feeds on another (the host) eventually causing its death. Many insect larvae are parasitoids because they feed on the body of the host, which later dies.

Parthenogenesis Reproduction without sex. The new individual grows from an unfertilized egg. Seen for example in certain generations of aphids, and also commonly in stick insects.

Pedipalp Appendage on the head of an arachnid, used either for grasping prey, or sometimes in reproduction.

Pheromone Chemical signal used, for example, by moths to find a mate, by ants to follow trails and by honey bees in communication.

Pollination Transfer of pollen to the female parts of a flower, often aided by visiting insects.

Predator An animal that feeds by catching other animals.

Proboscis The tube-like extended mouthparts of an insect, used for sucking in liquid food.

Proleg Outgrowths of the body of an insect larva that act as legs but which are not true limbs.

Pronotum Tough, shield-like cuticle covering the first segment of the insect thorax.

Prothorax The first of the three segments of the insect thorax.

Rostrum Beak-like mouthparts of bugs (Hemiptera).

Royal jelly Substance produced by worker honey bees and fed to larvae. Those larvae fed on royal jelly only develop into fertile queens.

Saprophage Animal that feeds on decaying matter.

Scale A flattened modified hair typically covering the wings of lepidoptera.

Segment A repeating body unit, most clearly seen in arthropods with long bodies, such as myriapods.

Social Organisms that live together in colonies.

Soldier A caste of worker in termites and ants that has a defensive role in the colony, usually with greatly enlarged jaws.

Spermatophore A packet of sperm produced by some arthropods.

Spinneret The organ in a spider's abdomen that produces silk. Also found in some insect larvae.

Spiracle The openings of the tracheae, along the sides of the arthropod body.

Stridulation Production of sound by rubbing together ridged surfaces, normally legs or wing cases. Typical of orthopterans.

Stylet Sharp, needle-like organ used for piercing. Found for example in the mouthparts of bugs, mosquitoes and fleas.

Subimago Adult-like stage in mayflies just before the full adult is formed. The subimago is duller than the adult, but like the adult it can fly.

Symbiosis Relationship between two organisms in which both partners benefit.

Synchronous Appearing at the same time, as in mass hatching of flying ants, for example.

Tarsus Final section of an insect leg, with claws at the tip.

Thorax The central body region of an arthropod, between the head and abdomen. The thorax bears the legs and wings (if present).

Tibia The fourth section of an insect leg, between the femur and the tarsus.

Trachea Small tubes lined with cuticle allowing gaseous exchange and opening through the spiracles.

Trochanter The short second section of an insect leg, between the coxa and femur.

Tubercle A small bump on the surface of the cuticle.

Venation The pattern of veins in the wing of an insect.

Venom Poison produced in special glands and used in defence or to immobilize or kill their prey. Typically found in many hymenopterans and spiders.

Viviparous Giving birth to live young.

Above: A European wasp, Vespula germanica, *at rest.*

INDEX

Above: Palpares
libelluloides.

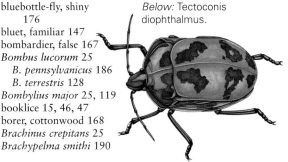

Below: Tectoconis
diophthalmus.

Above: Coccinella
transversalis.

Below: Palomena prasina.

Below: Zonocerus variegatus.

Above: Ixodes ricius.

Above: Lucanus cervus.

PICTURE ACKNOWLEDGEMENTS

Alamy Page 52br, 54tc, 61tr, 69br, 73bl, bc and br, 75br, 91tl, 99tr, 100bl, 104tr, 110cl, 115bc, 121br, 126cr, 133br, 140tr, 142br, 143bl, 156bl, 157tr, 166cl, 176cr, 200tr, 213cl, 245tr.

Corbis Page 23tr, 36bl, 37bc, 37br, 53tr, 96br, 159bl.

FLPA 143tl, 156tr, 160cl, 165tl, 171bc, 187tl, 190br, 192br, 205tr, 239tr.

Fotolia Page 17cr and bc, 22bl, 23tc, 26br, 27tl, bc, 28tr, bl and bc, 29bc, 30tr, 34tr, 44tr, 45br, 61bl, 67br, 70bl, 93tl, 111tr, 112cr and br, 118bc, 121br, 124bc, 134br, 135tl, 138tl and tr, 144tr, 146tr, 149br, 150br, 157tr, 172tr, 202br,

Istock Page 41tc and tr, 50tr, 51br, 52bl, 54tr, 55tl, 59tr, 60bl, 68tr, 69tl, 87bl, 88br, 89bl, 113cl, 118cr, 119tl, 122tr, 127tl, 129cl, 132br, 133tl, 138tc, 144br, 162tr, 170tc, 188cr, 196l and c, 222bl and tr, 246bl.

Nature Picture Library, Page 212cr, 224tr, 247tr

NHPA Page 35bl, 46bl, 47tr, 49bl, 50bl, 103tl, 131tl, 137br.

Photolibrary Page 146br, 148br, 194br, endpaper.